EDWARD CONDON'S COOPERATIVE VISION

s = const

EDWARD CONDON'S COOPERATIVE VISION

SCIENCE, INDUSTRY, AND INNOVATION IN MODERN AMERICA

THOMAS C. LASSMAN

UNIVERSITY OF PITTSBURGH PRESS

A John D. S. and Aida C. Truxall Book

Published by the University of Pittsburgh Press, Pittsburgh, Pa., 15260

Manufactured in the United States of America
Printed on acid-free paper
10 9 8 7 6 5 4 3 2 1

Cataloging-in-Publication data is available from the Library of Congress

ISBN 13: 978-0-8229-4534-5

Cover art: Photomontage of Edward Condon, Westinghouse Electric Corporation Photograph Collection, Thomas and Katherine Detre Library and Archives, Senator John Heinz History Center, Pittsburgh, PA, and the Westinghouse Atom Smasher, Allegheny Conference on Community Development Photograph Collection, Thomas and Katherine Detre Library and Archives, Senator John Heinz History Center, Pittsburgh, PA.

Cover design: Alex Wolfe

For Rebecca, Anna, and Jacob

CONTENTS

ACKNOWLEDGMENTS

THROUGHOUT THE WRITING of this book, I relied on the expertise and assistance of many individuals, whose generosity I am very pleased to acknowledge. I completed my graduate education under the direction of Robert Kargon in the Department of the History of Science and Technology at Johns Hopkins University. I am especially grateful to Professor Kargon for his wisdom and encouragement, not least for teaching me the history of American science. My training in the department would not have been complete without the guidance of Sharon Kingsland, Lawrence Principe, and Stuart Leslie. In the Department of History, Louis Galambos and Ronald Walters skillfully directed my fieldwork in US political, economic, and labor history. Working with them encouraged me to think broadly, outside my field of specialization in the history of the physical sciences.

At Carnegie Mellon University, David Hounshell, whose pathbreaking work on the history of US industrial research frames much of the analysis in this book, read the entire manuscript and provided a thorough and timely critique that significantly improved the final product. I owe him more than I can possibly repay for his friendship and generosity. Michael Neufeld, senior curator at the Smithsonian Institution's National Air and

Space Museum, also read every chapter, sometimes more than once. The book is much better because of his skillful editing as well as the feedback provided by the two anonymous referees the University of Pittsburgh Press tapped for reviews. Also for their support, encouragement, and thoughtful criticisms of my ideas as they evolved over the course of this project, I extend a special thanks to Glen Asner, Elliott Converse, Ronald Kline, Susan Morris, Sander Gliboff, Keith Barbera, and Gerard Fitzgerald. I am pleased to acknowledge the Johns Hopkins University Press and the University of Chicago Press, both of which published some of my research in *Technology and Culture* and *Isis* in the early 2000s. Journal editors John Staudenmaier, Margaret Rossiter, and Bernard Lightman skillfully guided my efforts to transform clunky draft essays into polished articles.

Research for this book took me far and wide to mine manuscript collections scattered across ten states and the District of Columbia. Generous support from the American Philosophical Society, the National Science Foundation, and, most recently, the National Air and Space Museum eased the financial burden significantly. I cannot acknowledge here every librarian and archivist who helped me along the way, but I thank all of them. Research is a collaborative effort, and without their assistance, historians who, like me, are at home in a sea of letters, reports, and memoranda, would be hopelessly lost. Dorothy Alke, vice president and senior counsel at the Westinghouse Electric Corporation (a subsidiary of the CBS Corporation), kindly granted me permission to cite and quote from archival documents preserved in the records of the Westinghouse Electric Corporation at the Senator John Heinz History Center in Pittsburgh, my hometown. I first met Carole Prietto, currently archivist at the Georgetown University Law Center, in the fall of 1990, my senior year at Washington University. More than anyone else, she taught me the craft of doing research in unpublished records and generously offered advice and support whenever I needed them most. At the University of Pittsburgh Press, Abby Collier expertly guided the manuscript through multiple revisions. Her patience, encouragement, and suggestions for improvement made the process much less stressful for me than it otherwise might have been. I am also grateful to Alex Wolfe, who managed the editing and preparation of the manuscript for publication.

Edward Condon died in 1974, but I did meet and learn from individuals who knew and worked with him. The late Frederick Seitz initially studied under Condon's direction at Princeton University in the early 1930s and went on to make major contributions to solid-state physics, some of which are discussed in chapter 3. They remained lifelong friends. Dr.

Seitz shared his recollections of Condon and his family with me through frequent letters and my occasional visits to his office at the Rockefeller University in New York City, and he also generously gave his time to read and comment on early drafts of the manuscript. So did the late Sidney Siegel, one of the first new PhD physicists Condon hired after he had moved to Pittsburgh in the fall of 1937 to work for Westinghouse. John Coltman, whom I met in 1998, joined Condon at Westinghouse in 1941 to work on high-power transmitter tubes for microwave radar sets crucial to the war effort. He taught me the ins and outs of this technology and also provided access to a rich cache of historical documents at the former Westinghouse Science and Technology Center outside Pittsburgh that has since moved to the nearby Heinz History Center. In 2004, a few years before he died, I had the privilege to conduct an oral history interview with Dr. Coltman that is now preserved at the Niels Bohr Library and Archives at the American Institute of Physics in College Park, Maryland.

Edward Condon and his wife, Emilie, had three children. In January 2016, I met Katherine Condon, the daughter of his youngest child, Paul (also a physicist). She kindly granted me permission to cite and quote from documents preserved in her grandfather's papers at the American Philosophical Society. By way of that personal connection and also through the writing of this book, I have come to recognize that Edward Condon's history is also, to some extent, my own. Born into a family of engineers whose lives followed the rhythms of Pittsburgh's metals and machine tool industries, I understood and appreciated from an early age the importance of science and technology in the local economy, not least by way of frequent visits to the engineering and manufacturing company my father owned and his father had founded shortly after World War I. I have always been interested in this history, more so now that I understand it through the life of someone who, despite a different background and professional perspective, experienced Pittsburgh in much the same way my family did, smoke and all.

Finally, it is with these more personal thoughts in mind that I acknowledge those closest to me. My parents lived through much of the history recounted in these pages, and I am deeply grateful to them for nurturing in me the confidence to pursue a subject that has interested me since childhood. I could not have done it without them. Heartfelt thanks go to my wife, Rebecca. Whenever I stumble or fail, she always gives me the strength and support to move forward. So do our children, Anna and Jacob. Watching them grow has been nothing less than delightful and also a constant reminder that history is always being made.

INTRODUCTION

IN THE FALL of 1927, Edward Condon, a twenty-five-year-old theoretical physicist who had just returned to the United States from Germany steeped in the new quantum mechanics, walked away from a promising academic career at his alma mater, the University of California at Berkeley. He went to work instead at the Bell Telephone Laboratories, the newly established research and development (R&D) arm of the American Telephone and Telegraph Company (AT&T), in New York City. Condon explained the rationale for this decision to Raymond Birge, a respected spectroscopist and his graduate advisor. "I hope to devote myself to theoretical physics in a broad way if my capabilities may permit it," he wrote Birge in November. "You probably smile at such a boyish hope in these days of great specialization[,] but nevertheless I intend to struggle to keep broad even if at a sacrifice of considerable depth."[1]

Birge did not understand Condon's logic. "There are certain types of very practically minded persons who may be . . . happier in industrial work than in University work," he responded flatly. "But in the case of a theoretical physicist like yourself, I cannot understand how there can be any question at all."[2] Condon resigned from the Bell Telephone Laboratories early in 1928 to pursue a university career, but his abiding interest in the industrial applications of physics persisted. His brief exchange with Raymond Birge evolved into an ambitious program of university-industry collaboration that directly challenged prevailing attitudes within the scientific and business communities about the legitimacy of academic science in a corporate setting. In the fall of 1937, after he had spent nearly ten years on the physics faculty at Princeton University, Condon moved to Pittsburgh to become associate director of research at the Westinghouse Electric and Manufacturing Company, the second largest electrical equipment maker in the United States. He left Westinghouse for Washington, DC, in 1945 to head the Commerce Department's National Bureau of Standards, a civilian government agency founded in 1901 to serve as the custodian of weights and measures.

This book explores the birth, life, and death of Condon's vision of cooperative research in industry through economic depression, world war, and the looming Cold War. It rose to prominence in Pittsburgh, a leading hub of heavy manufacturing in the United States. For years, Westinghouse and other large firms in the electrical and chemical industries had maintained elaborate laboratories stocked with university-trained scientists and engineers to protect their established product lines and diversify into new markets. This type of institutionalized invention proliferated after World War I, in no small part because of the synergies that existed between R&D and manufacturing. Some firms, however, also adopted a longer-term perspective; they exploited what contemporary practitioners called fundamental or basic research. This type of research aimed at product development, but it lacked the same institutional pedigree. More speculative studies sometimes proceeded without the internal guidance that seasoned managers, established markets, and core scientific and technological strengths readily afforded.[3] In 1935, Westinghouse plunged headlong into nuclear physics, a new field still confined to university laboratories, on a scale unknown in the electrical industry. This commitment to cutting-edge academic research lacked precedent among Pittsburgh's other smokestack industries, where R&D owed its legitimacy as an essential corporate function to the new knowledge gleaned from more established fields in the scientific and engineering disciplines.[4] When he arrived at Westinghouse two years later, Condon found a research environment uniquely suited to his temperament and sensibilities.

Condon's central role in the growth of fundamental research at Westinghouse adds a new and largely unexplored dimension to the history of industrial R&D in the United States. Much of the historical literature—

especially the pioneering studies of AT&T, the General Electric Company (GE), and the E. I. du Pont de Nemours and Company—highlights the strategies that these and other firms employed to reconcile the professional standards of academic research and the commercial imperatives of the corporate laboratory.[5] The extent to which Westinghouse absorbed and put to use Condon's knowledge of quantum theoretical physics deepens this historiographical tradition. Fields that straddled both communities and yet remain largely unexplored in the historical literature, such as solid-state physics (broadly known as the physics of solids before World War II), populated from the outset Condon's vision of cooperative research.[6] Crucially, however, Condon also conceptualized his new job in Pittsburgh in strikingly different terms than his predecessors and contemporaries elsewhere. Rather than see the diffusion of science into industry as a process of accommodation between academic and corporate interests, he articulated a wholly new strategy that transcended professional boundaries and rigid categories of knowledge production.

Condon often used terms such as "fundamental" and "pure" science and "basic" and "applied" research in his writings and public pronouncements to explain how new knowledge could be put to practical use, but he did not privilege any one of them at the expense of the others. Instead, he employed these terms interchangeably as symbols or rhetorical cues to converse in the languages of science and engineering, to translate ideas and skills from one community of practitioners to another to solve practical problems.[7] That conceptual breakthrough structured his understanding of and prescription for scientific and technological progress in industry. It emerged inchoate at the Bell Telephone Laboratories in the late 1920s, reached full maturity at Westinghouse Electric a decade later, and collapsed under the weight of military priorities at the National Bureau of Standards after World War II.

In the opening address he delivered at the dedication of the Charles Benedict Stuart Laboratory of Applied Physics at Purdue University on June 19, 1942, Condon proclaimed, "We must resolve not to neglect the cultivation of the basic science which we hope some day to apply. More and more, industry in America is recognizing the debt it owes to fundamental science—a debt it can hope to repay by fostering more basic research in its own research laboratories and by working in close cooperation with the universities." Elaborating further, he dismissed the notion of a research hierarchy in this collaborative process. "I feel sure," Condon declared, "that those who are interested with furthering scientific research at colleges see this problem of applied physics in all its broad implications. They recognize, as we do in industry, that all physics is applied physics—so- called pure physics being simply that part whose application is to satisfy the curiosity of the physicists."[8] Statements like this one guided Condon's approach to R&D at Westinghouse and within industry as a whole.

Shortly after he arrived in Pittsburgh, Condon established a postdoctoral fellowship program to recruit, on a rotating basis, young PhD scientists to conduct research on topics of their own choosing. On the face of it, he did not introduce anything radically new. Consulting arrangements, sponsored research projects, and other modes of collaboration among academic and industrial scientists and engineers had originated in the United States during the second half of the nineteenth century and proliferated after World War I.[9] The fellowships reinforced these established patterns of behavior, but they also broke new ground. A first for the electrical industry and for a firm reared in a long tradition of practical research, Condon's organizational innovation made the systematic introduction of open-ended studies in disparate and seemingly esoteric scientific fields, such as nuclear physics, a vital corporate asset. He expected Westinghouse to benefit from cutting-edge academic expertise and the universities to acquire, through the employment of former postdoctoral fellows, a workforce specifically attuned to industry's scientific interests. World War II abruptly ended Condon's local experiment in university-industry cooperation as the demand for military hardware crowded out fundamental research, and it also permanently restructured the political economy of science in the United States. Universities no longer relied so heavily on wealthy individuals and corporate sponsors for financial support after the war.[10] As the Cold War deepened, they more often looked to the federal government, especially the defense establishment, for resources to diversify and expand academic research in the physical sciences and engineering disciplines.[11]

Against this backdrop, Condon left Westinghouse to head the National Bureau of Standards in Washington, DC. Despite their pervasiveness as the primary units of analysis in the Cold War historiography, universities did not experience the impact of new patterns of patronage in isolation while the Bureau of Standards and other government research agencies languished passively on the sidelines as repositories of scientific and technological backwardness. Historians have reinforced that perception by way of uncritical analysis or outright omission.[12] Several possible scenarios of what the federal scientific research establishment might look like in the future began to emerge at the end of World War II.[13] The postwar revitalization of the Bureau of Standards that Condon spearheaded on behalf of business and industry competed directly against other initiatives designed to optimize the relationship between science and the state. Perhaps the best known among them is the contentious debate that culminated in the establishment of the National Science Foundation (NSF). The two main protagonists, West Virginia senator Harley Kilgore and Vannevar Bush, wartime director of the Office of Scientific Research and Development (OSRD), articulated opposing visions of public support for science. The delayed enactment of the NSF's founding legislation, which languished until 1950, allowed the military departments to fill the void left by the dismantling of the government's temporary wartime R&D contracting agencies.[14] Always present but rarely observed in this historical narrative is the role played by a third protagonist—Secretary of Commerce Henry Wallace, scion of a progressive Iowa farming family and an ardent New Dealer who ran the Department of Agriculture in the 1930s and occupied the vice presidency during the war. Wallace allied himself with Kilgore in the NSF debate.[15] However, he also acted independently to remake the Commerce Department—by way of the Bureau of Standards—into a complementary source of science-based economic growth.

The programmatic charge that Wallace handed to Condon at the Bureau of Standards in 1945 embodied a broader rethinking of late New Deal economic policy. World War II had pulled the United States out of the Great Depression, as industrial production and factory employment surged to meet the matériel demands of the armed forces. The cancellation of military contracts at the end of the war, however, stoked fears of another downturn, a possibility that alarmed Henry Wallace and other liberal reformers who had witnessed firsthand the economic dislocations of the 1930s. Always suspicious of but not inherently opposed to big business, Wallace believed that large corporations, which had fulfilled the bulk of production quotas for the wartime mobilization, leveraged their patent holdings and in-house R&D capabilities to manipulate markets and restrict competition at the expense of smaller firms without similar technical resources. Unlike some of the radical elements of the New Deal who continued to favor direct intervention in the economy to break up large firms and prevent a return to prior patterns of corporate behavior, Wallace and other like-minded advocates of national planning—while not unsympathetic to the thinking behind these measures—understood the problem of postwar reconversion in terms of stimulating consumption rather than regulating production. Opposed to the wholesale restructuring of capitalist institutions, they envisioned sustained peacetime economic growth and full employment as the natural outcomes of fiscal policies designed to increase purchasing power and consumer demand.[16] In his new role as secretary of commerce, Wallace anticipated the same economic dividends from public investment in R&D at the National Bureau of Standards. The improved competitive performance of small firms promised to raise the productive capacity of industry as a whole and to stimulate the consumption of goods and services in an expanding national economy.

Hired to rebuild the Bureau of Standards scientifically and overcome its reputation for second-rate research, Condon faced the added challenge of pulling off this transformation in a restructured political economy in which national security programs consumed an increasingly large share of the federal R&D budget. The onset of the Cold War recast Wallace's conception of an activist state in strictly military terms. It also set limits on Condon's ability to exploit, on the bureau's behalf, the type of university-industry collaboration he had established in Pittsburgh to broaden the scope of research at Westinghouse. Public investment to stimulate technological innovation in small businesses became a military subsidy for weapons R&D that no longer prioritized the importance Wallace had initially placed on firm size. His prescription for postwar prosperity quickly faded as Condon obtained seemingly limitless defense resources to diversify R&D at the bureau and bypass chronic bureaucratic constraints on institutional growth.

Condon's career at Princeton, Westinghouse, and the National Bureau of Standards is analyzed here in seven chapters that combine biographical and institutional history. Chapter 1 sets the stage for Condon's industrial career at Westinghouse. It traces his life from humble beginnings in New Mexico Territory at the turn of the twentieth century to the pinnacle of professional accomplishment in quantum theoretical physics at Princeton University in the mid-1930s. The origins of quantum mechanics in Europe, the transmission of this new knowledge to the United States, and the rapid growth of the discipline within the academic physics community structured his intellectual and professional development during this period.[17] So did a brief interlude at the Bell Telephone Laboratories, where Condon observed firsthand how scientists and engineers in a corporate setting applied their knowledge and skills to the solution of practical problems.

The technical resources manufacturing firms in Pittsburgh exploited to develop new commercial products did not incorporate the type of physics research that Condon cultivated at Princeton. Chapter 2 explains why

Westinghouse challenged this conventional wisdom and how it prepared to take advantage of Condon's scientific skills. Pittsburgh's history is inextricably linked to the rise and fall of the materials processing and machine tool industries. Condon's career at Westinghouse broadens that historical narrative. Rather than focus on the familiar inputs of capital and labor, the chapter explores the pervasive but largely unknown role of science in the growth and diversification of the local economy.[18] Shortly before Condon left Princeton, Westinghouse added to its R&D repertoire a new research paradigm that incorporated to a far greater extent than before the trappings of an academic physics department. In this case, Condon's absence from the chapter story line is deliberate. It is necessary to understand how Westinghouse and the industrial community in which it resided and in many ways emulated accommodated the unique brand of industrial research that he brought to Pittsburgh in the late 1930s.

Chapter 3 puts Condon back into the narrative. Unlike many of his academic peers who did not consider the benefits of an industrial career sufficiently enticing to give up their university appointments, Condon relished the opportunity to expand fundamental research at Westinghouse. He anticipated a more satisfying professional experience that complemented rather than diminished his academic pedigree in theoretical physics. Through such novel innovations as the postdoctoral fellowship program, Condon transformed the Westinghouse Research Laboratories into a significant source of new knowledge in nuclear physics, mass spectrometry, the physics of solids, and microwave electronics that impressed industrial rivals and universities alike. An explicit expression of the narrow specialization he had always avoided, the fellowships also served a vital corporate interest. Condon appropriated the company's engineering and manufacturing expertise to translate the results of research in these disparate fields into products for sale to academic and industrial consumers.

Chapter 4 examines the technological outputs of Condon's strategic vision, as well as the extent to which they succeeded in the marketplace. On the eve of World War II, scientists had already begun to exploit radioisotopes produced in high-voltage accelerators to treat diseases in the same way their predecessors had taken advantage of x-ray tube technology to improve medical diagnostics. Condon tuned the nuclear physics program at Westinghouse to the same practical ends. His corporate handlers, however, saw limited potential for commercial growth in the field, despite initial enthusiasm for the in-house research that supported it. Their response exposed in stark terms a clear disjunction between management's tolerance for risk and Condon's optimistic predictions for new business opportunities. His ability to converse in the languages of science and engineering succeeded up to a point; knowledge produced in the laboratory did not seamlessly transition into commodifed technologies for a mass market. Mass spectrometry research followed the same abortive trajectory. Only the research Condon started in the physics of solids and microwave electronics, two fields that bore some connection to the company's core technological strengths, moved quickly from the laboratory to the factory floor.

Chapter 5 focuses on the extent to which military requirements narrowed the focus of fundamental research on problems that dovetailed nicely with Westinghouse's expertise in engineering and manufacturing. Condon spearheaded the growth of microwave R&D, which propelled the company to fifth place (in dollar sales) among all US makers of radar equipment during the war.[19] He also served briefly as J. Robert Oppenheimer's handpicked deputy to organize the Manhattan Engineer District's secret atomic weapons laboratory at Los Alamos, New Mexico. A longer assignment at the University of California helped Westinghouse meet short-term production goals to scale up the manufacture of weapons-grade uranium for the bombs assembled at Los Alamos. It did not, however, become the permanent corporate R&D presence in Berkeley that Condon had envisioned and pressed his superiors to establish as a logical extension of the prewar fundamental research program in Pittsburgh. Already less sanguine about the prospects of a bright future at Westinghouse, Condon turned his attention to the politics of postwar atomic energy policy. This abrupt shift in priorities during the summer of 1945 culminated in a meeting with Secretary of Commerce Wallace, who selected Condon to head the National Bureau of Standards.

Condon's six-year tenure at the Bureau of Standards is the subject of chapter 6. From the outset, he modeled the bureau's transformation into a nationally recognized technical resource for the small business community on the founding principles of the fundamental research program at Westinghouse. This initiative foundered on restrictive civil service requirements, low staff salaries, a reputation for useful but unimaginative research, and the competition for resources in a federal R&D establishment increasingly dominated by the military departments. Unable to remake the bureau in Westinghouse's image, Condon charted a more pragmatic course that betrayed his instincts. He staked the bureau's future on research in select fields that catered directly to the technological requirements of the armed forces.

This compromise—born out of necessity to accommodate the broader shift toward permanent military preparedness—wrecked the cooperative vision of academic and industrial research that Condon had carefully nurtured. It also undermined the stability of his career, a theme that is discussed in the concluding chapter. Condon did not pull punches, especially when he judged someone to be misguided or wrong. He had already acquired a reputation for impetuous and combative behavior. These personality traits and a commitment to progressive-left politics clashed with the rising tide of anticommunism after the war. No longer interested in public service under such conditions, he resigned from the Bureau of Standards in 1951. The political fallout and ongoing harassment from the Un-American Activities Committee in the House of Representatives (HUAC) cut short his brief tenure as director of research at the Corning Glass Works in 1954 and precluded a seamless return to the academy. He finally obtained a permanent position at Washington University in St. Louis in 1956, but the appointment marked a steady withdrawal from active research and brought to a close a long career in industry and government. Thirty years earlier, Condon had set out to pursue an abiding interest in industrial research that nearly a decade on the leading edge of theoretical physics at Princeton did not diminish. To the contrary, by way of his introduction to quantum mechanics in the 1920s, Condon glimpsed for the first time the practical bent that structured his scientific outlook and career ambitions.

EDWARD CONDON'S
COOPERATIVE
VISION

CHAPTER 1

Rise of a Theoretical Physicist

TRAINED AT THE University of California and reared on the new quantum mechanics in Europe, Edward Condon played a leading role in the growth and diffusion of modern theoretical physics in the United States. Before he accepted a faculty appointment at Princeton University in 1928, however, Condon worked briefly at the Bell Telephone Laboratories in New York City. Despite the brevity of this industrial experience, it left a lasting impression on his intellectual and professional development. In a corporate environment where research served no other purpose than to fulfill the technological requirements of the telephone system, he witnessed firsthand the limited commercial appeal of theoretical physics. At the same time, Condon savored the opportunity to broaden the scope of his research interests beyond quantum mechanics to include more practical subjects, even in the engineering disciplines. Rather than derail a promising career, which his mentors at Berkeley had feared, the introduction to corporate R&D showcased a unique blend of intellectual versatility that blossomed at Princeton and later at Westinghouse.

ORIGINS

The details of Condon's youth and upbringing are recounted in a partially completed, unpublished autobiography that he began to write in 1948. Condon, born in Alamogordo, New Mexico Territory, on March 2, 1902, spent most of his childhood on the West Coast. He had no siblings. His father, William Edward Condon, an engineer and surveyor, had migrated west from Missouri—his birthplace—and briefly settled in Alamogordo to work for the local railroad. Edward's mother, Caroline Uhler, hailed from Ohio. William and Caroline divorced in 1908, after which young Edward's life entered a period of constant upheaval. Estranged from his father, who had relocated the family from Alamogordo to San Francisco, Condon and his mother moved frequently to make ends meet. By the time he completed the eighth grade, Condon had already attended fourteen different grammar schools in four states—California, Washington, Wyoming, and Colorado. In 1914, just before the outbreak of World War I, they returned to Northern California and settled permanently in Oakland, across the bay from San Francisco.

In 1919, one year after he graduated from high school, Condon prepared to enter the freshman class at nearby Berkeley—the main campus of the University of California. He intended to major in journalism, an interest that he had developed in high school, where he served as editor of the school newspaper. College life did not suit him, however, and he withdrew from the university in September, shortly after classes had started, to go back to work full time. Condon had gotten a start in the newspaper business at age sixteen, initially working part-time for the *Oakland Daily Post*, and he returned to that vocation, adding jobs at other local news outlets for the next two years. Although the work generated a good, steady income, his commitment to journalism as a long-term career objective soon faded, and in the fall of 1921 Condon reenrolled at Berkeley as a freshman. This time, however, he focused on the hard sciences, which, like journalism, had always been a source of interest. After dabbling in chemistry and astronomy, Condon decided in 1923 to major in physics. He attended classes year round and advanced rapidly through the curriculum.[1]

Condon completed all of the requirements for the undergraduate major in the summer of 1924, but he remained on campus to obtain a doctorate in theoretical physics. In the fall of 1925, he enrolled in a graduate course on the band spectra of diatomic molecules taught by Raymond Birge, a

1.1. Berkeley, 1925. Edward Condon is on the far left. Raymond Birge and Leonard Loeb are standing next to him. Emilie Condon is standing fourth from the right (*in front*). Emilio Segrè Visual Archives, American Institute of Physics

leading molecular spectroscopist who had transformed Berkeley into a nationally recognized center for research in the field. Condon completed his doctoral dissertation under Birge's direction in the summer of 1926 (the *Physical Review* published it later that year) and then, like many of his peers at the end of their academic training, prepared to study quantum mechanics abroad with Max Born on a postdoctoral fellowship at the University of Göttingen.[2] Condon had already been introduced to the new physics earlier that year when Born visited Berkeley, but his working knowledge of quantum mechanics remained rudimentary to the extent that his dissertation did not include a discussion of the subject. Birge lacked the requisite skills to close the gap. "[Condon] is working on the general field of band spectra, and so is under my direction, although I am able to give him little assistance on the purely theoretical part of the work," he wrote the fellowship selection committee at the National Research Council (NRC). Now fresh out of graduate school, Condon, his wife Emilie—a former student at Berkeley whom he had met through his newspaper connections and married in the fall of 1922—and their three-month-old daughter Caroline Marie left Berkeley by rail for New York City, the last stop stateside before sailing for Europe. They expected to reach Göttingen by the end of September.[3]

ACADEMIA VERSUS INDUSTRY

New York City initially overwhelmed Condon, who had never traveled east of Denver. "[New York City] is just like Oakland or [San Francisco] . . . except that there is so very much more of it," he wrote his mother on September 10, the day before the family set sail for Germany. "It simply extends for miles and miles. . . . I shouldn't care to live here," he added.[4] The Condons arrived in Göttingen on September 26, a month before the university opened. Condon and the other American physicists already there used the free time to read up on the latest research. "Every morning we get together to read the Schrödinger articles," he wrote Leonard Loeb, a physics professor at Berkeley and a lifelong friend. "I keep working away at the thesis band spectra problem by Schröd[inger]'s [wave] mechanics, but so far with but few results." Within a short time, however, Condon decided to leave Göttingen, which he found to be "disappointingly dull." He did not get along with Born. "I always feel hurried and ill at ease with him," Condon complained to Loeb. "He does no lecturing on the modern things, and the great theoretical seminar[,] which only meets twice a month, is addressed mainly by students."[5] Condon requested and received permission from the Rockefeller Foundation to transfer his fellowship— awarded by the International Education Board (IEB)—to the Institute for Theoretical Physics at the University of Munich, where he planned to study with Arnold Sommerfeld.

Condon liked Sommerfeld, and he quickly acclimated to the research environment in Munich. Only his correspondence suffered as a consequence of the improved conditions. "I wrote frequent peevish letters from Gottingen, where I was unhappy," he told Leonard Loeb half-jokingly in July 1927. "Here [in Munich], where I get lots of stimulation, I hardly write at all." Raymond Birge got that impression too, though Condon conveyed it without the same wit. "This has been a very pleasant semester," he told Birge, "and I have learned a great deal." Condon considered the option to remain in Europe for one more year, but IEB guidelines prohibited renewal of his fellowship.[6] Unable to work out a suitable alternative, he applied for and received an NRC fellowship and prepared to return to the United States for further postdoctoral study. Birge and Loeb welcomed Condon's decision to take the fellowship at the California Institute of Technology (Caltech) because they had expected from the outset that he would come back to Berkeley as a full-time member of the physics faculty. Birge in particular saw in Condon's return the opportunity to give the department a much-needed boost in theoretical physics. "I feel rather discouraged as to the possibility of our getting any of the younger, really brilliant, men at California," he wrote physics department chairman Elmer Hall after canvassing universities on the East Coast for potential recruits in the spring of 1927. "I think that our only hope at California is to develop brilliant men of our own, like Condon, and then keep them."[7]

The Condons arrived in New York City on October 9, but they did not proceed to Pasadena. Instead, they settled into a small apartment near Columbia University on the city's Upper West Side. Before he returned to the United States, Condon had declined the NRC fellowship and accepted an editorial job in the publications bureau at the nearby Bell Telephone Laboratories. Founded in 1925, the laboratories originated in the engineering department of the Western Electric Company, the manufacturing arm of the Bell Telephone System.[8] Condon had first pondered this career option, which rekindled a prior interest in journalism, in February 1927, before he left Göttingen for Munich. He answered an advertisement in the *Physical Review* for a PhD physicist with writing experience to provide editorial assistance to members of the research staff preparing papers for publication in professional journals. The job also called for the writing of science articles for the popular press, a requirement that prioritized the ability to explain complex technical concepts to a lay audience. Condon felt at ease in this milieu, and now, at the Bell Telephone Laboratories, he prepared to go further and explore at the working level the synergies between the scientific and engineering disciplines.

Condon also looked forward to a bright and financially secure future. Living in Europe had been expensive; he and Emilie had depleted their small savings and relied on cash advances from the International Education Board and the Berkeley physics department. The company promised to exceed Condon's minimum salary requirement of $3,500—already 30 percent more than Berkeley had agreed to pay upon his return—and granted him the freedom to pursue his own research interests, a rare privilege in an industrial R&D organization. "I am assured of a light program, which will enable me to keep on doing research work in pure physics," Condon wrote Loeb in late August, before returning to the United States. "Also they have colloquia there on physics and I am informed that the atmosphere of the place is much like that of a university." Condon singled out the significance of the recent discovery of electron diffraction by Clinton Davisson and Lester Germer—an unintended by-product of their research on thermionic emission in vacuum tube amplifiers—but, nonetheless, "real hot stuff for the new developments in quantum theory."[9] Condon's enthusiasm, however, shocked Loeb and Birge, both of whom questioned his judgment and dismissed industrial research as a suitable career option. "I feel you have made a grave mistake in throwing up a perfectly good National Research [Council] Fellowship," Loeb told Condon flatly. "All of us [at Berkeley] feel that you are selling yourself pretty cheaply to an organization which is far below where you belong."[10]

Loeb and Birge feared that Condon might not return to Berkeley in the fall of 1928. Their disappointment gave voice to a personal distaste for industrial research, but it also rested on a misunderstanding of Condon's intellectual development that had nothing to do with the future of theoretical physics at Berkeley. "I hope to devote myself to theoretical physics in a broad way if my capabilities may permit it," Condon wrote Birge after less than two months on the job at the Bell Telephone Laboratories. "You probably smile at such a boyish hope in these days of great specialization[,] but nevertheless I intend to struggle to keep broad even if at a sacrifice of considerable depth." While gaps in the documentary record preclude a definitive explanation of the motives underlying these statements, Condon's postdoctoral training in Munich may have reinforced the affinity for industrial research that nonplussed Loeb and Birge. Arnold Sommerfeld's conception of and pedagogical approach to theoretical physics included practical subjects that also incorporated engineering knowledge. The details of Condon's stay in Munich are unknown, but the glowing review of Sommerfeld's lectures he communicated to Loeb and Birge before returning to the United States suggests that the experience had some impact on his thinking. Elaborating further on his commitment to maintaining a broad perspective, Condon told Birge, "Applied physics seems to be just on the eve of adequately using things in theoretical physics that have been known a long time, and the place where I now am seems to be taking quite an active part in this trend."[11]

Few industrial R&D organizations in the United States possessed the institutional resources necessary to support a research environment that appealed to someone like Condon. Although it shared the research spotlight with other well-known, science-based firms such as General Electric, the Bell Telephone Laboratories nevertheless remained in a class by itself given its unique status as an operating unit of a regulated monopoly.[12] "If any industrial laboratory can afford a nice working atmosphere for a physicist," Condon wrote Loeb, "then this one should." At the same time, however, Condon clearly understood and accepted the limits of the company's tolerance for what he called "pure physics," a reference to the type of research otherwise conducted in a university setting, free from commercial priorities. "These laboratories are strictly commercial—as all industrial labs are," he explained to Birge. "It was a sheer accident that led Davisson [and] Germer to their great discovery [of electron diffraction], and it's not really appreciated here as a great one. But they do have problems which are not . . . less interesting than those of pure physics." Condon's "desire to take a rest from high specialization in physics" fully accommodated the commercial priorities that a corporate R&D organization such as the Bell Telephone Laboratories imposed. He relished the opportunity to "learn a little of the engineering viewpoint . . . [and] [d]o a little broadening." These statements confirmed the extent to which he rejected the privileged status Loeb and Birge ascribed to theoretical physics. "Industrial physics work is the only thing which has ever rivaled academic work in my inclinations," Condon told Loeb, "and if that issue is settled this year once and for all I'll be a much better university man for having settled it by direct experience."[13]

A light work schedule in the publications bureau afforded Condon plenty of time to tour the laboratories and sample the latest research. "My principal interest during office hours," he wrote Isidor Rabi, a postdoctoral student from nearby Columbia University whom he had met in Munich, "has been in the statistical work of the laboratories and I have learned a lot about [the] theory of probability and such stuff as is basic to [the] new statistics in physics." In-house research in this field, which had come to rely more heavily on quantitative methods of inquiry, focused on the solution of engineering and manufacturing problems and organizational bottlenecks that impeded the efficient operation of the telephone system. "They are making some very interesting applications," Condon told Rabi, "to problems of inspection engineering in large manufacturing plants."[14] In this case—his first foray into industrial research—Condon demonstrated an innate ability to converse in the languages of science and engineering. That proficiency made an impression on physicists elsewhere, among them William F. G. Swann, Ernest Lawrence's doctoral advisor at Yale University, who had moved on to become director of the Franklin Institute's new Bartol Research Foundation in Philadelphia. Swann had invited Condon to lecture in Philadelphia and later identified the source of the intellectual versatility he demonstrated at the Bell Telephone Laboratories. "[Condon] is endowed with a strong physical instinct," Swann wrote late in 1928. "He is one of those who subordinate[s] the technique of the analysis to the physical content of the subject, and is the type of man who would be likely to see deeply into the content of a theory without having to wade through a great deal of analysis."[15]

The resourcefulness Condon demonstrated at the Bell Telephone Laboratories exacted a high price. "I am beginning to feel myself slipping on wave mech[anics]," he confided to Raymond Birge one week before writing to Rabi, "because papers are coming out by leaps and bounds." Condon looked outside the company for intellectual sustenance. He joined a theoretical physics working group that met weekly at Columbia to stay abreast of the latest research. Participation in this group, however, did not fully resolve the underlying tension between his academic interests and predilection for industrial research as an equally attractive career choice. On December 1, 1927, Condon received a 10 percent pay increase that raised his annual salary to $4,000: "That makes it awfully hard to go back to university work[,] for the prospects of continued advancement here are excellent." Determined not to remain in the publications bureau indefinitely, Condon obtained a pledge from his boss to be placed anywhere in the company he chose. "These are awful days," he told Rabi anxiously. "The financial prospects are so good with the Bell Company and the work is highly interesting and pleasant. But to continue would mean giving up research in the broad and carefree mood of university life and I am very reluctant to give it up." Awakened to the plentiful opportunities now available for professional advancement in theoretical physics, especially on the East Coast, Condon reconsidered his long-standing allegiance to Berkeley.[16]

Condon first expressed doubts about Berkeley after the Christmas holiday, when he attended a meeting of the American Physical Society in Nashville. "I like the Eastern folks very much and am tempted to look upon the great divide across the U.S. as a disadvantage in California," he explained to Loeb in January. Only twenty-five years old, Condon had already begun to receive employment inquiries and firm job offers. New

York University, the University of Wisconsin, and the University of Minnesota all expressed interest. "There is a great boom of interest in theoretical physics," he told Rabi, who had yet to return stateside, "and every institution in the U.S. has suddenly realized that it just must have some young fellow who can talk the [wave mechanics] stuff." Like Berkeley, the Bell Telephone Laboratories lost some of its earlier luster as Condon acquired more self-confidence in academic circles. He resigned from the company, effective February 1, to sort out career options and get back to research full time.[17]

Pleased that Condon had cut short his industrial career, but no doubt bothered by the appeal of the East Coast academic elite, Loeb still anticipated a bright future for Condon at Berkeley. He promised to set aside funds to offset the cost of travel to professional physics meetings, and he also assured Condon of rapid advancement through the academic ranks with commensurate increases in pay. On January 25, Loeb wrote Condon, "Think it over and, whatever you do, give us full information of your other offers and give us a chance to meet them. That is the only request we make of you and the only obligation you have." Condon, who had formally accepted Berkeley's offer of an assistant professorship the previous spring, backpedaled. "The Berkeley offer I am somewhat reluctant to accept because I do not think it a good policy for a man to remain where he received his training, [and] also because it is so far west that visits in the East are troublesome and costly," he had written to Henry Erikson, chairman of the physics department at the University of Minnesota, five days earlier. Prepared to accept Erikson's invitation to come to Minnesota, Condon suddenly declined when Karl Compton intervened on Princeton's behalf.[18] The opportunity to go to Princeton permanently knocked Berkeley out of the running and put Condon at the center of the burgeoning growth of American theoretical physics.

Condon had met Compton, a rising star at Princeton who performed the first experimental verification of Albert Einstein's quantum interpretation of the photoelectric effect, in Göttingen and immediately took a liking to him. Before he settled on Caltech for a second year of postdoctoral study, Condon considered Princeton, but he told Loeb disappointingly early in 1927 that "there is no theoretical physics at Princeton." Compton had yet to fill the gap that prompted Condon's assessment. Princeton boasted a strong research program in experimental physics; four new assistant professors, all Princeton-trained experimentalists, had joined the faculty in 1925.[19] Edwin Adams, a Harvard-trained PhD who had come to

1.2. Karl Compton. Emilio Segrè Visual Archives, American Institute of Physics

Princeton in 1903, carried primary responsibility for teaching and research in theoretical physics. A respected expert on quantum theory—he wrote a commissioned survey of the field for the National Research Council in 1920—Adams shied away from the new quantum mechanics, preferring instead to cultivate his abiding interest in the classical subjects of electricity and mechanics. Initially, Compton tried to recruit John Van Vleck, one of Condon's better-known contemporaries at the University of Minnesota. "While our experimental activities are in a very prosperous state, we need someone like yourself on the theoretical side," Compton wrote Van Vleck one month after Condon struck Princeton from his list of candidates for postdoctoral study. "[Adams] is extremely valuable . . . but he is not so active in productive work along the newer lines in theoretical physics. . . . We have a very strong Mathematics Department who are [*sic*] extremely anxious to cooperate with us. . . . The fact remains, however, that the mathematicians are on somewhat too high a plane for most of us and someone like yourself would be an invaluable bond between the two departments." The negotiations with Van Vleck stalled in March when Compton had to withdraw the offer because of financial constraints.[20]

Circumstances quickly improved when the university received the first of several large donations from the family of a wealthy Chicago alumnus. That infusion of cash, in addition to the promise of matching funds for research from the Rockefeller Foundation, put Compton's plans to strengthen theoretical physics back on track. "Princeton is suffering from an attack of ambition, to become the American center for theory," wrote Francis Bitter, a newly minted PhD in physics from Columbia who had visited Princeton in October 1927 on his way to Caltech to start an NRC postdoctoral fellowship. Now flush with funds, Compton opened a dialogue with Condon, who had just arrived in New York City. Late in October, before Bitter departed for Pasadena, Compton invited Condon to present to the department some of his research on the band spectra of molecular hydrogen, which had grown out of discussions with the theoretical physics working group that met weekly at Columbia.[21] On November 18, the day after his presentation, Condon toured the Palmer Physical Laboratory. "The experimental people at Princeton are the liveliest group I have yet seen anywhere. . . . [They] are . . . making that place hum," he wrote Raymond Birge. Impressed, Compton queried Condon about a possible move to Princeton during the Nashville meeting of the American Physical Society in December. On January 31, 1928, a few days after Minnesota's Henry Erikson came up empty handed, Compton offered Condon an assistant professorship. The position paid more than Berkeley had put forward and focused exclusively on graduate instruction and research. Two weeks later, Condon resigned from Berkeley.[22]

Stunned, Leonard Loeb also felt personally slighted. He accused Condon of acting in bad faith because he had not given the department a chance to match Princeton's offer. "I must confess," he wrote, "that these incidents have spoiled the very high opinion which I had of you." Also taken aback, department chairman Elmer Hall complained to Loeb that "it isn't the fact that Condon left us, but the way in which he handled the affair which has disappointed me in him." Raymond Birge expressed his disappointment more candidly. "Princeton thinks very highly of Condon," he wrote a colleague at Harvard, "but I am very mad." Condon insisted that he had not violated Loeb's trust. He discussed the matter with Compton, who "thought that it would not be improper to resign provided this were done early in the spring." Loeb dismissed Condon's explanation: "You should first have obtained a definite release from Professor Hall . . . before accepting an offer from Compton."[23] Either way, Berkeley no longer served as the most attractive academic alternative to an industrial career. "There are some features in regard to which Berkeley couldn't equal Princeton," Condon wrote Loeb five days after he resigned, especially the $10,000 Compton now had on hand to pay for an endowed professorship in mathematical physics. "This will either provide me with a valuable older associate . . . or will be split up and used to get two or three younger men so that we can have a regular institute of theoretical physics." Loeb's anger subsided and their relationship recovered. Birge's bitterness persisted. "Birge in particular has never forgiven me for not coming back [to Berkeley]," Condon later recalled.[24]

PRINCETON UNIVERSITY

"I am very glad that this [appointment] has gone through and we all hope that it will be the first step in bringing together a stimulating group of men interested in theoretical physics," Karl Compton wrote Condon on February 3, 1928. That fall, Condon taught a new graduate seminar on quantum mechanics, one of a series of special courses and lectures offered for the first time on mathematical physics in collaboration with the mathematics department.[25] The seminar, Condon wrote, covered "recent developments in the quantum theory with emphasis on the work of Schrödinger and its application to the interpretation of the spectra of atoms and molecules." Its early success prompted both departments to expand the series into a permanent program in mathematical physics for undergraduate and graduate students that began in the 1931–32 academic year.[26]

Research occupied Condon's time outside the classroom. "That first year at Princeton," he later recalled, "was the most productive of my life." He wrote a longer and more extensive theoretical treatment—now grounded in the new quantum mechanics—of his doctoral dissertation on the intensity distribution of band spectra, which the *Physical Review* published in December 1928. Collaboration with Ronald Gurney, a postdoctoral fellow at Princeton, culminated in the introduction of a quantum mechanical explanation of the spontaneous emission of alpha particles (positively charged helium ions) from radioactive nuclei. Their quantum interpretation of alpha decay, discovered independently by George Gamow at Niels Bohr's Institute for Theoretical Physics in Copenhagen, explained the transit of insufficiently energetic helium ions through the potential energy barrier that stabilized heavy nuclei. Gamow had gone further in his analysis than Condon and Gurney; he applied the same quantum mechanical explanation of alpha decay to the absorption of charged particles by the nucleus, a phenomenon Ernest Rutherford first observed experimentally in 1919. This work, especially Gamow's contribution, had important implications for experimental nuclear physics. Gamow had gone further in his analysis than Condon and Gurney; he applied the same quantum mechanical explanation of alpha decay to the absorption of charged particles by the nucleus. Drawing on Gamow's calculations, Cambridge physicists John Cockroft and Ernest Walton built a low-voltage accelerator and bombarded lithium atoms with protons to record the first artificially induced nuclear disintegration in 1932.[27]

These research accomplishments only confirmed Karl Compton's initial impressions of Condon's ability. "I think that one or two other young mathematical physicists . . . are probably abler mathematicians," he wrote after Condon had been on the job for only three months, "but I do not believe that there is Condon's equal in this country in tying together his mathematics with sound physical intuition." Condon's productivity also extended to teaching—"I consider . . . Condon to have done more to stimulate interest in theoretical physics among our graduate students than any other influence here since I have been connected with this department," Compton added—and to the writing of textbooks. In the fall of 1927, just after he started working at the Bell Telephone Laboratories, Condon had begun writing what became the first American textbook on quantum mechanics.

Compton helped Condon obtain a contract from the McGraw-Hill Book Company, which used the forthcoming text to launch a new international physics series.[28] In early June 1928, he pressed Condon "to do everything possible to get [the book] out promptly. . . . There are others who are also working on the same line and the first really good English treatise on the subject will probably get the lion's share of the credit." Condon enlisted the editorial assistance of Philip Morse, one of Compton's students, to finish the manuscript. Even before the final product appeared on bookshelves, Condon distributed copies of the manuscript to the participants who attended his lectures at the second summer symposium on theoretical physics at the University of Michigan, the first major yearly gathering of specialists in the field convened in the United States. McGraw-Hill published *Quantum Mechanics*, which Condon and Morse dedicated to Compton, in the fall of 1929, shortly after the Michigan symposium.[29]

After he returned from Ann Arbor, Condon got back into research full time, but not at Princeton. In the spring of 1928, more than a year before *Quantum Mechanics* debuted, he called on Henry Erikson at the University of Minnesota to inquire about the status of the faculty appointment he had been offered a few months earlier but rejected out of deference to Princeton. When the teaching obligation at Princeton turned out to be heavier than Compton had initially promised, Condon reopened negotiations with Erikson on the assumption that the original terms of employment—a much lighter course load and a higher salary and rank— remained in force and could be applied to the 1929–30 academic year. Erikson responded enthusiastically. To ward off the competition, however, Compton arranged a promotion, which Condon accepted.[30]

Then, in March 1929, two months after he had turned Erikson away for the third time, Condon backtracked, but for reasons that had nothing to do with his professional ambitions. Separately, he confided to Compton that his wife, Emilie, did not like living in Princeton and wished to leave. "Mrs. Condon is a very nice young woman and I think would fit in very nicely almost any place better than in Princeton, for she has been used to the greater freedom and lack of conventionalities of the West," Compton wrote Erikson confidentially after their conversation. He hoped the most recent offer Erikson had tendered, in January, could be salvaged to alleviate the Condons' unpleasant circumstances.[31] When he learned that Compton had intervened on his behalf, Condon quickly accepted Erikson's invitation. "I am quite enthusiastic about the prospects and am glad to be joining the group there," he wrote Erikson on March 25. Emilie, who had felt isolated and unwelcome in Princeton's social circles, immediately took a liking to Minneapolis.[32] The bright future she and Condon anticipated, however, quickly faded after the fall semester got under way. "I may as well blurt out the truth at once and tell you that I am scientifically lonesome and unhappy," he wrote Compton, who had just become physics department chairman. "The situation here is not impossible, it is just that there is not one percent of the research spirit that Princeton has." Assured that Emilie had agreed to give Princeton another chance, Compton welcomed Condon's return in 1930, but not before he recorded some research successes at Minnesota.[33]

Mass spectrometry, a technique physicists used to measure the mass and abundance of isotopes, had attracted Condon's interest at Minnesota. Research in this field also enjoyed robust growth at Princeton, initially under the leadership of Henry Smyth—one of the four locally trained experimentalists appointed to the faculty in 1925—and then under Walker Bleakney, whom Condon had singled out at Minnesota and brought back to the Palmer Physical Laboratory to work with Smyth on a postdoctoral fellowship.[34] Bleakney's pioneering studies of molecular structure, to which Condon contributed, and precise measurements of isotopic mass and abundance—including the first experimental verification of the existence of deuterium (heavy hydrogen) and tritium (the third isotope of hydrogen)—sealed his academic reputation and further solidified Princeton's leading position in the field.[35] Also at Minnesota, Condon had returned to the subject matter of his doctoral dissertation and started a "new line" of study "on the application of [quantum mechanics] to atomic spectra." This research extended into the writing of another textbook that got under way in 1931 with the assistance of George Shortley, an electrical engineering major who, like Bleakney, had accompanied Condon back to Princeton for graduate study in theoretical physics.[36] In both of these cases, the brief interlude in Minneapolis provided fertile ground for some of the research fields that Condon later cultivated more aggressively at Westinghouse.

In the spring of 1934, Condon and Shortley submitted their completed book manuscript to the publisher. No longer burdened by writing deadlines, Condon once again prepared to return to research full time. He told Shortley, "I am going to work on nuclear theory in earnest this summer."[37] The high-voltage accelerators that produced the data for his calculations had only just moved into the mainstream of physics research in the universities. Moreover, the cyclotron and electrostatic generator acquired legitimacy as academic research tools at the same time industry expressed interest in their usefulness as potential sources of practical applications.[38] R&D managers at Westinghouse justified the corporate investment in nuclear physics on expectations of rapid growth and an abundance of new research findings that skilled practitioners like Condon inculcated at Princeton.

Princeton had gotten off to a seemingly auspicious start in nuclear physics in 1929. Robert Van de Graaff, an Oxford-trained NRC postdoctoral fellow, built inside the Palmer Physical Laboratory a 1.5-million-volt (MeV) electrostatic generator—the modern variant of a technology first introduced in the late nineteenth century—to accelerate charged particles. Van de Graaff's fellowship ended in 1931, at which point Karl Compton, who had resigned from Princeton the previous year to become president of the Massachusetts Institute of Technology (MIT), hired the young upstart to continue developing the generator technology on a much larger scale in an abandoned airship hangar at the Round Hill research station near Cambridge.[39] Van de Graaff's departure left nuclear physics at Princeton in the hands of Rudolf Ladenburg, who had just come to the United States from Germany, where he previously held joint appointments at the University of Berlin and the Kaiser Wilhelm Institute for Physical Chemistry and Electrochemistry. Supported by a grant from the Rockefeller Foundation's General Education Board, Ladenburg acquired a 400,000-volt transformer-rectifier set from General Electric in 1932—the same year James Chadwick discovered the neutron, Cockroft and Walton bombarded lithium atoms with accelerated protons, and Harold Urey published the results of his research confirming the existence of deuterium (heavy hydrogen). Ladenburg intended to raise the maximum potential to 600,000 volts, but his work progressed slowly, initially because of the faulty equipment GE supplied and insufficient testing of the apparatus prior to delivery. A weak ion source and leaks in the accelerating tube also "prevented steady work" to the extent that Ladenburg had to disassemble the equipment and start over.[40]

Ladenburg finally got his accelerator to operate smoothly in May 1935. "I have succeeded in getting new and interesting results on nuclear disintegrations of different targets," he wrote in his annual report to new physics department chairman Henry Smyth that year. By then, however, a more ambitious research program had been approved, and it quickly overshadowed Ladenburg's progress. Just a few weeks earlier, in late April, at the spring meeting of the National Academy of Sciences (NAS) in Washington, DC, Ernest Lawrence learned that the National Research Council had just awarded a postdoctoral fellowship to Milton White, one of his

1.3. Robert Van de Graaff. Emilio Segrè Visual Archives, American Institute of Physics

most promising graduate students. White had plied his trade, conducting proton and deuteron scattering experiments, on Berkeley's eleven-inch cyclotron—the first practical accelerator of this type to produce more than 1 MeV.[41] Initially, before the NRC announced the fellowship awards, White had indicated a preference to continue his research at either Harvard or Cornell, home to the first cyclotron built outside Berkeley. He explored local opportunities as well, visiting Caltech and Stanford University, but these options lacked sufficient appeal given the preference for research using high-tension accelerators, such as the Van de Graaff generator. Lawrence discussed White's uncertain future with Condon, a longtime friend (they had first met in Göttingen in 1926) also present at the NAS meeting.[42] The details of their conversation are unknown, but the available evidence confirms that Condon, whose initial enthusiasm for Ladenburg's high-voltage research had ebbed, now envisioned a more ambitious program; he encouraged Lawrence to persuade White to come to Princeton to build a cyclotron.[43]

After the meeting in Washington, Lawrence visited Princeton and discussed Condon's proposal with Henry Smyth. In early May, Smyth set aside $5,000 to begin construction of a 3 MeV cyclotron contingent upon White's participation. He also offered an instructorship to Malcolm Henderson, who had come to Berkeley from Cambridge University in 1932 to work with Lawrence as an unpaid research assistant. Although he prioritized Henderson's ability to teach—a more pressing requirement— Smyth highlighted the encouraging prospects for nuclear physics research, which pleased Lawrence. Henderson, who had worked closely with White under Lawrence's supervision, accepted the offer on May 7.[44] Buoyed by these developments, Smyth and Lawrence pressed White to join him. Like Ladenburg, Smyth had also recently pursued low-voltage studies—in this case, analyses of the products of deuteron-deuteron collisions by way of Walker Bleakney's mass spectrometers. "These two projects represent our start in the field of nuclear physics," Smyth explained to White one week later. "Nevertheless, we feel that the advantages of really high energy bombardment are so great that we would like to go in this direction." That fall, Louis Ridenour, a promising young recruit from Caltech's high-voltage laboratory, came to Princeton on a postdoctoral fellowship to help the Berkeley duo build the cyclotron. Given his new administrative responsibilities as department chairman, however, Smyth did not expect to devote much time to the work, but he did promise White faculty support. He immediately singled out an assistant professor who had previously worked with Ladenburg. "Condon [is] also much interested," Smyth told White, "and I think would probably actually be of more assistance to you than I would."[45]

Lawrence's negotiations with Smyth at Condon's behest put the department's commitment to high-voltage nuclear research on a much firmer footing. Smyth relied on Henderson and White to establish the operating parameters for the new cyclotron, the first of the larger accelerators to be built outside Berkeley. "Plans . . . are still in a very unformed state and will be left very much to you and White," Smyth wrote Henderson in Berkeley in early June. Meanwhile, Condon had already expressed confidence in their progress. "I am terribly pleased with White [and] Henderson," he told Lawrence. "They have done a lot already to stir up the nuclear situation." White and Henderson scrapped the original voltage limit—3 MeV—that the physics department had approved on May 1. Their more ambitious design, which now relied on a supplemental cash infusion from the university's reserve research funds, called for the construction of an electromagnet capped by pole faces thirty-five inches in diameter. The magnet in Lawrence's largest cyclotron at Berkeley measured twenty-seven inches across at the poles (increased to thirty-seven inches in August 1937). Henderson expected the Princeton cyclotron to produce a maximum voltage of 10 MeV, more than twice the potential—at 4.2 MeV—that he had seen recently on the twenty-seven-inch cyclotron at Berkeley. "The cyclotron boys are doing nicely," Condon wrote George Shortley in June 1936, "and the [cyclotron] should be operating by fall so there should be some nuclear physics to do around here interpreting data and thinking of experiments by next spring."[46]

Forced to wait for locally produced high-voltage data to interpret, Condon relied in the interim on the results of experiments conducted elsewhere to further his research. He collaborated with Gregory Breit at the Institute for Advanced Study and Richard Present at Purdue University to analyze and interpret the high-voltage proton scattering data previously obtained by Milton White at Berkeley and Merle Tuve's group—Robert Van de Graaff's major rival in the use of electrostatic generator technology—at the Carnegie Institution of Washington. Their calculations based on this evidence explained the source of nuclear stability. In this case, Condon, Breit, and Present showed that the stabilizing effect of a charge-independent nuclear force acting between protons in atomic nuclei offset the Coulomb electrostatic force pushing them apart. Condon and Breit presented the preliminary results of their research at the spring 1936

1.4. Annual summer symposium on theoretical physics, University of Michigan, 1936. Condon is standing in the front row, fifth from the right. Cornell physicist Hans Bethe, who had fled Nazi Germany in 1933, and Gregory Breit (*wearing glasses and jacket*) are standing next to him. Ernest Lawrence is seated on the left in the front row. Isidor Rabi is standing next to him. Edward U. Condon Papers, American Philosophical Society

meeting of the American Physical Society in Washington, DC.[47] Condon's work along this line got an additional boost later that year at the annual summer symposium on theoretical physics at the University of Michigan. Collaboration with Benedict Cassen, a Caltech-trained nuclear physicist and former NRC postdoctoral student in mathematics whom Condon had known at Princeton, culminated in the joint authorship of a paper on "the various types of exchange forces that are being used in current discussions of nuclear structure," which the *Physical Review* editorial office received on August 10, nearly two weeks into the symposium.[48]

Almost to the day, Ernest Lawrence, who also lectured with Condon in Ann Arbor that summer, watched the first particle beam exit Michigan's new cyclotron—larger (the pole faces measured forty-two inches across) but comparable in size to the one that neared completion at Princeton. "Two microamperes at two megavolts!" Condon wired Milton White during the initial startup.[49] In late October, White and Henderson coaxed the Princeton machine into operation and got 4.5 MeV. "Grand News. Heartiest Congratulations," Lawrence replied elatedly to their announcement. Within a few days, they had increased the voltage to 5.5 MeV and doubled the beam current to one microampere. Further refinements to the cyclotron continued into 1937, at which time White's NRC fellowship expired and he joined Henderson and Ridenour on the physics faculty at Princeton as a full-time instructor. By mid-January, White had managed to increase the current for 5.5 MeV alpha particles to 2.6 microamperes. Despite spreading of the beam, a recurring problem, he confirmed progress on the experimental program. "I don't believe we will do any more about the beam for a while," White wrote Lawrence in March, "since there are many things that can be done in nuclear physics as it stands."[50]

While White, Henderson, and Ridenour prepared to turn the cyclotron loose on problems of nuclear structure, Condon returned to industry temporarily to pursue a wholly different line of study that also provided an opportunity to earn some extra income during the summer. He agreed to consult for six weeks as a "special investigator" at the New Jersey Zinc Company's central research laboratory in Palmerton, Pennsylvania, sixty miles north of Philadelphia. How Condon obtained the appointment is not known, but the company, which owned major interests in ore extraction and zinc production, hired him to broaden the scope of in-house R&D. He worked in the laboratory's metallurgical department, where, *India Rubber World* reported in a 1935 feature article, "the varied equipment and metallurgists . . . are addressed mainly to ore reduction processes and studies of metals and their alloys." He diversified this research tradition, "tutoring one of the metallurgists in theoretical physics so he can try to read papers on [the] quantum mech[anics] of metals." Condon also relished the opportunity to acquire a working knowledge of the company's metallurgical problems. "My work here with the metallurgists . . . is proving extremely interesting," he wrote in late July, after just one week on the job. "I am . . . learning a great deal about the problems of theory of metals . . . [and] am working especially on problems of plastic flow in single crystals and talking with the [metallurgists] about everything else."[51]

The location and content of Condon's research had changed, not the motivation behind it. In his nuclear physics investigations at Princeton, Condon used theory to make sense of new experimental data. Studies of zinc materials in Palmerton proceeded in the same fashion but with the added expectation that they also had to serve practical purposes. Condon demonstrated the continuity of his research style—in this case, the ability to converse in the languages of physics and metallurgy—in a research report that analyzed, by way of a "Theoretical Discussion," the results of a three-week investigation of heat flow in zinc slabs. "The following notes," he wrote in the report's introduction, "represent the development of a certain point of view regarding heat flow in [zinc] briquetes [*sic*] undergoing reduction which may be of use in discussing data obtained experimentally."[52]

Condon's brief appointment at the New Jersey Zinc Company rekindled an interest in industrial research that had lain dormant for nearly a decade. On the other side of the state—in Pittsburgh—the Westinghouse Electric and Manufacturing Company had watched closely for more than two years as physicists at Berkeley, Princeton, MIT, and elsewhere built cyclotrons, Van de Graaff generators, and other types of accelerators for high-voltage research. Initially, nuclear physics drew a measured response from industry. Even among the large electrical equipment manufacturers that stood out as the most likely candidates to invest in this type of research, interest remained tepid beyond satisfying specific market demands—for example, the development at General Electric of x-ray equipment for medical diagnostics and therapy. Only Westinghouse broke ranks and upended this pattern of behavior. When Condon moved to Palmerton in the summer of 1937, Westinghouse had already begun to modify its once-dominant engineering-based research tradition to include a new R&D strategy that publicly extolled the commercial value of nuclear physics and other seemingly esoteric scientific disciplines on a scale unmatched in the electrical industry. This radical transformation and the diverse research opportunities and technological breakthroughs expected from it—all discussed in the next chapter—set the stage for Condon's arrival later that year.

CHAPTER 2

Science in the Steel City

AT THE NEW JERSEY Zinc Company, Edward Condon briefly glimpsed patterns of R&D behavior that persisted on a much larger scale among manufacturing firms located in Pittsburgh. In this smoke-filled city, where capital-intensive materials processing industries dominated the local economy, corporate laboratories typically shunned scientific investigations that strayed too far from the technological requirements of established product lines. A similar outlook structured the content and scope of R&D at Westinghouse until 1935, when nuclear physics debuted as a legitimate field of research to be exploited. The decision made the following year to build a ten-million-volt pressurized electrostatic generator—a much larger variant of the accelerator Robert Van de Graaff had recently introduced at Princeton—turned this local R&D tradition on its head. It also put Westinghouse ahead of rival General Electric as the standard-bearer of cutting-edge research in the electrical industry.

The only sign of nuclear technology in Pittsburgh in 1936 could be found at the United States Steel (US Steel) Corporation's sprawling steelworks in the nearby town of Homestead, situated along the banks of the Monongahela River a few miles east of the city center. In February, Milton

White visited Homestead to inspect the forty-two-ton electromagnet just built for the Princeton cyclotron.[1] Given the absence of sufficient homegrown expertise in nuclear physics, even among Pittsburgh's major educational institutions, R&D managers at Westinghouse solicited the advice and services of academic specialists farther afield, such as White and his colleagues at Princeton.[2] This strategy delegated programmatic decisions and assessments of commercial applications to outside experts who prioritized from the outset the academic quality of the company's research investment. Later on, when Westinghouse expanded the nuclear physics program into a more elaborate scheme aimed at broadening the scope of R&D even further, negotiations between company representatives and prospective academic recruits foundered on similarly mismatched expectations. These patterns of behavior are significant, because they determined some of the technological priorities that Condon inherited late in 1937 and subsequently molded into a more coherent vision for science-based innovation at Westinghouse.

R&D IN THE WORLD'S WORKSHOP

The quiet and idyllic lifestyle to which Condon and his Princeton colleagues had grown accustomed in suburban New Jersey contrasted sharply with the bustle of industrial Pittsburgh, a noisy, dingy city whose steel mills, foundries, and machine shops had prompted one visitor in December 1866 to call it "Hell with the lid taken off." The local economy had long thrived on the region's navigable rivers and railroad lines, access to plentiful natural resources, and the vertically integrated manufacturing firms that turned these raw materials into finished goods. Metals production led the way. By the late 1920s, on the eve of the Great Depression, factories operating within a thirty-mile radius of Pittsburgh shipped one-quarter of all the iron and steel produced in the United States.[3]

Homegrown companies became national giants. The Jones and Laughlin Steel Corporation, founded in Pittsburgh and the third-largest domestic steelmaker in 1929, had built all of its production capacity in the region. Other local firms had similar lineages. The Aluminum Company of America (Alcoa), the Gulf Oil Corporation, the Mesta Machine Company, the United Engineering and Foundry Company, and the Pittsburgh Plate Glass Company—to name just a few—controlled large, and in some cases the largest, segments of their respective industries, even during the economically depressed 1930s.[4] So did Westinghouse Electric. Founded in 1886 by George Westinghouse—inventor of the railroad air brake and pioneer of alternating current electric power transmission—the company grew into the second-biggest manufacturer of electrical equipment in the United States, behind rival General Electric. Manufacturing firms founded elsewhere also maintained a significant presence in the region. The United States Steel Corporation, heir to Andrew Carnegie's industrial empire and the world's largest steelmaker, with headquarters in New York City, operated production facilities—equivalent to 35 percent of the company's entire steel ingot capacity in 1936—at locations in and around Pittsburgh.[5]

Among these large firms and many others located in the region, laboratory investigations prioritized product development and often incorporated concepts and practices drawn from the scientific and engineering disciplines. Research at the big integrated steelmakers, specialty metals and glass producers, and machine tool and mining companies consisted of process and quality control, materials testing, and physical, chemical, and metallurgical analysis of steels and alloys, typically conducted in laboratories sited at major factory locations.[6] Some of these local commodity producers—US Steel, Pittsburgh Plate Glass, and Alcoa, for example—and also firms in the electrical industry, such as Westinghouse, established separate, independently located research departments staffed with professional scientists and engineers to support product development in the factory laboratories.[7] Still other companies tapped local R&D expertise through extramural contracts, called industrial fellowships, at the Mellon Institute of Industrial Research, founded at the University of Pittsburgh in 1911 and bankrolled by local industrialists Andrew Mellon and Richard Mellon. Fellowships typically focused on classes of materials—including food and paper products, textiles, and glass—and manufacturing processes, such as petroleum refining, chemical extraction from coke by-products, and gas purification.[8] In some cases, they evolved into new R&D programs at the sponsoring firms. In 1929, Gulf Oil—the world's third-biggest oil producer and the second largest in the United States behind the Standard Oil Company of New Jersey—spun off part of its petroleum refining fellowship, one of the earliest established at the Mellon Institute, into a separate laboratory near the University of Pittsburgh dedicated to petroleum production problems. Six years later, the company moved the laboratory to a new and much larger R&D campus located on the city's outskirts.[9]

Given the dominance of large process industries in Pittsburgh, local corporate R&D managers more often than not extolled the value of engineering experience in the laboratory. "It has been my experience that we secure the most desirable results from men who have been given a preliminary engineering training," wrote Samuel Kintner, director of research at Westinghouse Electric, in the *Journal of Engineering Education* in 1929. "Such men seem to have a better vision and a truer appreciation of what they are undertaking to do." A visitor to the company's research laboratory that year confirmed Kintner's preference; he noted that "all research seems to be of [the] engineering type." Even PhD scientists employed at the laboratory quickly lost their academic gloss. "Though Ph.D.'s and the badges of scientific fraternities are . . . common," Kintner had acknowledged two years earlier, in 1927, "they are not at all the vague and whiskered type sometimes thought of as scientists."[10]

Kintner's training in electrical engineering and many years of practical experience—he had worked in the industry, mostly at Westinghouse, since 1903—no doubt informed his approach to industrial research.[11] PhD physicists who occupied research directorships at other large, technologically diversified manufacturing firms nearby shared the same outlook. It figured prominently in the management style of Lars Grondahl, a Johns Hopkins PhD who had taught physics at the nearby Carnegie Institute of Technology. He resigned in 1920 to become director of research at the Union Switch and Signal Company, founded by George Westinghouse in 1881 and the largest producer of railroad signaling and traffic control equipment in the United States. "I like to get a man who has had a good thorough training in electrical engineering with graduate work in physics toward the Ph.D. . . . [someone] who has had the practical point of view," Grondahl declared to a gathering of academic and industrial physicists that the American Institute of Physics (AIP) convened at the University of Pittsburgh in mid-November 1935. Grondahl relied on this practical perspective to guide the work under way in the company's research department. "It is often said that industry needs more fundamental scientific principles in order to solve its problems," Grondahl had written in *Science* six years earlier, in 1929. "This is undoubtedly true and everyone will agree that the more thoroughly we know nature the better. However, I think it is also true that the science that is already known has only begun to be exploited."[12]

Austin Cravath, a Berkeley-trained PhD just off an NRC postdoctoral fellowship at Princeton and one of three new physicists added to the research staff in 1930, had no formal training in engineering, but his working knowledge of vacuum tube electronics, an established field directly relevant to Union Switch and Signal's product lines, likely appealed to Grondahl's sensibilities. "I don't know just what my work will be," Cravath wrote Leonard Loeb, his graduate advisor, in May, shortly after moving to Pittsburgh, "but it will be something in the electrical field. . . . It won't be pure science." In this corporate environment, where engineers occupied more than half of the research positions in the department in 1929, investigations at the leading edge of academic science—in quantum theoretical physics, for example—lacked institutional legitimacy as a source of technological innovation. "There are no quantum [physicists] among us," Cravath told Loeb in January 1932.[13] A similar scenario played out in the new research laboratory Gulf Oil established in 1929 to study problems of petroleum production. Although it employed first-rate scientists schooled in relevant disciplines such as geophysics, the laboratory did not indulge their purely academic interests. Research remained tightly focused on commercial objectives. "There is simply no future here for a theoretical physicist," Morris Muskat wrote his PhD advisor at Caltech in November 1930, after less than two years on the Gulf payroll. Muskat enjoyed the job—"the work of the laboratory is quite interesting as well as varied"—but he did not want to remain there permanently. "I have become familiar with a number of problems of classical theoretical physics," he explained. "However," he added, "I have been . . . falling behind in *modern* theoretical physics [original emphasis]. I am afraid unless I return soon to the academic field I may never be able to 'catch up.'"[14]

Cravath's and Muskat's experiences notwithstanding, industrial laboratories did occasionally produce significant scientific breakthroughs. Irving Langmuir at General Electric and Clinton Davisson at the Bell Telephone Laboratories both won Nobel Prizes, for chemistry and physics, in 1932 and 1937, respectively. Their contributions to science, however, originated as the by-products rather than the intended consequences of corporate research programs that—like their counterparts in Pittsburgh—focused on the solution of technical problems in electric lighting and long-distance telephony.[15] Shortly after Davisson received the Nobel Prize, the Bell Telephone Laboratories began to absorb, on a limited basis at first, the new quantum mechanics to help reconcile contradictory theories of solid, especially metallic, behavior. This diversification of research added to the company's existing knowledge base a fundamental understanding of the materials used in vacuum tube amplifiers and other core technologies required for long-distance communication. Despite their outwardly academic flavor, such studies supported a vital corporate function—improvement of the telephone system.[16]

General Electric had followed a similar strategy. The company expressed limited enthusiasm at the outset for a headlong rush into cutting-edge, academic-style research. Three veteran GE scientists—Langmuir, Saul Dushman, a specialist in high-vacuum research, and Albert Hull, an electronics pioneer who, like Dushman, also served as assistant director of the company's central research laboratory—questioned the industrial utility of what their contemporaries called modern physics during a visit to the MIT physics department in the spring of 1935. A weak economy and a tight academic labor market had left MIT unable to place many of its physics graduates, an outcome that drew attention to job opportunities in industry. Put off by students "being taught too much wave mechanics and nuclear structure," Langmuir, Dushman, and Hull favored more training in established subject areas. "What they want," wrote one observer in the department, "is a knowledge of classical physics plus an appreciation of the problems that interest industry. They say the physicists turn up their noses at such things."[17]

Hull, Dushman, and their contemporaries from other firms—including Lars Grondahl from Union Switch and Signal and Westinghouse's Samuel Kintner—converged on the University of Pittsburgh in November 1935 to discuss the practical uses of physics in industry with a larger audience of corporate R&D managers. They gathered at the inaugural meeting of the AIP advisory council on applied physics, held the day after the university convened a separately planned but co-located conference on industrial physics that drew more than four hundred participants.[18] This follow-on gathering of corporate luminaries marked the culmination of a scheme the American Institute of Physics had hatched in 1934, just three years after its founding, to facilitate as broadly as possible the application of physics research to industrial problems through new publication outlets—such as the *Journal of Applied Physics*—joint conferences, and educational reforms in the universities. The council's mandate acknowledged a widening professional schism between practically minded industrial physicists, whose numbers had swelled since World War I, and their academic counterparts.[19]

"I have often emphasized," declared Gulf Oil research director Paul Foote during a council discussion about physics instruction in the universities, "the importance of more training in mechanics, hydrodynamics, electricity and magnetism, elasticity, and other subjects of classical theoretical physics." Elaborating further, he complained that "not enough of such courses are required, and with the elective system these classical subjects are often subordinated for the newer developments of modern physics." Foote had already institutionalized the thinking behind these pedagogical preferences in the Gulf research organization. At the company's new research laboratory five years earlier, Morris Muskat had singled out the importance of investigations in classical rather than modern theoretical physics. Saul Dushman and Albert Hull, still fresh off their visit to the MIT physics department a few months earlier, agreed with Foote. "It seems to me highly desirable," Dushman asserted in a prepared statement (he did not attend the council meeting), "that the physicists who intend to go into industrial organizations should receive some training of a more 'practical' nature." Hull specifically highlighted the value of engineering: "The physicist who, having finished his training, decides to go into industrial physics but does not find an opportunity at once should not take more physics but should take engineering. . . . Physicists who have failed recently to find jobs should have done that."[20] At a December 1934 meeting of the American Institute of Physics and the NRC's Division of Physical Sciences, Oliver Buckley, director of research at the Bell Telephone Laboratories and a member of the new AIP advisory council, had called out training in classical physics and "an aggressively practical mind" as essential prerequisites for a productive career in industry. So did the chief physicist of the Eastman Kodak Company.[21]

Approved shortly before the AIP advisory council convened in Pittsburgh, the nuclear physics program at Westinghouse directly challenged these prevailing corporate attitudes about the practical utility of cutting-edge academic research. The headlong rush into this new discipline turned an otherwise historically productive R&D tradition in Pittsburgh on its head. Nuclear physics debuted in East Pittsburgh, a working-class community across the river from Homestead and the site of Westinghouse's largest manufacturing facility. Established in the engineering department in 1906, the research division initially comprised laboratories for materials testing and development and the control of manufacturing processes. Headed by Charles Skinner, an accomplished electrical engineer who had joined the company in 1890, the research division added a new laboratory—directed by a PhD physicist recruited from Eastman Kodak—in the nearby borough of Forest Hills in 1916. The laboratory initially produced scientific knowledge to support the factory laboratories, but this more diversified research program quickly ran its course. In 1920, management split the factory laboratories from the research division and consolidated them in a new materials and process engineering department. Only the laboratory in Forest Hills remained within the newly downsized and renamed research

2.1. Lewis Chubb. Smithsonian Institution Archives, Image No. SIA2008-0099

department. A deliberate shift in priorities back to more practical aims that Skinner favored—the development of radio equipment—prompted the director of the laboratory to resign the following year.

When Charles Skinner stepped down in 1922 to become assistant director of engineering, Samuel Kintner, who had worked with him previously, took over as director of the research department and the laboratory in Forest Hills. That outcome reinforced management's preference for engineers rather than scientists to manage in-house R&D.[22] Kintner himself acknowledged in 1927 that "much of the work of the laboratories is not spectacular, though tending to make Westinghouse output more consistent and satisfactory, improving performance, and reducing costs." This research ethos persisted into the 1930s. Kintner's promotion to vice president of engineering in 1930 left the research department without a director. Lewis Chubb, an engineering graduate of Ohio State University and veteran of the research department who had joined the company in 1905, replaced Kintner and served as head of the newly named Westinghouse Research Laboratories until 1948.[23]

NUCLEAR PHYSICS

The persistence of an engineering-based research tradition at Westinghouse Electric did not preclude studies in scientific fields related to the company's core expertise but without clear and immediate practical objectives. Some high-voltage research of this type occurred fleetingly prior to 1935. In 1927, Joseph Slepian, a Harvard-trained mathematician with a practical bent—he pioneered the development of lightning arresters, circuit breakers, and mercury-arc rectifiers at East Pittsburgh—obtained a patent for an x-ray tube he designed to accelerate electrons by means of magnetic induction.[24] Isolated instances of high-voltage research like this one at East Pittsburgh set the stage for the more deliberate, large-scale investment in nuclear physics that followed after 1935. These early initiatives did not, however, proceed far enough to establish an in-house technical capability sufficient to execute the enlarged program. The company turned instead to the academic physics community for much of the requisite expertise.

Nuclear physics research originated in the company's patent department at East Pittsburgh. R. P. Jackson, an engineer who had followed developments in the field, drafted a preliminary report in May 1935 comparing Ernest Lawrence's cyclotron to the induction accelerator Joseph Slepian patented in 1927. Engineering vice president Samuel Kintner instructed research director Lewis Chubb to study Jackson's findings and discuss with Slepian "what, if anything, should be done in following up his [Slepian's] earlier work." Elaborating further, Kintner told Chubb that "if, in your judgment, there is a real field for this and it will be to our disadvantage if we don't get into it promptly, a statement should be made to the management so that we will not be accused later of passing up something of real importance." Chubb, who had concluded that "the Lawrence scheme . . . is considerably better than [Slepian's]," stopped short of requesting funds to establish a separate research program. "When we are in a position to do fundamental work without seeking any direct application," he wrote Kintner, "I believe we should consider work of this kind. . . . At present there is nothing of direct interest to us. Possibly in five or ten years chemical processes, medical treatments, etc., will use high velocity particles and there should be a market for electrical equipment to practice these processes."[25]

Financial troubles also limited access to resources and manpower. The Depression had already hit some of Pittsburgh's industrial laboratories especially hard. In October 1932, Paul Foote confirmed a hiring freeze at Gulf Oil's four-year-old research laboratory near the University of Pittsburgh. Also in a move to cut costs and preserve cash, Union Switch and Signal had dismissed 40 percent of its research staff the previous year. Those still employed endured a 20 percent cut in pay and work hours. Then in May 1932, the company executed a fresh round of layoffs that further reduced the staff of the research department by another 40 percent. A similar retrenchment played out at Westinghouse. The company posted record earnings of $27 million in 1929, but a sharp decline in sales had produced cumulative losses in excess of $21 million by 1933. Management cut factory production, slashed the labor force, and reduced expenditures for research and development. The number of salaried research personnel (engineers and scientists) employed at East Pittsburgh dropped nearly 50 percent in the twelve months to June 1934. "Conditions are worse than ever in the Pittsburgh district," Gulf's Paul Foote told AIP director Henry Barton later that year. Although Kintner had offered to obtain funding for nuclear physics research on his behalf, Chubb could not justify the expense. "Until we can see something," Chubb concluded, "we are hardly justified and do not have the man power to work in this pioneering field."[26]

Despite the bleak financial picture Chubb disclosed to Kintner, Jackson forged ahead. He visited Berkeley that summer. Jackson did not have the opportunity to observe the twenty-seven-inch cyclotron accelerate deuterium nuclei to the high velocities required to produce radioisotopes, but he interviewed Ernest Lawrence at length "and got the drift of how it operates." In his trip report, Jackson also confirmed Lawrence's expectations for practical applications in medicine. Harvey Rentschler, director of research at the Westinghouse Lamp Company (a wholly owned subsidiary located in Bloomfield, New Jersey), enthusiastically embraced Jackson's findings. "This work on atomic nuclei is taking the field of physics by storm," he wrote O. H. Eschholz, manager of the patent department and Jackson's boss. "I cannot help but feel that some commercial applications must come out of investigations of this character. Of course the application of induced radio-activity in hospitals, in itself, is an illustration of the correctness of my belief." Elaborating further, Rentschler told Eschholz about a potentially useful alternative to the cyclotron. At the University of Rome in 1934, Enrico Fermi had shown that neutrons emitted from a combination of naturally occurring radon and beryllium, and slowed on passage through paraffin, transmuted a wide range of stable elements into radioactive isotopes. The prospect of commercial applications prompted Fermi to obtain patent protection, first in Europe. Rentschler learned from Benedict Cassen, Condon's colleague from Princeton now on the research staff at Westinghouse's new x-ray subsidiary, that Fermi "was willing to dispose of his [patent] rights in this country under a very reasonable arrangement provided he was assured that the method was being developed."[27]

Cassen, meanwhile, put forward a separate proposal to develop a wholly different type of neutron source more robust than Fermi's low-yield radon-beryllium combination and less technologically complex than the high-voltage accelerator Lawrence had perfected. Cassen discussed the technical details with Rentschler, who advised Eschholz that "a satisfactory supply of neutrons could be produced using a discharge tube which required only approximately 200,000 volts instead of several million volts or its equivalent as Lawrence uses." Jackson had reached the same conclusion about the productive limits of Fermi's method, but he still favored the cyclotron. "If artificial radioactivity is to come into use," he wrote Lawrence after returning to Pittsburgh from Berkeley in August, "it must be produced by power rather than from known radioactive substances. The scheme you have employed appears to be the most effective that has so far appeared, and I have been trying to get our people interested in the subject."[28]

Dayton Ulrey, manager of the physics division at the research laboratories in East Pittsburgh, acknowledged the potential for practical applications "in the medical field," but he assumed that "the X-Ray Co[mpany] will follow up this line." Although he stood out as the most qualified internal candidate to direct such an effort, Cassen had just resigned from the Westinghouse X-Ray Company, a manufacturer of medical diagnostic equipment, to accept a research position elsewhere.[29] In September, nearly one month after Cassen's departure, Chubb and Rentschler drafted and submitted to Kintner a proposal for nuclear physics research. The opportunity to exploit Fermi's research added a financial incentive to the discussion that had not existed when Chubb and Kintner first considered options for a new program back in May. "We have looked over the patent application . . . and it is our feeling that this field of research is basic in nature and may result in apparatus and products which will be of great commercial importance in the electrical, chemical, and medical fields," Chubb and Rentschler wrote. "Without the patent rights," they explained, "our opinion is as in the past, that we are not justified in doing more in nuclear physics than possibly keeping up with the literature, the possible applications, and probably part-time experimental work of one man. With patent rights, we feel that the 1936 budgets should include an activity of at least two men—one to work at Bloomfield, particularly on neutron production, and one at [East] Pittsburgh on general nuclear physics work." In mid-October, one week after Fermi's business agent in New York City filed the application for a US patent, Kintner, in consultation with Jackson, Eschholz, and Westinghouse Electric president Frank Merrick, approved Chubb and Rentschler's recommendation to support the work of two physicists: one to develop a neutron generator at Bloomfield along the lines Cassen suggested and the other to initiate at East Pittsburgh the production of neutron-activated radioisotopes "somewhat similarly to the method now employed by Lawrence" but based on a modified version of Joseph Slepian's induction accelerator.[30]

While the patent department at East Pittsburgh continued to investigate the legal basis of Fermi's claims, Charles Slack, a veteran electronics expert at Bloomfield, began preparations to build a neutron generator "capable of producing radioactivity in usable quantities." Slack, who had obtained a doctorate in physics from Columbia, expressed a genuine interest in nuclear physics, but he did not share Harvey Rentschler's optimistic predictions for commercial applications. "It is a very interesting field," Slack wrote Princeton's Rudolf Ladenburg, whose assistance he had sought to get started, in mid-November, "and we hope to do some work in it even though the commercial possibilities are extremely remote." By then, Chubb's sanguine assessment of Fermi's patent options had run its course. "My understanding from the Patent Department," Chubb wrote, "is that Fermi disclosed his work in the literature prior to its being invented in America . . . and that this precludes any possibility of his getting a valid patent in America." Without patent protection, Eschholz concluded, Westinghouse had nothing to gain commercially from an investment in Fermi's research. Kintner agreed. Negotiations between Fermi's agent and Westinghouse ended in January 1936 along with preliminary plans to expand Charles Slack's research program at Bloomfield. "The likelihood that we will have any funds available for development along the[se] lines . . . is extremely remote," Eschholz wrote. "What funds may be allocated, it is our intention to apply to working out further [at East Pittsburgh] the principle covered by the Slepian patent."[31]

In October 1935, after Kintner approved his research proposal, Chubb instructed Dayton Ulrey to contact academic physics departments directly "to see what we might get in the way of a good research man for the Nuclear Physics work." Ulrey reached out to Princeton, Caltech, Columbia, Berkeley, MIT, and the Carnegie Institution of Washington, all established or newly emerging centers for high-voltage research. Despite its prominent position in an industry whose electrical products nuclear physicists now used to build accelerators, Westinghouse seemed an unlikely candidate to join this elite group. Even archrival General Electric got a warmer reception. When he visited Berkeley to see Lawrence's twenty-seven-inch cyclotron for the first time, in the summer of 1935, R. P. Jackson learned that GE had initiated a small nuclear physics program and planned "to take up the subject actively." Lawrence's enthusiasm for GE's interest in the field—he provided consulting advice to the company—did not extend to Westinghouse. To the contrary, his reaction to Jackson's inquiry combined surprise and a tinge of condescension. In the notes he wrote immediately after their first meeting, Jackson recalled that "Lawrence was under the impression that our research work was pretty well abandoned and that we could not very well be interested in anything of the kind [that] the [General Electric Company] would be."[32]

This attitude toward Westinghouse as a second-rate latecomer to nuclear physics colored Lawrence's response to Ulrey's inquiry about potential job applicants. "It is a little difficult for me to suggest a suitable man," Lawrence wrote Ulrey in early November, "without more detailed information as to the extent of the nuclear program you are undertaking. There are several men in our laboratory that might be interested in joining your group, although at the present time there are open to these men attractive university positions." Ulrey's obfuscation, however, had been deliberate. "Our plans for carrying on investigations in the Nuclear Physics field," he responded later that month, "are not very definite and we have purposely kept them indefinite until we select a specialist whose judgment we wish to take into consideration in mapping out our program." Referencing Kintner's preference for research "along the general lines which you [Lawrence] have developed" as one possible alternative, Ulrey did not exclude other options: "There are, of course, a number of lines of attack and we realize that each of these will probably require considerable fundamental research unless we merely copy what others have done."[33]

Milton White, whom Condon had handpicked to build the cyclotron at Princeton, answered Ulrey's call, but he expressed at the outset only marginal interest in an industrial career at East Pittsburgh. "Right now I am not at all interested in going there, unless conditions and salary were really attractive," he wrote Lawrence on November 1. White also requested from Ulrey "some assurance that nuclear research would be supported for at least five years." Inchoate plans and the absence of a firm budget slowed Ulrey's progress.[34] The delay also prompted Louis Ridenour, White's collaborator on the Princeton cyclotron, to withdraw his candidacy, despite a promising visit to East Pittsburgh in December. White's reservations about job security and Ridenour's fading interest as the negotiations languished pointed to a broader criticism of the company's plans that crippled the entire search. "Practically all of the prospective candidates and some of the authorities contacted for recommendations," Ulrey complained to Chubb two days before Christmas, "apparently have some anxiety as to the permanence of a position at Westinghouse." Chubb confirmed the absence of concrete research objectives in February 1936, nearly four months after Ulrey had started to canvass universities for prospective candidates. "We are still somewhat undecided just what we should do in the way of Nuclear Physics here," he wrote a company consultant, "and so far, have not made any real attack on the problem, as we have not decided upon a man for the work."[35]

The logjam broke in June when Ulrey recruited William Wells from the University of Minnesota. Wells's appointment prompted a major reorientation of the nuclear physics program at East Pittsburgh. Early interest in the cyclotron all but disappeared in favor of a rival and operationally distinct accelerator—Robert Van de Graaff's electrostatic generator. Wells had cut his professional teeth on this technology under the direction of

2.2. William Wells. Emilio Segrè Visual Archives, American Institute of Physics

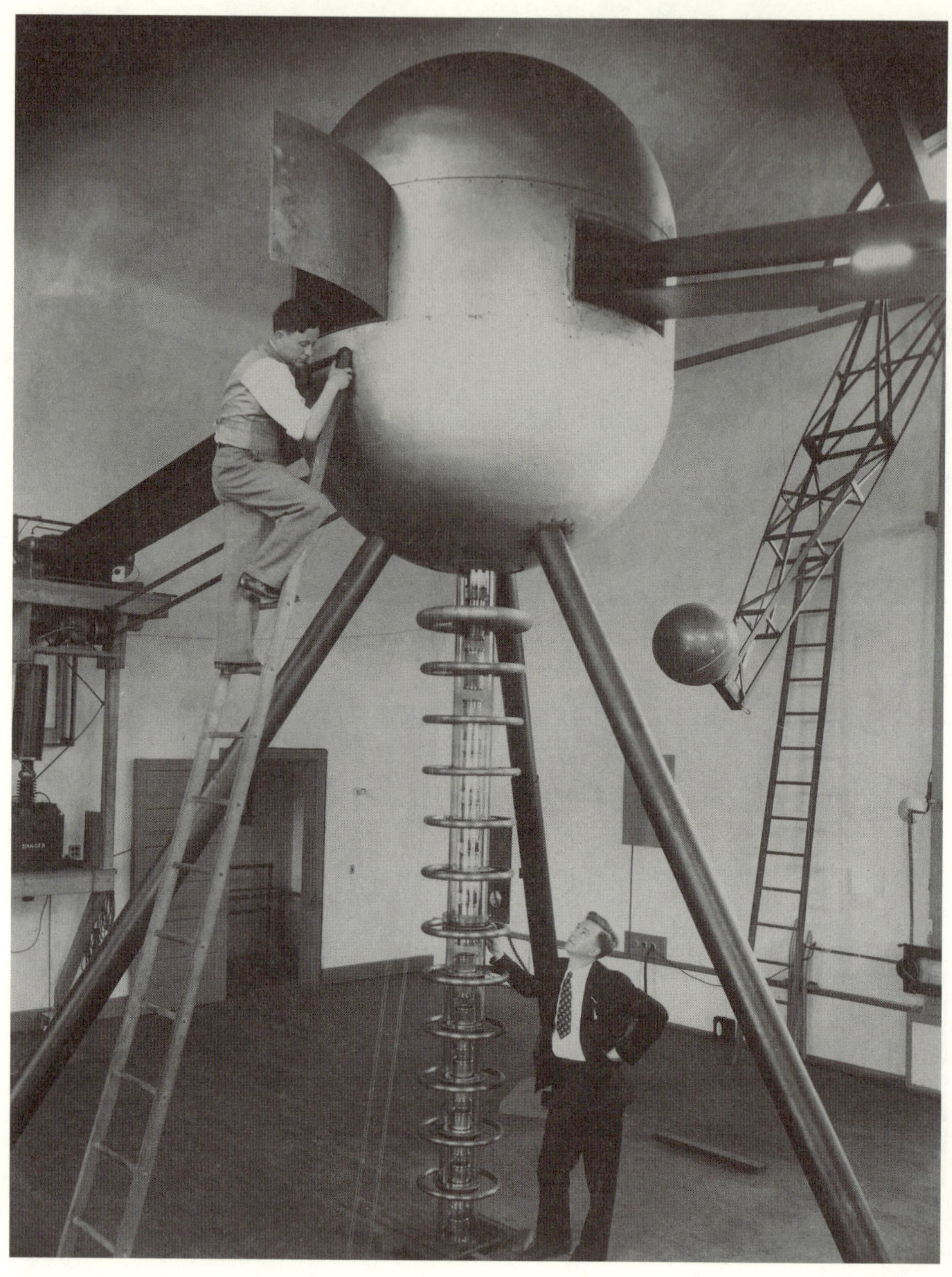

2.3. Department of Terrestrial Magnetism, 1935. Merle Tuve stands next to the accelerating tube of the 1.2 MeV Van de Graaff generator that William Wells used for his proton scattering research. Carnegie Institution of Science

Merle Tuve in the Department of Terrestrial Magnetism (DTM) at the Carnegie Institution of Washington. He performed a series of proton scattering experiments on Tuve's 1.2 MeV generator, and they provided the data for his PhD thesis, completed at Johns Hopkins University in 1934.[36] That summer, the physics department at the University of Minnesota, flush with funds from the Rockefeller Foundation, hired Wells to initiate a high-voltage research program built around the same technology.[37] Unlike White and Ridenour, on whose behalf Henry Smyth quietly labored to keep at Princeton during the negotiations with Westinghouse, Wells faced an uncertain future by way of the impending expiration of the Rockefeller Foundation grant at the end of the 1936–37 academic year. Westinghouse offered more job security, even if the long-term outlook for the research program at East Pittsburgh remained unresolved. It certainly helped to have the confidence of Merle Tuve and DTM director John Fleming, both of whom gave Wells high marks in response to Ulrey's initial inquiry for job candidates. "We regard . . . Wells . . . as the best fitted of the available candidates, most of whom are known personally to us," Fleming had written Ulrey in October 1935.[38]

Shortly after he accepted Ulrey's job offer, Wells attended a symposium on nuclear physics at Cornell University to survey the state of the field and evaluate the most promising accelerator designs. After the symposium, Wells visited Columbia and Princeton "to further determine the stage of development of their cyclotrons." Later that month, he called on Charles Slack at Bloomfield to learn the details of his neutron generator research and also visited the Carnegie Institution of Washington to obtain an update on the DTM's high-voltage research program, especially Merle Tuve's ambitious plan, proposed the previous year, to build a 20 MeV pressurized electrostatic generator (scaled down and reduced to 10 MeV just before Wells arrived).[39] In the trip report he initially wrote for Kintner but subsequently submitted to Chubb in his absence (due to illness), Wells reviewed and compared the major accelerator technologies then in use. "Of the various methods . . . for the production of high energy particles," he wrote, "only two, the cyclotron and the pressure electrostatic machine, seem at the present time to be indicated for use here [at Westinghouse]. I recommend that we aim towards 10 M.V. [million-volt] deuterons . . . as it seems that to even stay in the pond of nuclear physics let alone swim around that such a voltage is to be aimed at."[40]

Wells preferred the electrostatic generator, which is not surprising given his training and experience. It also manifested certain operational advantages that the cyclotron could not match, namely, the ability to mete out reliable quantitative data based on precise measurements of nuclear phenomena, a functional attribute that Wells identified "as one of the fundamental necessities in the design of nuclear disintegration apparatus." He expected practical applications to follow naturally from the discovery of first principles. "It is only by an extrapolation of our past experience that new fields of knowledge are in general always utilized in very practical ways sooner or later, that one can predict with some surety that there will be commercial applications of nuclear physics in the near future. Personally, I believe that there will be as we have already some indications to that effect." At the end of his report, Wells highlighted the commercial prospects for chemical and medical research, but he cautioned in the case of the latter that "the major question . . . is that of determining whether the M.D.s [medical doctors] can think of successful uses [of radioisotopes] and they in turn must wait until they have large enough quantities." At Berkeley, the cyclotron had already demonstrated its superiority over naturally occurring radioelements as a reliable, cost-effective source of appreciable quantities of radioisotopes, an advantage that prompted Lawrence's first sponsor—the Research Corporation—to patent accelerator improvements and test the viability of a commercial market for medical and biological applications.[41]

In early August 1936, one month after Wells had completed his fact-finding mission, Chubb invited Harvey Rentschler and Charles Slack to East Pittsburgh to finalize planning for the nuclear physics program. Having abandoned the option to acquire or license Fermi's patent rights, which had not yet been granted in the United States, Chubb informed Rentschler of the decision just made "to go ahead on Nuclear Physics with the idea of doing advance work as a speculation and also to obtain certain prestige and publicity. . . . It is going to be an expensive development and one which should be carefully considered from all angles before spending the money." The meeting, which also included representatives from the patent department, convened on August 11. Chubb confirmed the unanimous approval of Wells's recommendation to build a pressurized Van de Graaff generator. "The use of the high-voltage direct method," Chubb wrote in a memorandum for the record, "will enable us to study matter; to gain experience of value to the Electric [Company] in high-voltage work; and [will] be equally good for developing useful applications in the medical, chemical, and power fields."[42]

Although he acknowledged that in the case of the last option—power conversion—the likelihood of obtaining practical applications remained "very speculative," Chubb nevertheless staked a claim alongside one contemporary precedent.[43] At the Massachusetts Institute of Technology four years earlier, in the summer of 1932, Robert Van de Graaff had filed a patent application for a novel system of electric power production and transmission based on the design of his 10 MeV electrostatic generator under construction at the Round Hill research station. On the lookout for new sources of income to fund research in physics and engineering, MIT president Karl Compton tried to market the technologically ambitious concept to the Tennessee Valley Authority as an alternative to conventional power transmission for rural electrification. The plan fell apart in the spring of 1935 for reasons that had less to do with the complexity of the system itself than the legal and contractual disputes between MIT and the federal government that accompanied efforts to turn it into a practical device.[44] Chubb did not mention Van de Graaff's research or Compton's plans to commercialize it in any of his surviving correspondence, but he may have known something about it from the publicity it received in the scientific literature and the news media, not to mention his familiarity—through the reconnoitering of Jackson, Ulrey, and Wells—with high-voltage investigations under way elsewhere.[45]

Potential practical applications notwithstanding, Chubb confirmed the operational versatility of the electrostatic generator vis-à-vis the cyclotron. He specifically referenced the advantages of doing "quantitative work," a requirement that Wells initially highlighted: "This arrangement gives the possibility of using protons, alpha particles, deuterons, or electrons[,] while the Cyclotron can use only positive ions and would not give us the experiences in high-voltage generation and control which we will get by the direct method." Finally, Chubb affirmed for the first time the significance of corporate publicity, an outcome not considered previously. "The publicity possibilities through work with the recommended set-up will be outstanding," he explained, dismissing in turn the only alternative to the electrostatic generator Wells, and especially R. P. Jackson before him, had recommended earlier. "The building of another Cyclotron," Chubb wrote, "would only duplicate activities at other laboratories." Chubb and his cohort at East Pittsburgh justified their decision to build an electrostatic generator—the first large-scale industrial atom smasher of its kind in the United States—on the basis of its uniqueness as a new but unexploited experimental tool for industrial R&D. Moreover, they envisioned the atom smasher—as popular accounts and company news releases later called it—

to be the new public face of Westinghouse's scientific and technological prowess, a marketing strategy already replicated elsewhere for a similar purpose. For more than a decade, General Electric had advertised for public consumption the contributions of its research laboratory, referred to in the national media as the House of Magic, and of the eminent scientists and engineers—such as Irving Langmuir and Charles Steinmetz—who worked there.[46]

Chubb traced an abiding interest in the accumulation of publicity to the spring of 1935, when planning for nuclear physics research first got under way, but the documentary record does not substantiate that claim. Instead, it captures his initial skepticism and reluctance to proceed. Publicity received no mention as an added incentive to authorize the program until Chubb called the final planning meetings in East Pittsburgh in August 1936. While it is possible that Chubb articulated himself or adopted from someone else this new programmatic role and simply did not mention it in any of his correspondence, it is more likely that the value placed on public relations as a desired end product coincided with the appointment of Marvin Smith as Samuel Kintner's replacement in the negotiations. Kintner, who had been ill, died in September. An electrical engineering graduate of the Agricultural and Mechanical College of Texas who had started working for Westinghouse in 1916, Smith cut his managerial teeth in the power engineering department at the factory in East Pittsburgh. Between 1926 and 1930, he ran the department's alternating-current (AC) generator section, then moved up to division engineer in charge of the design of all large rotating AC machinery, a position he held until Kintner died.[47] Like Kintner, Smith had no direct knowledge of nuclear physics, but he supported the program on a far more ambitious scale.

Prior to his absence from East Pittsburgh, Kintner had encouraged measured progress leading to modest goals—for example, the reduction of Joseph Slepian's induction accelerator patent to practice for radioisotope production. "This field of work [nuclear physics]," Smith wrote shortly after filling Kintner's post, "has the advantage that it is developing quickly. It is of a spectacular nature which attracts the attention of the general public, so that the publicity is valuable." Smith did not, however, deliberately relegate to a secondary role the likelihood of commercial products based on this research. To the contrary, he recognized that obtaining the necessary corporate funds to pay for the program required some rationale for practical applications. "It is our feeling," he told Ralph Kelly, the vice president in charge of the company's East Pittsburgh manufacturing operations, whose

2.4. Marvin Smith. Courtesy of Hagley Museum and Library

approval Smith needed to proceed, "that having an extremely high voltage source of this kind available will lead to investigations that will provide valuable information in other subjects such as porcelain insulators and dielectrics of various kinds." Smith's justification for support, a combination of publicity and optimistic expectations for new product innovations, likely reflected a higher tolerance for programmatic risk than Kintner had deemed prudent. At the same time, however, Smith focused on improvements to conventional technologies—insulators and dielectrics—found in the AC electric power systems Westinghouse had pioneered. His conception of applications to be derived from nuclear physics research, while more ambitious than Kintner's, still signaled an extension of, rather than a radical departure from, established product lines. This distinction matters because it highlighted the persistence of an engineering-based research tradition that managers like Smith favored as the nuclear physics program got under way.

The rapid progress in East Pittsburgh quickly overshadowed Charles Slack's neutron generator research at the lamp division in Bloomfield. Chubb had pondered its fate at the August 11 meeting and concluded that further work should focus on immediate applications, "particularly in the medical and chemical fields . . . by cooperating with some outstanding medical research organization, hospital, or clinic." Commenting on the planned disposition of Slack's research, to be completed by the end of 1936, Smith informed Ralph Kelly that "the useful field of [Slack's] apparatus is somewhat limited and cannot be used for a broad and extensive investigation of the type we believe our Research Laboratories at East Pittsburgh should enter into." This decision effectively ended the lamp division's direct involvement in the nuclear physics program.[48]

In the meantime, Wells put to rest the last remaining element of Kintner's original agenda for nuclear research—the modification of Slepian's induction accelerator into a proton source suitable for neutron-activated radioisotope production. At Chubb's request, Wells completed a detailed analysis of this option and concluded that Slepian's design lacked the flexibility to accommodate the required modifications. It did, however, show promise as a source of high-energy electrons that might "have possibilities in the X-ray field"—one of Westinghouse's established businesses—although the projected development costs and time commitment precluded further work along this line, given the substantial investment already planned for the electrostatic generator. In early September 1936, Chubb prepared a detailed construction and operating budget for the generator, which Smith and vice president Ralph Kelly approved later that month. The estimate for materials plus assembly—slightly more than $40,000—came in at twice the maximum projected cost of reducing Slepian's electron accelerator patent to laboratory practice. In mid-October, they got authorization to proceed from the company's comptroller, and work commenced shortly thereafter.[49]

The decision to build a 10 MeV electrostatic generator enclosed in a steel tank filled with compressed air or gas promised to extend the operational limits of accelerator technology, but it also exposed Wells and his staff to significant technical challenges already faced by physicists working on similarly ambitious nuclear physics projects elsewhere. The obstacles encountered at Round Hill, where Van de Graaff's huge open-air generator threw the first high-voltage sparks—"man's closest approach to the voltage of nature's lightning," declared *Science News Letter*—in late November 1933, helped set the stage for the design choices Wells made at East Pittsburgh.[50] The production of high voltages had not proved especially difficult; Van de Graaff's machine could hold 6 MeV by January 1934. That voltage, however, had to accelerate a tightly focused particle beam of sufficient current through a forty-foot vacuum tube mounted horizontally between the ion source and the target housed separately in two large fifteen-foot diameter spheres. Each sphere sat atop a twenty-eight-foot support column, one of which contained the belt-driven charging apparatus for the ion source.

Despite the rapid development of the technology in recent years, existing high-voltage tubes could not meet these stated performance requirements. That limitation indicated the extent to which Van de Graaff sought to extend the state of the art.[51] Shortcomings in the design and construction of the vacuum tube forced Van de Graaff to abandon the horizontal configuration between the spheres in favor of a simpler arrangement—a shorter tube mounted vertically inside one of the support columns—that limited usable voltages for nuclear bombardment experiments to one-quarter of the originally stated objective of 10 MeV. Wells had firsthand knowledge of the technical troubles at Round Hill as early as the summer of 1936, more than a year before Van de Graaff scrapped his original design: "Concerning the Round Hill machine, they are now constructing the accelerating tube but due to the magnitude of every step in their development it is safe to say that it will be six months or a year before they have particles under control between the two spheres." In the spring of 1937, Karl Compton confirmed Wells's findings. "The original design of the Round Hill apparatus," he wrote, "is not as well adapted for nuclear disintegration experiments as later modifications of the electrostatic generator which involve . . . a single high voltage terminal with the working end of the tube at ground, and . . . operation in a tank of compressed gas."[52]

Operational (without the vacuum tube) in the spring of 1938, the new electrostatic generator at East Pittsburgh produced 1.3 MeV at atmospheric

2.5. Robert Van de Graaff's electrostatic generator at Round Hill, near the MIT campus. Note the accelerating tube mounted horizontally between the spheres. Emilio Segrè Visual Archives, American Institute of Physics

pressure. When the pressure inside the specially designed steel tank increased to seventy-five pounds per square inch, however, the voltage nearly tripled, to 3.7 MeV. Westinghouse did not introduce the concept—academic research along this line had originated elsewhere earlier in the decade by way of smaller vessels filled with compressed gas—but the company took advantage of the recent industrial development of large spherical tanks for gas storage under pressure to scale up the technology. This push to a higher voltage range also narrowed the field of competitors, which Westinghouse had watched closely since the program's inception in 1936. Based on the field survey Wells completed that summer, Westinghouse maintained a decisive advantage over the Carnegie Institution of Washington, where Merle Tuve pressed ahead with his 10 MeV pressure installation. "Due to experience in this field," Wells wrote, "I feel they will make rapid progress if they start this but one can safely say that it will be about a year before they can hope to have such a machine in operation due to building and shop difficulties and man power available to push the work." Construction did not begin in Washington until late May 1937, at which point the atom smasher at East Pittsburgh had already produced high voltages under pressure and continued to advance toward completion.[53]

Although it carved out a place for itself at the leading edge of high-voltage research, Westinghouse avoided from the outset a public relations misstep that might undermine the justification for such a risky investment. Wells, the principal architect of the nuclear physics program, proceeded cautiously. In a direct reference to the advance publicity MIT put out that proclaimed operational benchmarks not yet achieved by Van de Graaff's high-voltage apparatus at Round Hill, Wells did not want to divulge to the public any technical details about the accelerator at East Pittsburgh until significant scientific research had been completed. "The first . . . time at which we should make a definite statement from a scientific point of view," he wrote in September 1936, "is when we have a beam of particles coming from the high voltage tube, and of such energy that we are well up among the leaders in the field." Wells also recommended that public statements about the generator's maximum voltage be limited to 5 MeV, half the value originally planned: "The problem of a vacuum tube to stand over 2.5 million volts has not been solved as yet, and although it looks straightforward, we had best not make any claims until we can make a tube to stand voltages of 5 million volts and over." He preferred to start even lower—at 3 MeV—"and advance from there as carefully as the experiments and the advance of the knowledge of nuclear physics indicates."[54]

2.6. Westinghouse Research Laboratories and the surrounding borough of Forest Hills, Pennsylvania. The atom smasher is visible on the left. The main research building (*on the right*) opened in 1916. Emilio Segrè Visual Archives, American Institute of Physics

Lewis Chubb also endorsed a modest publicity strategy at the outset, but for reasons unrelated to the likelihood of self-inflicted embarrassment that initially troubled Wells. Rather, Chubb did not want to release too many technical details that might be exploited by the company's competitors. Doing so "has the advantage of showing an advance in the solution of the problem in an early announcement from us," he wrote Smith, "but it has the disadvantage that it permits our chief competitor to go on from this point and take advantage of what has been worked out here, if they are inclined to compete in this work." Chubb's reference is surely to General Electric, but he overestimated the company's desire to stake out a competitive position. Despite its early interest, GE supported research in nuclear physics on a much smaller scale than Westinghouse. In 1933, Kenneth

Kingdon pressed GE research director William Coolidge for permission to expand his prior work on electron accelerators, which had already proceeded on a very limited basis, into a more elaborate exploration of nuclear physics. Coolidge agreed, but only if the results of his nuclear disintegration studies supported the development of x-ray tube technology, an established product line, for medical applications.[55]

Kingdon's experience illustrates the extent to which commercial priorities limited the scope of research. Saul Dushman, Coolidge's assistant director, queried Ernest Lawrence early in 1935 about potential recruits for the laboratory. "At the present time we have no openings available," he explained, "but I hope that economic conditions will improve sufficiently in the next year to enable us to add one or two more physicists to our organization." Even after the company lifted the hiring moratorium and began expanding the laboratory staff, the prospects for nuclear physics research remained bleak. "I think it very unlikely that nuclear physics will be taken up in the near future," wrote Frederick Seitz, a Princeton-trained theoretical physicist and one of the first new hires GE brought on board, in November 1937.[56]

A similar scenario played out at the Bell Telephone Laboratories. Kenneth Kingdon's research at GE had been known to Chubb and his subordinates at Westinghouse since the summer of 1935, but they never mentioned the Bell Telephone Laboratories during their preliminary discussions about nuclear physics, which suggests the absence of a corresponding program there. The program contemplated in the spring of 1940 added high-voltage research to studies of solid materials, which had already begun to incorporate concepts from quantum theoretical physics to improve the performance of critical technologies in the telephone system. Prospective research in nuclear physics fit this practical bent. "We are not interested in trying to crash the nuclear game in a big way, but rather in trying to make use of some of its techniques for our problems," wrote William Shockley, whose pioneering research spearheaded the invention of the transistor after the war. Shockley, who had contacted Tuve for technical advice, envisioned a much smaller version—in the range of 2 MeV—of the new 10 MeV pressurized electrostatic generator at the Carnegie Institution of Washington.[57] It is not known if the project got beyond the initial planning stage, but Shockley's reference to its limited scope at the outset points to a research strategy more akin to the one followed at General Electric.

EXPANDING FUNDAMENTAL RESEARCH

Early in 1937, after Westinghouse had entered into negotiations with the Chicago Bridge and Iron Works to obtain a welded steel tank to enclose the electrostatic generator for high-pressure operation, Marvin Smith, the vice president for engineering, began to map out a broader program of speculative scientific research modeled on the nuclear physics program. This ambitious turn to more open-ended studies proceeded on a scale that no other firm in the electrical industry ventured to match. Smith took advantage of the improved economic climate; sales and income had returned to pre-Depression levels. He also leveraged efficiencies that originated in a company-wide reorganization that had gotten under way earlier in the decade. Following the lead of other large, vertically integrated manufacturing firms, Westinghouse created a new decentralized management structure. In an effort to focus on long-term priorities, the headquarters staff relinquished daily administrative control of the operating divisions by way of newly formed business units built around specific product lines that operated autonomously with their own engineering (and other functional) departments. Decentralization relieved the workload in the research laboratories at East Pittsburgh and opened the door for a programmatic expansion of R&D along the lines Smith envisioned.[58]

Citing "changes in our Research organization and personnel with the idea of eliminating certain unsatisfactory conditions and obtaining more efficient and effective operation," Smith proposed—likely with input from Chubb and other R&D managers—the establishment of a separate "fundamental division" at East Pittsburgh, organizationally equivalent to the physics and other divisions that already populated the research laboratories. Smith acknowledged that the "present [research] organization is based largely on the consideration of practical or applied research." Such work focused on "finding the answer to a problem which already exists and is not of a pioneering nature. For some time we have felt the need of doing more fundamental and pioneering research," Smith wrote to Westinghouse president Frank Merrick in March. "This kind of work," he elaborated further, "is in the nature of a quest after facts about the properties and behavior of matter without regard to specific or commercial application possibilities at the time it is started." Smith did insist, however, much as he had in the case of nuclear physics the previous year, that "it must of course be limited to fields in which we have hopes of obtaining information that can be applied in the development and extension of new and improved products. It can be justified only by our belief that this kind of work ultimately pays dividends in the form of new ideas and information that can be practically applied to our industry."[59]

Lewis Chubb singled out the company's investment in nuclear physics research to reinforce the delicate balance that Smith tried to strike between knowledge production and commercial applications. "Although at present the only products of this type of research are artificial radioactive materials which may be used in [medical] therapy," Chubb declared in an address during the Fifth Annual Convention of the Edison Electric Institute in Chicago in June 1937, "we feel justified in going ahead as it seems today the best method of learning the secrets of nature, and feel confident that the new knowledge will bring forth new things or at least supplement our present activities which are becoming so diversified." A broader management directive to facilitate the transition of new knowledge from the laboratory to the factory floor provided the institutional leverage to meet Chubb's expectations and also those of Smith. "Responsibility for developing lines of activity which either are new in the art or which have not been previously undertaken by the company has been centralized . . . with the formation of a New Products Division," the trade journal *Steel* announced in August, two months after Chubb's public address in Chicago. The division laid the groundwork for the full-scale commercialization of new ideas coming out of East Pittsburgh. It identified potential market outlets, standardized production techniques, and trained sales personnel before the hand-off to an appropriately designated operating division—with its own engineering, manufacturing, and sales departments—for further development.[60]

Research in the separate fundamental division required a long-term, multiyear commitment, which Smith attributed to "a consensus of opinion of leaders in industrial research." He singled out the R&D policy at DuPont, where in 1927 chemical director Charles Stine had established a similarly structured fundamental research program. By the time Smith pitched his plan to Merrick, whose approval he needed to proceed, Stine's program, now in its tenth year of operation, had earned the respect of the scientific community, introduced neoprene—the first commercial synthetic rubber—and discovered nylon, a synthetic fiber that became a major source of corporate earnings in the years ahead.[61] Rather than bill research expenditures to the operating divisions—the standard practice at East Pittsburgh—Smith planned to adopt DuPont's policy and draw funds from a separate budget to avoid the short-term pressure to reduce program costs during downturns in the business cycle. Stine's training, R&D experience, and intimate knowledge of nitrocellulose (explosives) technology—DuPont's primary source of product diversification—guided the selection of appropriate research topics and the hiring of qualified laboratory personnel, such as the brilliant but mercurial organic chemist Wallace Carothers. Smith, however, lacked the same depth of expertise to articulate a like-minded programmatic agenda for fundamental research at Westinghouse. Instead, he simply recycled the strategy Chubb and Dayton Ulrey had recently used to recruit a specialist to run the nuclear physics program. "In most of our . . . engineering activities, including practical and applied research," Smith wrote, "management is capable of providing a reasonable amount of assistance in directing it, by reason of engineering knowledge and judgment of the concrete results obtained. On a fundamental research program of the type we are now considering, Management cannot offer very much in the way of concrete assistance in directing the work because of the lack of scientific knowledge and the difficulty of evaluating accomplishments. We must depend largely on the vision, judgment and inspiration of the man selected to lead this activity."[62]

The search for a candidate began a few weeks before Smith briefed Merrick. Merle Tuve, William Wells's mentor at the Carnegie Institution of Washington, got the first call from Westinghouse. The offer highlighted the fundamental division's elite status—initially to be populated "with outstanding scientists" and "expanded carefully by the selection of 'top notch' scientists as they are found"—and its organizational independence from other corporate R&D functions "in such a way as to free it from the current interruptions and the many urgent calls for specific investigations that customarily come to many research workers." This flexibility also promised broad discretion in the choice of research subjects, which included physics, chemistry, metallurgy, and the "border-line sciences." Already content in a state-of-the-art research facility devoid of commercial pressures, Tuve expressed no interest. "At present I can hardly picture to myself any possibility of abandoning the program of nuclear physics investigations I have underway here," he wrote the personnel director at East Pittsburgh.[63]

Westinghouse turned next to Lee DuBridge, a respected authority on photoelectricity and the ambitious new chairman of the physics department at the University of Rochester. It is not known why or how the company identified DuBridge as a candidate for employment. Unlike Tuve, DuBridge had cut his professional teeth on the cyclotron—Rochester

2.7. Lee DuBridge. Courtesy of the Archives and Special Collections, California Institute of Technology

had built one in 1935—rather than the electrostatic generator. Tuve may have suggested his name. Either way, DuBridge used the offer to solidify his own position at Rochester and garner more resources for the physics department. "I am submitting to Westinghouse a proposal for a program involving a $200,000-a-year budget, increasing later," he wrote Rochester president Alan Valentine. "If they accept, I can not turn it down." Marvin Smith recoiled at DuBridge's proposal. "He outlines a fundamental research program which is somewhat more extensive than we had in mind," Smith told company president Frank Merrick, "but a discussion of this general subject with him has helped to crystallize our ideas and bring us to the realization that it is necessary to build this activity up to a reasonable degree before we can hope to accomplish satisfactory results." DuBridge's lavish demands helped Smith arrive at a realistic assessment of the company's commitment to fundamental research. He promised to set aside $150,000 annually for five years and assured Merrick that "our future policy [will] be determined on the basis of the results accomplished at the end of this time." This smaller sum still equaled one-fifth of the total research budget at East Pittsburgh in 1937.[64]

The negotiations ended when Valentine offered DuBridge a higher salary and additional resources for physics research at Rochester. DuBridge also felt uneasy about an industrial career. "It was the uncertainty about how I would like it," he recalled that summer, "which was a big factor in my decision."[65] Presumably without other interested candidates, Westinghouse made one last attempt to hire Merle Tuve. Joseph Slepian, Chubb's associate director who participated in the job search, pressed Tuve to accept a five-year contract in early May. Promised more money for research at the Carnegie Institution, Tuve again declined.[66] However, he recommended an alternative strategy that put Slepian on track to hire Edward Condon later in the year. "Why not postpone the more general [fundamental research] program for a year or two, or select a man to guide it without reference to special qualifications in nuclear physics," Tuve asked, "and with yourself as guide and advisor simply get behind [William] Wells, the two of you to push the nuclear physics program together? . . . At least with [Wells] on hand the problem of finding a 'director' . . . is less urgent." The evidence indicates that Slepian followed Tuve's advice, only on a shorter schedule. DuBridge had first met Condon in 1929 at the second summer symposium on theoretical physics, at the University of Michigan, and they became close friends. Now out of the running, DuBridge urged Slepian to put Condon in charge of the new fundamental division at East Pittsburgh.

Slepian, DuBridge later recalled, "had some misgivings about theoretical physicists, but apparently has got over them." Condon expressed no reservations. He quickly arranged a leave of absence from Princeton and agreed to serve as a consulting physicist for one year at the Westinghouse Research Laboratories, effective September 1.[67]

Joseph Slepian's willingness to hire Condon signaled the extent to which Westinghouse, in the space of only a few years, had extended the scope of R&D beyond the engineering-based research tradition that had long guided technological innovation at East Pittsburgh and other manufacturing firms in the region. Quantum theoretical physics, Condon's specialty, resided at the periphery of the company's established technological strengths. Unlike Merle Tuve and Lee DuBridge, both of whom declined to give up academia for industry, Condon envisioned a merger of equals. The bump in income—his annual salary nearly doubled—facilitated rather than precipitated the transition from Princeton to Westinghouse. Prior stints at the Bell Telephone Laboratories and the New Jersey Zinc Company had honed the skills—a unique combination of scientific and engineering knowledge—that Condon began to put to work on a much larger scale at Westinghouse in the fall of 1937.

CHAPTER 3

Atom Smashing at East Pittsburgh

"**I AM VERY INTERESTED** to see the new program which you are instituting at Westinghouse," Leonard Loeb wrote Condon on New Year's Eve 1937, four months after he had come to East Pittsburgh. "If [the company's] directors will stick to the program and let you manage it," he added, "Westinghouse may even show General Electric a thing or two." Commenting on "Condon's desertion to the ranks of industrial physics" in October, Princeton's Milton White told Ernest Lawrence, "Commercial research is to a large extent what the director of research makes it, and if Condon were to become Westinghouse's director I am sure that many of the harsh words now being uttered against Westinghouse would die out."[1] These statements captured some of the anxieties that prompted the company to commit to fundamental research on a scale unequaled elsewhere in the electrical industry. Burnished credentials within the scientific community dovetailed nicely with public proclamations from engineering vice president Marvin Smith and research director Lewis Chubb about the commercial value of such a large strategic investment.

Condon embraced the same rhetoric for a different purpose; he exploited their insecurities and self-proclaimed inexperience to pursue research goals that often prioritized his own academic interests rather than the company's traditional technological strengths. The new fundamental division Joseph Slepian hired him to organize and direct lacked the common lineage of product development that the other R&D divisions at East Pittsburgh shared. On Condon's watch, the division adopted—in terms of social organization, professional standing, and scientific output—the conventional trappings of a university physics department. Condon did not, however, intentionally mislead his corporate patrons to achieve narrow academic objectives nor did he come to Westinghouse simply to escape financial hardship at Princeton, although pecuniary considerations informed the decision to resign. Unlike Merle Tuve and Lee DuBridge, who used job offers from Westinghouse to extract concessions from their employers, Condon took advantage of the opening to execute a fully matured conception of how science might better serve the technological requirements of industry. He brought to Westinghouse a unique vision of academic-industrial cooperation that did far more than put a public face on management's claims about the commercial value of fundamental research. It upended established patterns of technological innovation at East Pittsburgh.

BUILDING A RESEARCH STAFF

In mid-July 1937, Joseph Slepian invited Condon to visit the Westinghouse Research Laboratories. "It would be very valuable for us to hear your ideas and views on the possibilities of fundamental research in a large industrial laboratory," he wrote. A brief stopover in Pittsburgh on July 16, just before he left Princeton to take up summertime residence at the New Jersey Zinc Company, convinced Condon to accept Slepian's offer, but on a one-year trial basis to begin on September 1. This temporary commitment merely extended the sabbatical he had already planned to take in California during the second semester of the 1937–38 academic year. Condon preferred to test the waters first, a decision that Lewis Chubb supported. "The idea of coming on a year's leave of absence . . . is a good one," he wrote, "as it will give a trial period for you to see how you like research work in industry and without the risk of losing continuity of service and position at Princeton." Chubb offered Condon a $7,500 salary plus a bonus based on a percentage of corporate earnings, with the understanding that the amount would increase should he decide to accept permanent employment.[2]

From the outset, Condon expected to plan and manage the fundamental research program without any restrictions. "You will have the full backing of Slepian and [Marvin] Smith," Lee DuBridge assured him. "Smith is the key man as I see it and he is all for the fundamental research idea. You may have to sell him the idea as to what fundamental research is, but since he is sold on the nuclear physics project, most anything else ought to be easy," he explained. Chubb, meanwhile, immediately solicited from Condon concrete recommendations for suitable research topics and the names of qualified personnel to recruit. "I would like to have suggestions," he wrote less than a week after Condon's first visit to East Pittsburgh, "with regard to research projects and possible outstanding young scientific candidates." Initially Condon adopted the preliminary plans DuBridge had drawn up during the negotiations with Slepian earlier in the year. In addition to nuclear physics, which already had a head start, DuBridge proposed "some work in electronics"—an established field at Westinghouse—but he also added the physics of solids, especially the quantum theory of metals, and mass spectrometry. "I thought the thing to do," he wrote Condon, "was to find a good man in each field [and] build the program in that field around him." In the preliminary plans he submitted to Chubb in August, which used DuBridge's recommendations as a guide, Condon split the fundamental division's work into three sections—nuclear physics, chemical physics, and the physics of metals. Mass spectrometry, infrared spectroscopy, and studies of semiconductors and other nonmetallic solids filled out the research program in chemical physics. The work he selected for the physics of metals section included physical metallurgy, electronics, superconductivity, and cryogenics.[3]

Condon's priority for nuclear physics focused on the hiring of additional personnel to help William Wells complete the atom smasher "and get that running as smoothly as possible." Mass spectrometry, by contrast, already showed promise as a candidate for industrial use. Walker Bleakney, whom Condon recruited from the University of Minnesota to complete postdoctoral study at Princeton in 1930, had improved the art to the point where it could identify trace amounts of light hydrocarbons in soil gases. The petroleum industry expressed interest in this line of research as the basis of a new and potentially valuable method of locating oil deposits. Bleakney first analyzed soil gas samples for the Humble Oil Company in 1936 and opened negotiations with the Phillips Petroleum Company the following spring for a more elaborate joint research program. Writing to Chubb a few months later, Condon recommended a similar research strategy at Westinghouse. He prepared to turn the mass spectrometer loose on molecular structure problems to obtain data "on fundamental molecular processes involved in the breakdown of insulating oils" and a host of other materials used in the electrical technologies the company produced. Condon also singled out for further study the type of micro gas analysis that Bleakney had pioneered on behalf of oil producers.[4]

Like-minded thinking guided the selection of infrared spectroscopy "as a general tool for molecular structure problems," Condon explained to Chubb. Here, too, Condon extended his abiding interest in the theory of atomic spectra that had originated at the University of Minnesota and blossomed at Princeton. In the spring of 1936, he interpreted on theoretical grounds the infrared data obtained by Ralph Barnes, a respected spectroscopist and instructor in the physics department who had just resigned to become the director of physics research at the American Cyanamid Company's new research laboratories in Stamford, Connecticut. Since 1933, Barnes had served as a consultant to American Cyanamid's recently acquired dyestuffs subsidiary—the Calco Chemical Company—working on infrared spectrophotometry. Fresh off the recent collaboration with Barnes, whose research program at Princeton he now supervised, Condon referenced some early work along the same lines also under way at the New Jersey Zinc Company. "Here . . . at the laboratory," Condon wrote Chubb from Palmerton, "they have worked out an important infra-red technique for measuring particle size in paint pigments. . . . So I think there would be plenty of scope for interesting fundamental work in the infra-red."[5]

Condon left unresolved the detailed plans for research in the physics of metals pending a survey of metallurgical R&D already under way at East Pittsburgh, but he deliberately left room for the establishment of a "fundamental program that fits in closely with the metals work that is now being done. . . . There has been in the past few years a great development of the basic physics of metals which I am sure will net us a great stimulus to the physical metallurgists." Starting modestly within days of his arrival at East Pittsburgh, Condon organized and ran a weekly seminar for the research staff on theories of the solid state "to consider recent developments in this field." The seminar replicated on a more structured level the individual tutoring he provided the metallurgists on staff at the New Jersey Zinc Company during the summer. Condon also delayed the selection of specific programming objectives for electronics research and studies of nonmetallic solids, two fields that "are so closely related to the present program of the [East Pittsburgh] laborator[ies] that I have felt it useless to think about plans in this direction until I learn more about the existing situation."[6]

With Condon less than a month into the job, Chubb and Slepian pressed him to remain at Westinghouse permanently. Given direct supervision of the nuclear physics program and access to the budget originally promised to Lee DuBridge, Condon told Princeton University president Harold Dodds that he "would be in a position to make a definite decision about the future by March 1 [1938]." In the meantime, Condon continued to refine his research plans. Rather than hire only accomplished experts to build fields of specialized knowledge, which DuBridge had originally proposed, Condon chose instead to recruit for the bulk of his staff—except for one additional permanent appointment—newly minted PhD scientists under the age of thirty-five on a rotating basis for a maximum of two years. Called the Westinghouse Research Fellowships, the program paid an annual stipend of $2,400, which compared favorably to the compensation Princeton doled out for similar postdoctoral appointments.[7] At any given time, ten fellowship recipients (five selected each year) would have the option to investigate topics of their own choosing within the broad subject areas Condon had previously identified. A first for the electrical industry, the program had few precedents elsewhere. In 1934, DuPont—building on its earlier pioneering venture into fundamental research and support of graduate instruction in the universities—spearheaded a pattern of institutional collaboration with academia similar to, but not as elaborate as, the novel approach Condon now proposed at Westinghouse. DuPont did not create a whole new in-house program; it distributed funds for postdoctoral study in organic chemistry to select universities.[8]

Condon presented the fellowship plan to management in October. "The main advantages to be gained from the plan," he explained, "are . . . [a] continuing source of the highest type of personnel for the Research Laboratories [and] a steady stream of fresh ideas from the best academic institutions . . . to give strength and vitality to a fundamental research program and to contribute new ideas to the rest of the laborator[ies] by direct contact with the men involved." The research fellowships helped mitigate the chronic difficulties that the company had previously encountered in trying to recruit qualified personnel to run the nuclear physics program and, more recently, to organize and manage the fundamental division. Condon specifically highlighted this advantage. "It does not need to be stressed," he wrote, "that risks involved in selection of men will be greatly reduced by drawing from this group [of fellows] for research and technical staff." Of the twenty-three investigators he hired between 1938 and 1942, more than half—fifteen in all—accepted full-time positions at the research laboratories after their fellowships expired. Finally, the fellowships promised to improve the company's image as a source of new knowledge, bringing it "the finest type of institutional publicity . . . [a] steady stream of research publications from the Laboratories . . . [and] the eventual development of a body of former Fellows on the faculties of leading American universities in whom will have been implanted a strong loyalty to [Westinghouse]."[9]

Condon got approval for the fellowship program on November 11. That same day he also received a firm job offer. "They are offering me a newly created position as [associate] director of the Research Laboratories," he wrote Harold Dodds, "at a starting salary of $9,000 a year, effective January 1, 1938." Condon welcomed the opportunity to earn more money; he had not received an increase in rank or salary at Princeton since 1930. Always mindful of his finances, Condon struggled to save money in a household that now included two more children. At the same time, Condon, fresh off his stimulating experience at the New Jersey Zinc Company, expressed a keen interest in the job. Tailored to suit his intellectual proclivities, it imposed no restrictions. "This position," he explained to Dodds, "enables me to work . . . in extending the use of modern atomic physics in all phases of the work now being done in the laboratories. . . . Assurance of the continuance of the work over a trial period of five years has been given by the management." Condon dropped the original plan to segregate fundamental research in a separate division in favor of a group comprising mostly postdoctoral appointees under his direct supervision. He expected them to work closely with other members of the research staff. Already convinced that "this opportunity seems to be more attractive than the future for me appears to be at Princeton," Condon nevertheless waited for Dodds to respond; he anticipated a counteroffer.[10] After consulting physics department chairman Henry Smyth, Dodds promised to promote Condon from assistant to associate professor and increase his salary 30 percent, to $6,500. Still smarting because seven years had passed since his last promotion, Condon replied, "The situation at Princeton is about what I expected it would be. . . . I see a better opportunity . . . here [at Westinghouse]." He resigned on December 3.[11]

Westinghouse launched the fellowship program and publicly announced Condon's appointment as associate director of research in mid-December 1937. The company distributed announcements to colleges and universities throughout the United States and also promoted the program by way of statements released to scientific journals, trade and industry publications, and major newspapers.[12] The media blitz generated results. Fifty-three applicants vied for the first five research fellowships, to be awarded the following spring. Condon's enthusiasm quickly waned, however, as the national economy tipped back into recession, prompting a decline in corporate profits and a sharp 46 percent drop in new orders booked for electrical equipment during the first six months of 1938.[13] Although Marvin Smith had promised at the outset that expenses for fundamental research would not be charged to the operating divisions "and thus be subjected to the usual pressure of the Divisions to have such charges reduced during business recessions," the company's financial woes mitigated any effort to protect the program. Condon chafed at Smith's backtracking. "Salaries have been cut ten percent and other economies are being put into effect in all directions," he told Lee DuBridge in May. "The fellowship program will survive but I am afraid that is only face-saving because we got so much publicity for it. It is going to be hard to get the money to back up their [the fellows'] work properly and quite impossible to get an experimental physicist for the regular staff as I had planned."[14]

Condon had also come to the conclusion—"quite apart from this unfortunate turn of circumstances"—that he had strayed too far from theoretical physics. "At first I thought this was a temporary phase," he wrote Henry Smyth in an attempt to get his old job back, "and did not notice it in the excitement of seeing and learning so many new things. . . . I have decided that I should return to academic work and regard this year's experience as valuable and broadening but not as a definitely new direction to my work." The same anxiety had prompted Condon to resign from the Bell Telephone Laboratories ten years earlier—"I am beginning to feel myself slipping on wave mechanics," he confided to Raymond Birge at the time— but now he faced a more competitive job market. In response to Condon's inquiry about employment opportunities at Rochester, DuBridge offered no more than personal encouragement for improved conditions at West-inghouse: "I am afraid the situation here is not particularly promising for the immediate future." After he wrote to Smyth, Condon appealed directly to President Dodds, hoping the promotion he had been offered prior to resigning in the fall might still be available. "Regarding the possibility of renewing our earlier offer to you," Dodds wrote Condon after consulting Smyth, "I have to report that it is not possible for us to do so. Arrangements have been made for next year involving commitments which must be maintained."[15]

Bleak economic conditions and a stated preference for academic research do not fully explain why Condon chose to leave Westinghouse after only a few months on the job. On the face of it, his uneasiness about the new surroundings at East Pittsburgh is consistent with prior patterns of behavior. The intellectual awakening and opportunities for professional advancement he first experienced somewhat anxiously in the late 1920s combined with pecuniary obligations and family priorities to justify brief stints at the Bell Telephone Laboratories and the University of Minnesota. His backpedaling in the spring of 1938, however, also revealed another source of insecurity that merits further explanation. Leonard Loeb had put a name to it in the summer of 1928—a stubborn restlessness that Condon initially dismissed but later acknowledged. "The fact is that I am always dissatisfied with what I am doing and would like to be doing much better than I can," he admitted to Loeb in 1933. "This leads me always into thinking about ways to change things for improvement. . . . So many people like to be deep in a well-oiled rut but I just can't stand it."[16]

This penchant for periodic change, which may have also borne some connection to Condon's itinerant upbringing, had grown more pronounced by the time Westinghouse called on him four years later, and it persisted throughout the remainder of his life. Physics department chairman Henry Smyth bristled when Condon declined what he considered to be a generous counteroffer to remain at Princeton. "He is in one of his moods of great enthusiasm for a new job," Smyth sniped after Condon resigned, "and apparently did not consider staying here at all." Milton White, builder of Princeton's cyclotron, rejected an offer from Condon to take a leading role in the nuclear physics program at East Pittsburgh early in 1938 for the same reason that upset Smyth. "A[s] much as I liked . . . Condon," White later recalled, "he had already acquired quite a bit of a reputation for being an enthusiast for a while and then dropping things. . . . I didn't want to get caught in that."[17]

Condon suppressed his restless proclivities and prepared to remain at Westinghouse in the interim. "There is nothing urgent about my situation," he told Lee DuBridge. "I can go along here alright for a while." His attention soon turned to the first group of fellowship recipients, who arrived at East Pittsburgh during the summer. Condon, Chubb, and Slepian—in consultation with the laboratory division managers—selected three nuclear phys-

3.1. William Stephens, William Shoupp, and Robert Haxby (*left to right*) check the instruments at the base of the atom smasher. William Wells is standing on the far right. Emilio Segrè Visual Archives, American Institute of Physics

icists two work with Wells: William Shoupp, Robert Haxby, and William Stephens, from the University of Illinois, the University of Minnesota, and the California Institute of Technology, respectively. After Walker Bleakney declined Condon's earlier offer of a full-time job, John Hipple, one of his most promising graduate students at Princeton, got the call for a fellowship. Hipple's appointment marked the beginning of an entirely new research venture at East Pittsburgh. It also signaled the extent to which Condon exercised authority over the selection process. "The mass spectrograph work which Hipple will do," he explained to Leonard Loeb that summer, "is a continuation of my interest in molecular structure, which grew out of my association with Bleakney." Sidney Siegel, a Columbia-trained PhD who planned to study the physics of metals, particularly the elastic properties

3.2. John Hipple. Emilio Segrè Visual Archives, American Institute of Physics

of single crystals, rounded out the group of young upstarts. Shoupp, the oldest, being thirty, raised their average age to a mere twenty-seven years.[18]

Condon dropped his original plan to hire another experimental physicist to work with Wells on the atom smasher. Disappointed that all of the candidates interviewed for the position had declined, he finally abandoned the effort altogether due to insufficient funds precipitated by the business slump. Initially lacking confidence in Wells's abilities, Condon had come to appreciate the work he had accomplished to date and subsequently recommended that Wells receive a promotion, making him "the top man in experimental physics in the fundamental research program. This would give us two of the three proposed permanent positions in Wells and myself, with the additional appointment put off until a suitable man appears." This decision put more pressure on the fellowship program to generate candidates for permanent employment at East Pittsburgh, a challenge that Condon welcomed. "We have had such a fine response to the fellowship plan," he wrote Chubb in August 1938, "that I feel that we are going to have in it an abundant supply of the best young research men in the country to draw on for the regular staff, and we can develop on sounder lines if we hold back to leave open more opportunities for some of these men."[19]

The fellowship program replicated at East Pittsburgh a working environment strikingly similar to the one Condon had cultivated at Princeton. He used quantum mechanics to reconcile competing theories of physical behavior through the interpretation of experimental data, a process that stimulated collaborative interest among practitioners within and outside the physics department. This intellectual proclivity also nurtured a broad perspective that his contemporaries did not always share. In one case, Condon praised the quality of J. Robert Oppenheimer's research to differentiate his own abilities and aspirations. "[Oppenheimer's] work is on the purest and most abstract side [of] theoretical physics and of the very highest quality," Condon explained to Leonard Loeb in 1933. "I don't have a flair for that sort of work. What I can do fairly well is to take the main equations of mathematical physics and apply them to the explanation of special phenomena. This is my professional line and I have tried to keep quite broad within it." Elaborating further, he set out a clear strategy to move forward. "It is my desire," Condon continued, "not to stay too narrowly within quantum theory but to use mathematical methods on the whole range of physical problems." Six years later, in 1939, Ernest Lawrence acknowledged Condon's strengths along similar lines. When he learned that Leonard Schiff, a recently minted PhD in theoretical physics from MIT who had just finished a postdoctoral fellowship split between Berkeley and Caltech, had made the short list for a job at the University of Rochester, Lawrence penned a glowing recommendation to Lee DuBridge. "Schiff is always actively interested in interpretation of experimental results and is a good fellow to have around for an experimenter to talk physics with,"

Lawrence wrote. "In this respect I suppose he might be classified as a Condon type."[20] The Westinghouse Research Fellowships—the industrial equivalent of an academic postdoctoral study group—institutionalized the research style that Condon explained to Loeb and that Lawrence later ascribed to him.

The fellowship program structured but did not limit the scope of Condon's intellectual activity at Westinghouse. To the contrary, the broad discretion he enjoyed as associate director of research afforded the opportunity to apply the same interpretive skills Lawrence later referenced to technical problems elsewhere in the company that relied on a knowledge of engineering rather than quantum theoretical physics. In the fall of 1938, a few months after the first group of postdoctoral fellows arrived at East Pittsburgh, Condon prepared a statistical analysis of fluctuations in the number of defective transformers manufactured for electric power distribution systems at the company's factory in Sharon, Pennsylvania, north of Pittsburgh. Ten years earlier, Condon had followed with great interest the pioneering research just under way on statistical quality control at the Bell Telephone Laboratories. The Western Electric Company put this knowledge to use to optimize the quality of manufactured telephone components. Condon, who had stayed abreast of research in the field, applied the same techniques to explain the failure rates in transformer production at Sharon. "The method of approach used in your [analysis] is one that reveals some fundamental facts," a grateful engineer in the transformer engineering department at Sharon wrote Condon in October. "I can readily agree with you that the high rates of failures . . . cannot be explained by chance, but are due to some changes in the manufacturing process."[21]

The ability to alternate seamlessly between the hard sciences and the engineering disciplines showcased the range of Condon's intellectual versatility. It also informed a broader conception of technological progress in industry that, from his unique vantage point, Pittsburgh stood poised to advance. "I honestly believe," Condon declared,

> that no other city in America is so well equipped as Pittsburgh to assume leadership in the next phase of scientific history, the period when an intellectually mature America gives to science a host of young, trained, inquiring minds, adequately backed with laboratories and material support, and encouraged by a philosophy which exalts the spirit of free inquiry as the surest means of promoting human welfare. Nowhere else is there such a well established group of technical men and industrial scientists. Nowhere

else is there a community whose wealth is so directly an outgrowth of the industrial progress which development of pure science brings.[22]

Condon rendered these lofty ambitions into concrete objectives, first through the postdoctoral fellowship program and then by way of a bolder initiative, national in scope, that he first outlined in detail in January 1939 and recounted again in a letter to University of Pittsburgh chancellor John Bowman in late 1940. "The . . . development of pure science in the universities and technical schools of this part of Pennsylvania is in no way commensurate with the standing of Pittsburgh as a . . . center of industrial progress," he wrote. Condon recommended the establishment of a new national institute of scientific research to stimulate the product-driven R&D already under way in the country's leading science-based industries. Representative examples of the intended beneficiaries included Alcoa, Gulf Oil, and Westinghouse. Condon expected these and other firms to provide the $25 million endowment he deemed necessary to support such a massive undertaking. He also envisioned an organization functionally akin to the Mellon Institute of Industrial Research but firmly grounded in the guiding principles of the fundamental research program at Westinghouse. In addition to the construction of new laboratory facilities in Pittsburgh, Condon called for the appointment of a small permanent staff of "scientists of international reputation," joined by a much larger contingent of temporary research fellows drawn from the universities who would be responsible for building, at the outset, "a strong program in the field of nuclear physics, in the physical aspects of organic chemistry, and in the fundamental parts of physical metallurgy." Elaborating further, Condon highlighted the importance of independent judgment in the selection of research topics: "It is principally desirable to keep the whole plan flexible, securing a staff of the best men possible and leaving the program of research entirely up to their judgment." Doing so provided added assurance "that the project shall be as effective as possible in producing the results desired and that it shall be protected against falling into senile decay."[23]

The proposal never got beyond the planning stage. Condon petitioned Westinghouse chairman Andrew Robertson for financial support to get the project started, but there is no evidence that he endorsed or acted on it. The University of Pittsburgh also expressed some interest by way of preliminary discussions within the administration about options to establish closer ties to local industries, in part to improve the quality and broaden the scope of scientific research in the region through an expansion of the science and engineering departments. University provost Rufus Fitzgerald invited Condon to campus for a meeting on March 8 to obtain more information about the proposed institute in preparation for a possible collaboration. "[Condon] thought that it might be worked out in such a way that Pittsburgh could do research in pure science that would be comparable to that which the Rockefeller Foundation has done in medicine," Fitzgerald wrote in his diary after their discussion. In the end, however, Fitzgerald opted for a less ambitious solution. Rather than get involved in such a huge undertaking, he favored the targeted solicitation of funds from local patrons to recruit new faculty—"outstanding men"—for the physics and mathematics departments. Without a receptive audience, Condon shelved the proposal until November 1940, when he appealed directly to Fitzgerald's boss, the university's chancellor, John Bowman. "I would like to present more fully the ideas for research in Pittsburgh . . . which I think [are] more timely than ever," he wrote Bowman.[24] There is no evidence of a response from Bowman or Fitzgerald. This outcome, however, should not obscure the significance of Condon's vision. The justification for a national institute of scientific research in Pittsburgh revealed in stark terms the full extent of his thinking about how academic science could most effectively serve the technological requirements of industry.

CONDON'S GROUP AT WORK

What Condon failed to achieve at the national level he pursued vigorously through the postdoctoral fellowship program. "The work which [Sidney] Siegel will do," he wrote Leonard Loeb in the summer of 1938, "represents a new venture for the [Westinghouse Research] Laboratories on what, at least, in this environment, seems like a rather academic approach to fundamental metallurgy." Condon did not privilege physics over metallurgy or assume that the company's accumulated expertise in the materials technology necessary for the development of electrical equipment lacked scientific rigor. William Johnson, a newly minted PhD in metallurgy from the nearby Carnegie Institute of Technology, joined Siegel on a postdoctoral fellowship in 1939 to study diffusion rates in silver alloys by way of radioactive tracers produced in the atom smasher. When Johnson inquired about a permanent job opportunity two years later, Condon praised his research skills, even his "very profound knowledge of the detailed metallurgy and physical chemistry of steel." He also solicited advice about Johnson's suitability for employment from the longtime head of chemical and metallurgical research at East Pittsburgh. "He knows much more about the practical side of metallurgy than I do," Condon confided to Loeb after the consultation, "[and] I have a high respect for [his] judgment in such matters." Rather than enforce a research hierarchy with physics at the top, Condon envisioned metallurgical knowledge and the diverse skills that Siegel and Johnson brought to Westinghouse in complementary terms. In this case, prior experience at the New Jersey Zinc Company served as a useful template. Condon broadened the scope of metallurgical research to incorporate new theories of atomic structure that contributed directly to the solution of practical problems.[25]

An article that appeared in *Research Progress*, a new in-house R&D publication Westinghouse first put out in 1938, explained the practical bent of Sidney Siegel's work: "Studies of the electrical and elastic properties of . . . [copper-gold] crystals may clarify the fundamental nature of interatomic forces that determine the properties of structural materials." Elaborating further, the article affirmed Siegel's prediction that "a workable theory . . . would help metallurgists to substitute exact knowledge for the usual trial-and-error methods of combining metals to produce alloys, for in principle all the properties of a metal should be calculable if only the position of its atoms and the forces acting between them were known." His studies of elasticity—the ability of a metal to retain its original shape after the removal of an external force—found immediate application in the testing of blades installed in the steam turbines Westinghouse manufactured. Investigations of elastic behavior in aluminum alloys containing copper that Siegel shared within the local research community, Condon later recalled, "aroused a good deal of interest in the research laboratories of the Aluminum Company [of America]." After his fellowship expired in 1940, Siegel joined the research staff at East Pittsburgh full time to work on the development of magnetic materials.[26]

The scope of Sidney Siegel's research illustrated the extent to which the physical understanding of solid materials had changed in the span of only a few years. In 1933, Frederick Seitz, a twenty-two-year-old graduate student whom Condon had recruited from Stanford University to study at Princeton, worked out a method to compute the band structure of metallic sodium. This breakthrough marked the first successful attempt to describe quantitatively the properties of real rather than ideal solids based on a quantum mechanical interpretation of the electron theory of metals. Further refinements culminated in a more extensive treatment of sodium and

3.3. Sidney Siegel. Bank and Stoller, Inc., courtesy of Emilio Segrè Visual Archives, American Institute of Physics

paved the way to a systematic analysis of the electrical, mechanical, and optical properties of other solids. "The new theories of atomic physics—quantum and wave mechanics—are clarifying knowledge of the structure of solids and thus bringing these seemingly 'impractical' playthings of the mathematical physicist into the realm of material objects which one can see and touch; objects which an engineer or layman could class as practical," proclaimed *Science,* paraphrasing comments Lee DuBridge delivered at a joint meeting of the Optical Society of America and the American Physical Society at Columbia University in February 1938, a few months before Siegel arrived at East Pittsburgh. After he worked briefly for DuBridge at the University of Rochester and then Saul Dushman at the General Electric Research Laboratory, Seitz reentered academia later that year. He accepted an assistant professorship at the University of Pennsylvania, where Gaylord Harnwell, a friend of Condon's recently recruited from Princeton to head the physics department, had promised him a free hand and plenty of resources to build a new research program in the physics of solids.[27]

Seitz's departure from General Electric pleased Condon. "I am glad you did this," he wrote, "because it will take a little of the stiffness out of our relations, which may have been imposed by our working for such definitely competing companies." Condon looked forward to cultivating closer professional ties with Seitz, who had just completed six weeks in residence at the University of Pittsburgh as a lecturer in the inaugural summer session on the physics of metals organized by Elmer Hutchisson, the newly appointed chairman of the physics department. "During the past few years," Hutchisson wrote in the program announcement, "much progress has been made in understanding the forces that hold atoms to-gether in solids. . . . The fundamental purpose of this summer session . . . is to review the results which have been obtained and to study their possible applications in the field of metallurgy." Condon, whom Hutchisson had appointed advisory professor in the fall of 1937 "to provide courses in . . . fundamental fields of physics which are especially applicable to industry and which will be taught by those actually applying physics every day," helped organize the first session. It reconvened every summer through 1941.[28]

Toward the end of 1938, Condon took advantage of the merging interests of physics and metallurgy in Pittsburgh and Seitz's new academic position in Philadelphia to expand the scope of fundamental research at Westinghouse. "The general plan of Westinghouse cooperation with the

3.4. Frederick Seitz in January 1937, shortly before he left the University of Rochester to take up industrial research at General Electric. Emilio Segrè Visual Archives, American Institute of Physics

University of Pennsylvania was fully approved," Condon wrote Seitz in mid-December, "and as a natural development of the close cooperation of the company with the universities here in Pittsburgh . . . I want to encourage you to think along general lines of getting some cooperative work started . . . in which we may be of some assistance in the field in which you are interested, namely physics of metals and in the physics of semi-conductors." This local arrangement embodied the broader objectives of the proposal for a national institute of scientific research that Condon prepared to put on paper. In January 1939, when he first spelled out the institute's functions, Condon invited Seitz to spend eight weeks during the summer on leave from the University of Pennsylvania as a paid research associate at East Pittsburgh: "The idea would be to have you come over and get acquainted with our problems, especially in the rectifier business and also in . . . ferr[o]-magnetism or any other branch which interests you, during which time we could figure out the plans for support of your work in Philadelphia."[29]

The freedom Condon imparted to Seitz that summer and the next (he returned in 1940) explicitly reinforced the academic flavor of the fundamental research program. In 1940, Seitz published *The Modern Theory of Solids*—the first synthetic, textbook treatment of the subject. He acknowledged in the preface Condon's input and "the privilege of spending a stimulating summer in East Pittsburgh in 1939." Also that summer, Seitz worked closely with Thomas Read, Sidney Siegel's classmate at Columbia whom Condon had brought to Westinghouse in the spring of 1939 on a postdoctoral fellowship to study the mechanical properties of solids, specifically internal friction—or energy loss—in metals. "This program of research on the internal friction of single crystals in metals," Read wrote in the research plan he submitted to Condon, "is designed to obtain data which are needed to test and elaborate the present theory of this phenomenon."[30]

Relying on the same experimental technique Siegel had used to grow copper-gold alloys and study their internal atomic arrangements, Read obtained data that attributed some of the observed energy loss to crystal lattice defects called dislocations, the theory of which had first been proposed in 1934 to explain plastic deformation—the tendency of a metal under load to retain new physical dimensions after the external force is removed—in solids. "Read's new evidence," Seitz later recalled, "prompted the two of us to prepare a series of review articles in which we attempted to describe, in a more or less systematic way, the roles that dislocations could play in affecting the various properties of crystalline materials."[31] Four pathbreaking articles, written with Siegel's assistance and published under the general title "Theory of the Plastic Properties of Solids," appeared in the February, March, June, and July 1941 issues of the new *Journal of Applied Physics*, which Elmer Hutchisson had helped found at the American Institute of Physics and he now edited at the University of Pittsburgh.[32]

The time Seitz spent at East Pittsburgh produced significant research results, and it also culminated in the establishment of a cooperative research program in semiconductor physics. Westinghouse already manufactured, through a license agreement with the Union Switch and Signal Company, copper-oxide rectifiers for power applications. Lars Grondahl, Union Switch and Signal's director of research, had invented this device more than a decade earlier, for use in railway signaling systems.[33] Temperamental and prone to chronic production problems, the enigmatic copper-oxide rectifier stymied the research personnel who studied its properties and behavior in the electro-physics division at East Pittsburgh. "In spite of its apparent simplicity," stated the division's quarterly research report in 1936, "the copper-oxide rectifier is one of our least understood devices." In January 1939, Condon grumbled to Philip Morse, who had coauthored *Quantum Mechanics* with him a decade earlier at Princeton, about "how little attention has been paid to the physics of semi-conductors in the universities' laboratories. Aside from the industrial importance of things like copper-oxide rectifiers, there is a lot of interesting physics there . . . [which] by no means has been thoroughly checked up experimentally." To help fill that gap, he told Morse, "I would like to get a man who really could make a good showing in the field of ferro-magnetism and another who would work on semi-conductors."[34]

Collaboration with Seitz gave Condon the opportunity to hire new recruits specifically suited to this task. In mid-January 1940, just before he agreed to return to East Pittsburgh for another summer appointment, Seitz reminded Condon of the offer he had initially made in December 1938 to "give us a starting push in the field of semiconductors." He asked Condon for a small stipend—$600 plus $200 for equipment—to pay for a graduate student fellowship for research "covering 'academic' materials as well as practical ones." Since it would be a direct extension of his own interests, Seitz agreed to supervise the research personally. "With your help on this matter for about two years," he assured Condon, "I think we could get valuable work started that might not otherwise be initiated for some time." Condon obtained approval from Lewis Chubb and Marvin Smith, while Gaylord Harnwell, Seitz's boss in the physics department, worked out the details of the agreement between Westinghouse and the University of Pennsylvania.[35]

Although "developments of industrial interest" remained a possibility once the research got under way, Harnwell emphasized the project's scientific merit, in full agreement with the priorities Seitz and Condon had established at the outset. "This work," Condon wrote the university's vice president for administration, "would be of a pure scientific nature and the Westinghouse Company would not look to any immediate industrial applications of the results." Chubb acknowledged that "in fundamental work of this kind the question of patents is not of any great importance," but the option to obtain rights based on the research completed remained with the company. In March, Seitz handpicked Stephen Angello, a graduating senior in the electrical engineering department, to receive the new fellowship. "He is a budding genius," Seitz wrote Condon, "and was so taken with physics and the possibility of applying modern physics to engineering problems that he could hardly apply . . . fast enough." After Angello accompanied Seitz to Pittsburgh "to get a proper introduction to our problems," he got to work studying electrical conductivity in copper-oxide rectifiers, the subject of his doctoral dissertation. When Angello finished in the spring of 1942, Condon redirected the junior physicist's research, now undertaken on a full-time basis at East Pittsburgh, to a decidedly more practical wartime objective—the development of silicon crystal detectors for microwave radar receivers.[36]

The research Siegel, Seitz, Read, and Angello conducted at East Pittsburgh diversified Westinghouse's accumulated expertise in metallurgy and materials technology. "In the last five years there has been a very great development in physicists' understanding of the nature of the metallic state," Condon explained to Chubb early in 1941. "It was my desire from the first that a program of fundamental research at the Laboratories should have an adequate development in this direction." Condon, however, never lost sight of the commercial priorities that justified the initial investment in personnel and equipment. "The field is also not so detached from present commercial interests, not so 'pioneering,'" he elaborated further, "and already finds close points of contact with already existing activities."[37]

The opportunity to build on these accomplishments and forge a closer research relationship with Seitz came later that year, when the longtime chairman of the physics department at the Carnegie Institute of Technology died. Condon got the call to replace him but declined; he recommended Seitz instead. "I really believe that there is a good opportunity for an aggressive fellow to make a fine thing of the job here," he told Seitz in March 1942, "and I am awfully happy you are interested in it." Seitz felt the same way, and he also looked forward to a continuation of the cooperative research Condon had cultivated on his behalf at East Pittsburgh. Seitz took the job, but the collaboration he and Condon envisioned did not come to pass.[38] Signs of an uncertain future first appeared in early 1941, when Seitz declined Condon's offer to return to Westinghouse for a third summer appointment. Consulting work for the army and the navy left no room in his schedule for the required time commitment. By the time Seitz arrived in Pittsburgh in November 1942, the defense mobilization had put civilian R&D on hold and monopolized the scientific and technological resources of Pittsburgh's industries and universities.[39]

Nuclear physics lacked the same practical perspective that guided the growth and diversification of research on solid materials before the war intervened. An inchoate commitment to commercial applications had justified the initial investment in high-voltage research. When William Shoupp, Robert Haxby, and William Stephens arrived at East Pittsburgh in the spring of 1938, the atom smasher had not yet been completed. The status of the partially finished accelerator troubled Marvin Smith, who closely followed developments elsewhere. After learning in February that the new cyclotron at Harvard University "may be in operation before ours," Smith instructed Chubb to work faster. "It is imperative," he wrote, "that we do everything possible to expedite the work on our unit." Pointing out that both accelerators "are hardly comparable," Chubb reassured Smith that the Harvard machine "will be in use sometime in the early summer. We expect our [electrostatic generator] to be working sooner than that."[40]

Smith raised a similar ruckus in mid-June after reviewing the conference program for the summer meeting of the American Institute of Electrical Engineers (AIEE) in Washington, DC; it announced a planned inspection trip to the Carnegie Institution to see Merle Tuve's 10 MeV pressure installation, still under construction. Referring to a brief description in the announcement that anticipated the accelerator "will be ready for use before June," Smith called on Chubb again to pick up the pace at East Pittsburgh. "If this statement is correct," he wrote worryingly, "their program and progress is considerably in advance of ours and the publicity which has been given our work may become a boomerang if we do not get our unit into operation at an early date." Smith's fixation on publicity and institutional competition had also been on Tuve's mind. In February, he had agreed to write for the AIEE journal *Electrical Engineering* a feature article on the 10 MeV atom smasher—"'counter-publicity' to compensate for Westinghouse," he told Department of Terrestrial Magnetism director John Fleming. Initially concerned following Smith's inquiry but unaware of Tuve's advertising campaign for the AIEE, Chubb dispatched Condon to Washington to obtain a status report; he "found that they are considerably behind us." Elaborating further on Condon's reconnoiter, Chubb told Smith, "What they mean by the generating equipment being ready for use before June is that they will be able to run the belts and have voltage [without the vacuum tube], such as we have done for several months."[41]

When Smith first queried Chubb about Tuve's progress, major construction at East Pittsburgh had already turned to the assembly of the vacuum tube, which measured more than thirty feet in length. "The big generator is finished . . . and work is going ahead on vacuum tube assembly," Condon wrote Leonard Loeb on June 8. "One of eight 4-foot sections has been assembled; the others will come along as fast as possible," he explained. By early September, Wells, now aided by Shoupp, Haxby, and Stephens, had finished off the remaining sections and prepared to hoist them into final position inside the pressure tank. The company had a showpiece set against a drab industrial landscape full of factory buildings that churned out motors, generators, switchgear, and other types of heavy electrical apparatus. Sited next to the main research building stood a new structure capped by an inverted, pear-shaped steel tank that weighed more than one hundred tons. It measured forty-seven feet in height and thirty feet in diameter at the widest point. The total height of the new facility reached nearly seventy feet. Inside the tank, which had been designed to operate at a working pressure of 120 pounds per square inch, an electrode fifteen feet in diameter sat atop four metal-reinforced porcelain insulating columns lined with two charging belts, leaving sufficient space in the center for the vacuum tube that would soon run half the length of the entire structure. No one who passed by the Westinghouse Research Laboratories could miss this latest incarnation of the company's scientific and technological prowess. Rival GE acknowledged the transformation, confirming Leonard Loeb's earlier prediction that Westinghouse "may even show General Electric a thing or two." Just off a guided tour of the new facility late in 1938, after the vacuum tube had been assembled and mounted inside the steel tank, GE research director William Coolidge returned to Schenectady "thoroughly impressed." Frederick Seitz, who recorded Coolidge's comment in a letter he wrote to Condon shortly afterward, added, "He [Coolidge] being very conservative, you can regard this as a compliment of the first order."[42]

On November 24—Thanksgiving Day—a one-million-volt potential (at atmospheric pressure) accelerated the first beam of protons through the vacuum tube. Satisfactory operation of the tube at higher voltages, however, remained elusive. Work to improve its performance proceeded slowly. In July 1939, Chubb reported to Smith, "The problem of getting a suitable ion beam focused down the 30 ft. long vacuum tube is quite a difficult one and has presented several unexpected problems. With the ion beam we now have[,] some nuclear physics research could be done but we have felt that it is more worth while to concentrate efforts on getting a larger intensity before undertaking to use the machine for the fundamental problems which are scheduled."[43] Mitigation of ion scattering in the tube proved especially challenging. In a progress report to Westinghouse president George Bucher, who had replaced Frank Merrick the previous February, Smith acknowledged that "the delay and consequent expense have been much greater than we anticipated." The original budget estimate for materials and construction had more than quadrupled—to nearly $200,000—by the summer of 1939.[44]

By September, the atom smasher had produced, under a tank pressure of sixty pounds per square inch, a one-microampere current at 4 MeV. "There is a tremendous amount of work to be done at voltages lower than 4 million," Chubb wrote Smith just after Christmas, "and in order to proceed with the scientific work, we decided some months ago to be content with what we have and make further improvements to bring it up to the rated t[en] million volts as opportunity affords." Westinghouse now occupied a leading position in the field. The accelerating tube in Robert Van de Graaff's generator at MIT could handle only 2.5 MeV, while the new pressure installation at the Carnegie Institution of Washington achieved a better result at 3 MeV. To obtain higher currents—between four and six microamperes—in the vacuum tube at East Pittsburgh, Wells had to drop the maximum voltage to 3.6 MeV, still well above the operational limit reached at the Carnegie Institution. Just off a visit to Tuve's group "to secure information on the operation of the atom smasher," Wells confirmed plans there to scrap the original vacuum tube in favor of a new one to reach higher voltages.[45]

While they fine-tuned the atom smasher for routine operation, Wells and the new postdoctoral appointees—Shoupp, Haxby, and Stephens—

also began to produce small quantities of radioisotopes (nitrogen, sodium, and silicon) and to measure the minimum threshold energies required to emit neutrons from light elements bombarded by protons. Target materials included beryllium, boron, carbon, and lithium and, later, much heavier elements, such as uranium. At a meeting of the American Physical Society held in New York City early in 1940, Wells proclaimed the scientific value of this research. "All of these reactions, or thresholds, occur at a very sharp and definite voltage," he declared, "and will consequently serve as a good inter-laboratory voltage standardization table." Similar data recorded elsewhere, *Science* reported after the meeting, collectively "serve as valuable calibration points for the giant electrostatic generators with which scientists smash atoms." Commenting on the results back in Pittsburgh, Chubb told Smith, "Although we cannot point to any commercial accomplishments, we feel that the scientific work has already been quite worthwhile and will add to Westinghouse prestige." Then, in May, Wells, Shoupp, Haxby, and Stephens recorded what Condon later called "a truly important scientific discovery—one which is a new phenomenon rather than simpl[y] a working-out of details on a known process." They discovered photofission, the process whereby uranium and thorium atoms are split, not by neutrons or deuterons, the only known methods at the time, but rather by high-energy gamma rays. Their findings appeared in the July 1 issue of the *Physical Review*.[46]

Professional recognition in the scientific literature dovetailed nicely with the company's ongoing publicity campaign. William Wells had helped prepare a layout of the atom smasher for the August 30, 1937, issue of *Life* magazine, which pleased George Pendray, a former science writer and editor with public relations experience who had recently become assistant to Westinghouse president George Bucher. "*Life* . . . is one of the most sought-after magazines on the newsstands these days," Pendray wrote Chubb when the article appeared. "Advertising men estimate it is read by about [ten million] people." Condon, whom *Fortune* had dubbed "Atom Smashing E. U. Condon" in February 1938, broadened the public appeal of the company's investment. "During the last three years," Condon reported to Chubb early in 1941, "I have given several dozen lectures to popular and engineering groups on our nuclear physics work everywhere from Boston to Los Angeles and from Milwaukee to Atlanta."[47] Nationally distributed periodicals and newspapers profiled his public appearances and also highlighted the company's burnished credentials as a source of new knowledge for the electrical industry.[48]

The publicity Westinghouse reaped by way of the nuclear physics program depended, in large part, on Condon's ability to recruit fresh talent and produce novel research results. He personally pressed colleagues in the academic physics community to solicit applications for the postdoctoral fellowship program. "We have a splendid bunch of men for the first batch of Fellows," he wrote Philip Morse at MIT in January 1939, "and will really have a marvelously successful project if each year we can get as good a group as we started off with." Appealing to Morse for help in finding applicants for the next round of openings, Condon added, "I am surprised that the people at M.I.T. haven't had more candidates interested in these fellowships because I think that the opportunities for work that are presented to young men in physics, chemical physics, and physical metallurgy are superior to any in the country right now." Condon executed what amounted to a sales pitch for new research recruits, and it worked. In the spring, a newly minted PhD from MIT who planned to study thermionic emission from metals got the call to report to East Pittsburgh for a two-year appointment.[49] Such tactics improved the company's reputation among academic scientists, and they conferred institutional legitimacy to the strategic vision for fundamental research in industry that Condon had carefully nurtured at East Pittsburgh. The academic-industrial collaboration he proclaimed on Westinghouse's behalf also exacted a high price. It ran afoul of more pragmatic elements in upper management who questioned the market potential of the research results Condon and his acolytes in the press championed as the United States edged closer to war.

CHAPTER 4

New Products for New Markets

INCHOATE REFERENCES TO commercial applications did not undermine the justification for a large-scale investment in fundamental research at East Pittsburgh. That initial vagueness only reinforced the autonomy Marvin Smith and Lewis Chubb granted to Condon to direct research along scientific lines he deemed worthwhile. The unrestricted scope of the postdoctoral fellowship program reflected the same priorities. The absence of preselected research fields explicitly tied to the company's technological and market strengths—a key incentive designed to boost the academic appeal of the fellowships—yielded on an annual basis a large pool of potential recruits whose research interests cut across a broad range of topics. The selection of five finalists for the second group due to arrive at East Pittsburgh in 1939 proved especially difficult. "We are having a hard time making up our minds about the Fellowship[s]," Condon wrote Frederick Seitz in March, "because there is such a variety of applications in non-comparable fields."[1]

Once Condon began to turn out product ideas based on the studies just under way in the disparate fields of nuclear physics and mass spectrometry, his corporate handlers began to assert their influence and react more cautiously. The mass production of radioisotopes for medical research and analytical laboratory instruments for the chemical and oil industries strayed too far from Westinghouse's established markets and core technological competencies. Especially in the case of nuclear physics, Condon identified seemingly ripe but unproven commercial outlets that did not take advantage of the program William Wells had spearheaded prior to his arrival. Rather than exploit the existing setup he had inherited from Wells, Condon handpicked a rival technology—the cyclotron—to establish new product lines, but uncertainties in the marketplace combined with the prospect of higher capital investment costs to thwart his ambitions. Although he had built a first-rate scientific research program at East Pittsburgh, Condon failed to translate its output into salable goods and services for academic and industrial consumers.

RADIOISOTOPES

Early in 1939, while his staff fine-tuned the electrostatic generator for routine operation, Condon opened discussions with the medical school at the University of Pittsburgh to establish a cooperative research program that focused on the medical applications of nuclear physics. This proposal replicated on a larger scale the collaborative arrangement he established later that summer with Frederick Seitz at the University of Pennsylvania to study the physics of solids. In both of these cases, Condon invoked the same vision of academic-industrial cooperation that structured the postdoctoral fellowship program. However, unlike the physics of solids, which complemented the company's long-standing expertise in metallurgy and materials technology, nuclear physics had no clear-cut correlation to an existing knowledge base at East Pittsburgh. The atom smasher had not yet produced any research results, and the speed with which Condon moved to collaborate with the University of Pittsburgh prompted Marvin Smith and other interested parties to question the legitimacy of his research objectives.

On the face of it, Condon did not propose anything radically new. Westinghouse had already established a precedent for collaboration with the University of Pittsburgh. Late in 1937, the company awarded the first grant—for $50,000—to the medical school's new department of industrial hygiene for a three-year research program to assess, according to a *Journal of Applied Physics* article in January 1938, "the value of artificial fever as a weapon against rheumatism, arthritis, venereal disease, the common cold, influenza, heart disease, tuberculosis, and brain disorders." Although practical objectives—such as improved productivity of factory workers—may have provided the initial impetus to fund the research, Westinghouse proceeded with a broader goal in mind: "to make the results available to all medical authorities as a contribution to public health." It is likely that Condon knew about these prior developments when he approached the medical school to propose a joint research program in nuclear physics. T. Lyle Hazlett, a physician who had served since 1920 as medical director at the factory in East Pittsburgh, also sat on the industrial hygiene department's advisory committee, a consulting body comprising university and business representatives and recently convened to provide overall direction and solicit local financial support for medical research on behalf of the industrial community. Condon drew inspiration from Hazlett, who had shown "an enthusiastic interest in developing an arrangement of this kind."[2]

Condon discussed research options with Hazlett, as well as physics department chairman Elmer Hutchisson and representatives from the medical school. Two fields of investigation showed promise: the use of radioisotopes as tracers to study chemical and biological processes and the direct application of neutrons to treat diseases. To build a stronger case, Condon referenced similar research programs under way elsewhere—at the University of Michigan, the University of Rochester, and, most notably, at Berkeley, where a new sixty-inch cyclotron, the largest one then in existence and dedicated exclusively to medical problems, had just been completed for radioisotope production and neutron therapy. MIT followed suit and broke ground for a cyclotron for the same purpose. In June 1939, Merle Tuve's group at the Carnegie Institution of Washington started building a duplicate of the sixty-inch cyclotron at Berkeley for research in medicine, biology, and also physics. Unknown to Condon, Rochester had already prepared to move upmarket; it drafted plans to acquire a sixty-inch cyclotron, also based on Lawrence's design, to use in an expanded medical research program.[3]

Condon recommended, pending the approval of the advisory committee to the industrial hygiene department, that the medical school—itself replete with funds for cancer research and additional resources under the committee's direct supervision—fill three new faculty appointments in physics, biochemistry, and physiology. Supported on an annual budget of $25,000 for five years, the new recruits, Condon wrote, "would devote themselves to the full-time study of medical problems using radio-active materials." The plan also leveraged related research under way in the

4.1. The new sixty-inch cyclotron at the University of California, Berkeley, August 1939. Milton White, whom Condon had recruited to Princeton four years earlier, is standing next to the magnet. Courtesy of Lawrence Berkeley National Laboratory

chemistry department. In 1936, a private foundation established ten years earlier by local retail magnate Henry Buhl awarded $15,000—the first of several grants the university received—to the department for research on the biochemical foundations of vitamin C. This work, which had since grown to include broader studies of animal nutrition, combustion reactions in cells, and the molecular structure of synthetic carbon compounds, stood poised in early 1939 to incorporate radioactive tracers, a natural point of entry for the enlarged program Condon and Hazlett envisioned. To get started, Condon expected the university to obtain radioisotopes produced in the atom smasher at East Pittsburgh. "If the work began to show results as important as I feel it should," he wrote, installation of additional high-voltage equipment at the university or at a nearby "cooperating hospital" would likely be necessary—"probably through Westinghouse engineering assistance."[4]

Condon deliberately emphasized the minimal corporate outlays in money and staff time required to initiate the program. "On the Westing-house side," he told Chubb, "the program does not call for expansion beyond the present plans for . . . fundamental research." One of the postdoctoral appointments scheduled to begin in 1940 had already been set aside for a specialist to work on the chemistry of radioisotopes. Adding in Wells and three of the five postdoctoral fellows brought on board in 1938—William Shoupp, Robert Haxby, and William Stephens—Condon confirmed that "we will have provided a staff of five men and . . . large equipment principally devoted to fundamental studies in the physics and chemistry of this field." He also highlighted the recent arrival of an old friend and collaborator from Princeton days. Benedict Cassen had returned to Westinghouse, this time as a member of the research staff at East Pittsburgh, to develop "high-voltage therapy equipment" for the Westinghouse X-Ray Company. Finally, Condon emphasized the likelihood of generous publicity. "I feel that the plan can be quite valuable in definitely identifying us with an important field of medical progress," he wrote, "in a way which can react favorably on our general scientific prestige." Marvin Smith, who had received a copy of the proposal, shared Condon's enthusiasm "for the idea," but only to the extent that "it does not interfere with our program on industrial research work." He told Chubb, "I am not agreeable to diverting any appreciable amount of our funds or the time of our people to the medical research problems at this stage of development."[5]

Condon shelved the proposal while Smith pulled out all the stops to get the atom smasher up and running. Condon also discontinued negotiations with the University of Pittsburgh, even though, as physics department chairman Elmer Hutchisson wrote a few months later, "Condon has offered to supply the University with artificial radioactive material for experimental work as soon as the generator [at East Pittsburgh] is finished." In the meantime, Hutchisson proceeded on his own to secure a reliable source of radioisotopes for medical research. Backed by the advisory committee to the industrial hygiene department, Hutchisson drew on funds the medical school made available to hire Alexander Allen, chief physicist at the Franklin Institute's Biochemical Research Foundation, founded in Philadelphia in 1935. Allen, who had recently built a cyclotron there, began sketching out designs for a new one—incorporating forty-four-inch pole faces—at Pittsburgh.[6] In a related move to exploit the investigative potential of radioactive tracers, Hutchisson also tapped into a new $45,000 grant from the Buhl Foundation for cooperative studies in chemistry, biology, and physics to establish a separate research program in biophysics headed by Samuel

Simmons, a newly minted PhD from Berkeley who helped Allen draw up plans for the cyclotron.[7] Even after it began operating in the fall of 1939, the atom smasher at East Pittsburgh quickly fell out of favor as a temporary source of radioisotopes for the new biophysics program, while the lack of funding for the cyclotron and a building to house it on the University of Pittsburgh campus delayed Allen's progress.[8]

The first tracer studies—of rat tissue—that Simmons carried out relied on radioactive copper shipped to Pittsburgh from Berkeley and the University of Rochester. "The Westinghouse type of equipment," quoted a January 1940 progress report to the Buhl Foundation, "does not operate at a high enough voltage to produce suitably active copper." Moreover, the postdoctoral fellows who worked for Condon at East Pittsburgh now consumed for their own research nearly all of the radioisotopes produced on site. Despite Condon's continued enthusiasm—"he has already given us much useful apparatus and . . . is pushing the cyclotron as if it were his own project," Alexander Allen wrote in October 1939—Marvin Smith pulled Westinghouse back from further participation.[9] The university labored unsuccessfully for more than a year to obtain financing for the cyclotron. Then, in December 1940, Sarah Mellon Scaife, the niece of industrialist Andrew Mellon, donated the funds Allen needed to proceed. Westinghouse expressed some renewed interest in nuclear physics at the University of Pittsburgh in the summer of 1942, but it had nothing to do with Condon's previously aborted proposal for cooperative research with the medical school; instead, the company agreed to underwrite the cost of stock electrical equipment for the cyclotron.[10]

Late in 1939, while his proposal for cooperative research lay dormant and the University of Pittsburgh embarked on the construction of a cyclotron, Condon revisited the subject of nuclear medicine. This time, however, he deliberately switched tactics. In a bid to attract a more favorable reaction from Marvin Smith, Condon explicitly connected his prior arguments about the prospects for growth in biological and medical research to corporate strengths in manufacturing, sales, and marketing—an approach he believed would be more likely to receive a sympathetic hearing. "This report," Condon wrote in the opening statement of the proposal he submitted to Smith, Chubb, and other executives responsible for new product development at East Pittsburgh and also at the lamp and x-ray subsidiaries, "initiates the analysis of the business of manufacturing artificial radioactive materials, together with auxiliary measuring equipment, for sale to the medical profession and to research institutes." He solicited their approv-

al to build a cyclotron for quantity radioisotope production at East Pittsburgh and also to manufacture on a commercial scale at the lamp and radio
divisions in Bloomfield, New Jersey, and Baltimore the necessary detection
equipment—counting tubes, amplifiers, and recorders. In the case of the
auxiliary measuring equipment, he anticipated a markup of more than
100 percent. "We are now getting experience at Research [East Pittsburgh]
with various types of circuits in connection with the atom smasher work,"
he wrote. "This experience will enable us to offer a good design for such
equipment within the next few months without extra development costs."[11]

Although the Research Corporation held on Ernest Lawrence's behalf
the patents on the cyclotron, no commercial market for radioisotopes existed. Condon nevertheless anticipated rapid growth and the concomitant
requirement for an industrial supplier to meet rising demand. "A rough
index of the interest in this field," he elaborated, "is given by the fact that
there are now more than twenty large cyclotrons in university laboratories
in America, with many others projected." Local experience also guided
his thinking. The University of Pittsburgh obtained radioisotopes for the
tracer studies, just under way in the physics department, not from the
atom smasher at East Pittsburgh but from the cyclotrons at the University of Rochester and Berkeley. Condon held out hope for improvement at
Westinghouse, "to double our present total voltage [to 10 MeV] . . . and
multiply the current at least by 10, but this will take time." In the interim,
he acknowledged that "the cyclotron is superior as a machine for producing radio-active materials in quantity." Merle Tuve, who had pioneered
the development of high-tension accelerators at the Carnegie Institution
of Washington and now presided over the construction of a sixty-inch
cyclotron modeled on Lawrence's at Berkeley, acknowledged the same
advantage. When William Shockley inquired about the acquisition of
electrostatic generator technology for a limited nuclear physics program at
the Bell Telephone Laboratories in the spring of 1940, Tuve stated bluntly,
"If you want . . . equipment for producing artificial radioactivity, there is
nothing that even faintly begins to compare with a cyclotron, nor has there
ever been."[12]

Condon suggested that "steps be taken to get an idea of the market
possibilities in this field," and he also requested "an investigation of the
patent situation" in advance of negotiations with Lawrence. "The project
could be handled at [the] Research [laboratories]," he wrote, "perhaps as
an activity of the New Products Division jointly with the X-Ray Company,
whose sales organization would distribute the product." Lacking in-house

expertise, Condon recommended the construction of a forty-four-inch cyclotron—expected to produce beam currents between ten and one hundred microamperes at 10 MeV—based on cost and design data provided by Alexander Allen at the University of Pittsburgh. He obtained from Samuel Simmons, Allen's collaborator on the cyclotron and the new head of biophysics research at the university, data on estimated yields of "the most important products"—radioactive phosphorus, sodium, iodine, and sulfur—from which he calculated the projected costs of production.[13]

Condon began to execute his plan of attack early in 1940. He arranged a trip to California, which included a visit to Berkeley, in February to get the latest information about "the medical use of radio-active materials." Then, in April, Lawrence enlisted Condon's help to solicit support from Westinghouse—"engineering help and advice . . . and special consideration in regard to prices"—to acquire electrical equipment for his colossal 184-inch cyclotron. The Rockefeller Foundation had given Lawrence more than $1 million for the project. He asked Condon to arrange a meeting with Westinghouse chairman Andrew Robertson to discuss the possibility of collaboration during a brief stopover in Pittsburgh. It is not known if that meeting took place, but, in late March, Condon had told Lawrence that company president George Bucher planned to be in San Francisco on business in May, accompanied by his assistant, George Pendray, "one of my best friends in the company and a man with a keen and broad interest in all modern development in science." On Pendray's request, Condon asked Lawrence to arrange a tour of the cyclotron laboratory. Other executives on the same trip included Robertson and also Marvin Smith, who had received a copy of Condon's proposal for commercial radioisotope production. Lawrence briefed them all on plans for the 184-inch cyclotron, while Robertson, according to the laboratory's assistant director, seemed "most anxious to help in every way possible." Smith "saw the laboratory and our plans for the new project" on May 13, and the next morning Pendray arrived with the company's head of sales for the West Coast. "[Pendray] was primarily interested in the medical applications of the cyclotron," Lawrence's deputy wrote that afternoon after the tour, which suggests his visit may have also served as a fact-finding mission to help build internal support for Condon's proposal to manufacture radioisotopes in quantity.[14]

Pendray's involvement did not strengthen Condon's hand. That summer, after the Westinghouse contingent had visited Berkeley, Lee DuBridge participated in a special weeklong symposium on the physics of metals at the University of Pittsburgh. Proximity to Westinghouse gave DuBridge the opportunity to catch up with Condon, who briefed him on the status of the discussions at East Pittsburgh. In late August, shortly after he returned to Rochester, DuBridge told Lawrence, "[Condon] finds the Westinghouse officials conservative in regard to undertaking the commercial marketing of radioactive materials at the present time." What DuBridge paraphrased as "conservative" behavior, however, Marvin Smith and other executives likely justified as a prudent business strategy. Estimated costs for materials, construction, and overhead to build and operate the cyclotron Condon envisioned totaled more than $70,000, a sum that likely proved hard to justify given the company's substantial outlay—nearly three times that amount—for the just-completed electrostatic generator. The projected financial burden put Condon on the defensive. His associates also may have felt slighted at the prospect of the atom smasher's downgraded status given the new emphasis on radioisotope production. Condon later denied such claims. "Although it has been hinted at that I really had in mind simply a desire to round out the nuclear work by adding a cyclotron to our already quite expensive equipment," Condon recalled early in 1941, "I insist that I then foresaw the great interest that has developed in this field."[15] Either way, the large-scale manufacture of radioisotopes resided outside Westinghouse's established markets. Those executives tasked with rendering a decision on his proposal most likely considered the investment too risky without more evidence of strong consumer demand.

How to manage and commercially exploit this type of research had weighed heavily on Marvin Smith. Early in 1939, just before Condon petitioned him to tie radioisotope production at Westinghouse to medical research at the University of Pittsburgh, Chubb, Joseph Slepian, and Thomas Spooner—a veteran electrical engineer at the laboratories and Chubb's assistant—prepared a research planning report at Smith's request. It attempted to quantify the value of fundamental research, or what they called "pure" or "pioneering" investigations "in new fields without specific application in mind, but with the hope that information gained will be of value in future application." Acknowledging the difficulty and arbitrariness of determining "just what monetary value can be assigned to the results of research," they nevertheless singled out as successful products those commercial technologies derived from pioneering investigations that each generated at least $1 million in new business for the company. Research on the conduction of electricity through gases, for example, had yielded important technological breakthroughs, what the report called "nuggets"—lightning arresters, circuit breakers, and rectifiers—that satisfied these selection criteria.[16]

In each case, Chubb, Slepian, and Spooner invoked a research strategy that Smith likely used to evaluate Condon's forthcoming proposals for the commercial development of nuclear physics. "While the discovery of an individual nugget may in itself be unpredictable," they wrote, "we can, to some extent, control its occurrence by digging in localities where nuggets are likely to be found." Elaborating further, they constructed a simple heuristic device: "If we are too conservative in this respect, we confine ourselves to well proven and established fields and engage in routine mining operations. The other extreme would be wildcatting with long odds against making a strike. The ideal for pioneering is somewhere between these extremes. Attempt is made to direct this unapplied research in the fields where strikes are likely." Research on electrical conductivity in gases struck a productive balance between tapping an empty vein and striking gold on the first try. Chubb, Slepian, and Spooner did not mention nuclear physics research or its prospects for commercial development in their report. "It is too early to predict the value of current work," they concluded.[17] Condon adopted a more aggressive research strategy that rested on a different set of assumptions embodied in the postdoctoral fellowship program. Rather than exercise caution and reconnoiter the R&D landscape for commercial prospects that bore some relation to the existing knowledge base at East Pittsburgh, he favored from the outset the exploitation of cutting-edge academic research. It constituted wildcatting, the parlance Smith's advisors had used. In the case of radioisotopes, Condon promised to give Westinghouse the early lead, but no market for this product existed in the late 1930s.

In June 1939, one year after the physics department at the University of Rochester had received a $35,000 Rockefeller Foundation grant to support, in collaboration with the medical school, biological research using radioactive materials, Lee DuBridge raised the possibility of charging consumers the cost of producing radioisotopes in the cyclotron rather than continuing to provide them free upon request. A novel idea at the time, DuBridge's suggestion and Lawrence's reaction to it laid bare the challenges Condon faced as he tried to open up the field on an industrial scale. "On one or two occasions," DuBridge wrote Lawrence, "individuals connected with either universities or industrial companies have inquired about the possibility of purchasing from us samples of radioactive materials made in the cyclotron." DuBridge speculated that increased demand might constitute an added and unsustainable burden on laboratories like his own, which had just lost an annual appropriation from the Research Corporation—"our main source of funds for the running expenses of the cyclotron program." He had no "desire to go into commercial production" but asked Lawrence "whether it would be worthwhile for the cyclotron laboratories to agree on some policy."[18]

Although he agreed that laboratories in need of financial support—he mentioned Rochester by name—should "have no hesitancy" to levy a fee, especially on industrial firms, for their services, Lawrence dismissed outright the need for a broad policy to offset and regulate the cost of production. Universities should act individually as needed on a case-by-case basis. Berkeley already possessed abundant financial resources from the Rockefeller Foundation and the National Advisory Cancer Council to provide radioisotopes to users on request and free of charge, a service that gave Lawrence control over the selection of research projects that he deemed worthy of support. That institutional leverage remained a priority. Lawrence also highlighted an "undesirable psychological effect" of large-scale industrial production: "When voluntarily giving the materials, all of us in the laboratory have the feeling that our efforts are in the direction of furthering important work, while if we should sell the materials I am sure that some of the boys would have a little bit [of] the feeling that they were functioning as routine technicians." DuBridge concurred—"I think we shall follow your lead and supply such materials to the limit of our capacity for work which seems to us to be important"—and he dropped the matter for the time being. "The question of supplying such materials to the industrial companies," he wrote Lawrence in late June, "has not yet come up in any critical form and we shall deal with it as the need arises."[19]

Discussions with Condon in Pittsburgh the following year, in the summer of 1940, prompted DuBridge to revisit the subject. By then, Condon's proposal to build a cyclotron for radioisotope production had been shelved, but he continued to investigate possible alternatives. "Condon is considering ways and means of finding out just what the immediate market [for radioisotopes] might be," DuBridge told Lawrence in August. Citing as evidence the regularity with which the Rochester cyclotron produced radioactive copper for Samuel Simmons's biophysics research at the University of Pittsburgh, Condon proposed alternatively to DuBridge that Westinghouse obtain from his laboratory, and possibly others, a limited supply of radioisotopes produced at cost and then repackage and sell them to physics, biology, and medical laboratories at "a reasonable profit." Given that the cyclotron already operated on a full production schedule, Condon offered to pay the cost of hiring additional personnel to expand the department's program and also negotiate directly with the Research Corporation "in regard to patent licensing."[20]

Condon's plan intrigued DuBridge. "I feel that the time is surely coming when the demand for radioactive materials on the part of many laboratories which do not have cyclotrons will be much greater than the existing university cyclotrons can supply," he explained to Lawrence. "Sooner or later one or more companies will find it worth while to set up cyclotrons for the commercial manufacture and distribution of such materials." He also speculated that the demonstrated existence of a viable market for radioisotopes after a trial run "of a year or two" might convince "the officials of Westinghouse" to act favorably on Condon's initial proposal to build a cyclotron for commercial production. DuBridge still proceeded cautiously, because of renewed misgivings about "the whole question of policy in regard to a university cooperating in the commercial distribution of radioactive materials." Lawrence, whom DuBridge had again solicited for advice, liked Condon's plan, but only to the extent that broader industrial participation should complement rather than replace the institutional autonomy Berkeley and other universities already enjoyed as producers and distributors of radioisotopes. "If any undesirable consequences develop," Lawrence told DuBridge, "you can always cancel the arrangement after a reasonable trial."[21]

Lawrence's support buoyed DuBridge, who prepared to start negotiations. "In view of what you say," he wrote Lawrence in mid-September, "I think we shall study our program with a view to making a definite proposal to Condon." Early in 1941, Condon confirmed the receipt of "small supplies" of radioisotopes from the University of Rochester and that "some first steps toward developing a commercial outlet for these things have been taken." Except for this preliminary groundwork, however, sufficient internal support to proceed on the larger scale originally envisioned failed to materialize. Meanwhile, what Condon called the "commercial applications of radio-active materials" did not extend beyond the limited use of radio-phosphorus produced in the atom smasher at East Pittsburgh to test and analyze the properties of different types of steel manufactured locally. "I regret very much," he complained to Chubb in January, "that thus far I have not been able to state my case persuasively enough to get approval."[22]

Rather than follow Condon's lead and plunge headlong into an inchoate commercial market for radioisotopes, executives at Westinghouse sanctioned a seemingly less risky strategy more closely aligned to core competencies in engineering and manufacturing. As his plans for radio-isotope production foundered, Condon obtained their approval to test the market for cyclotrons, to be built on order for industrial and academic consumers. Marvin Smith had already given some indication of this preference in the spring of 1940, when Lawrence inquired about the acquisition of a power supply for the new 184-inch cyclotron under construction at Berkeley. Smith offered to sell more than stock electrical equipment to Lawrence. He recalled during the summer, "We were quite willing . . . to do a reasonable amount of development work that might be necessary to supply equipment of the proper rating and characteristics to meet the requirements of the project which you were undertaking and in the successful operation of which we were naturally interested." In the interim, rival General Electric had submitted a bid for the same power apparatus without the associated development costs. Lawrence queried Smith again to find out if "Westinghouse might be able to come forward with a more attractive proposal," but he declined to change the terms of the original bid. "Our interest and attitude have not changed," Smith wrote. "In view of the fact that the General Electric quotation has apparently been prepared on the basis of excluding development charges, as we had already agreed to do," he explained, "there is no reason to expect that our price would be any lower than theirs." Still favorably inclined to proceed as a supplier of custom equipment for cyclotrons—"We stand ready and anxious to assist you in any way possible so far as consultation and engineering service are concerned," Smith told Lawrence—Condon's handlers laid the groundwork for the more ambitious plan he executed to ramp up production of entire units commodified for a mass market that, like the one for radioisotopes, did not yet exist.[23]

CYCLOTRONS

Westinghouse's first inquiry about building a cyclotron came from the University of Pennsylvania after Condon and DuBridge met in Pittsburgh to discuss options for a cooperative radioisotope production program at the University of Rochester. In January 1940, William Donner, a Pennsylvania steel tycoon turned philanthropist whose son had died of cancer, pledged $200,000 to the university to purchase high-voltage x-ray equipment for the radiology department in the medical school. Plans to upgrade the department's facilities, which Donner had agreed to underwrite in the fall of 1937, called for a separate building to house the new installation and a dual track program of biological research and cancer treatment. Consultation with the physics department, however, raised concerns about the choice of available technologies in the desired 3 MeV range.[24] Clinical evidence also indicated that higher x-ray voltages did not improve remission rates among cancer patients.

Physics department chairman Gaylord Harnwell investigated the efficacy of neutron therapy via the cyclotron. For advice, he turned to Ernest Lawrence's younger brother John, a Harvard-trained physician who had spearheaded research in the field at Berkeley. Lawrence confirmed the limited benefits of high-voltage x-rays. Cautious optimism also guided his assessment of neutron therapy as a suitable alternative. "I would say that there is better than a fifty-fifty chance that neutrons are going to be of great value in therapy," he wrote Louis Ridenour, who had queried Lawrence on Harnwell's behalf. "The answer . . . will have to wait until a large group of patients are studied, treated, and followed over a period of several years." The younger Lawrence did, however, encourage Harnwell's group "to look at the whole field of the possible applications of nuclear physics in therapy." Lawrence touted the cyclotron's ability to produce large quantities of radioisotopes, some of which, he pointed out, had already shown promise in the treatment of leukemia and other ailments. With this information in hand, supplemented by additional cost and installation data obtained from Berkeley, Harnwell persuaded the university administration to use the funds from the Donner bequest to build a duplicate of Berkeley's sixty-inch medical cyclotron.[25]

Harnwell visited Berkeley on a fact-finding mission during the summer. Back in Philadelphia, the physics department prepared to hire the staff to operate the cyclotron for the medical school. The university solicited a bid from Westinghouse in September. "There now seems at last to be a definitely growing interest in commercial construction of a cyclotron," Condon wrote Ernest Lawrence on September 18. "The tendency is now to think along the lines of building one of which is essentially a direct copy of your 60-inch job." Condon obtained from Lawrence a complete set of drawings and specifications for the sixty-inch cyclotron and began preparing a formal bid in consultation with engineers at the Westinghouse X-Ray Company, the subsidiary tasked with building it. From the outset, they agreed to forgo income potential to accumulate the engineering and manufacturing experience required for quantity production. "Engineers of the Westinghouse Company . . . have expressed themselves as greatly interested in being given the opportunity to bid for the contract of constructing our instrument," the university's vice president for medical affairs wrote Lawrence in late September, shortly after he and Harnwell returned to Philadelphia from Berkeley. "They have expressed the greatest interest; appear to be willing to take a loss if necessary for the sake of the experience, so that we think it might be advantageous to have further conversations with them."[26]

Two months later, in late November, following discussions "with East Pittsburgh and particularly Condon," L. D. Canfield, the vice president and general manager of the x-ray company, who in 1937 had diversified the newly profitable medical apparatus maker to produce industrial-scale equipment, acknowledged "that to make a commercial unit will require a great deal of design work. . . . None of us would feel satisfied in building the unit as it is layed [sic] out unless further development was done." Canfield also highlighted a costly distinction between quantity manufacturing of commodity electrical equipment—motors, generators, steam turbines, and transformers—and what amounted to the batch production of a new technology not yet standardized for the commercial marketplace. "A home-made job may be satisfactory to [those] who build it," he wrote the manager of the x-ray company's Philadelphia office in late November, "but it will be necessary that purchased equipment should be a finished design and capable of continuous operation." Canfield anticipated a sale price in excess of $200,000 "for the unit installed . . . and with no allowance for unforeseen troubles in development." That figure still came in at twice the amount the university planned to spend based on a cost estimate Harnwell had obtained in March from Merle Tuve for the sixty-inch cyclotron already under construction at the Carnegie Institution of Washington. Put off by the high bid from Westinghouse—"This is in line with the expectations I entertained"—Harnwell waited on a response from General Electric, the only other vendor the university solicited for a bid.[27]

Unlike Westinghouse, General Electric proceeded much more cautiously. This distinction merits some explanation because it highlights the extent to which Westinghouse rather than GE—the electrical industry's leading innovator—tolerated risk to leverage cutting-edge research as a legitimate source of corporate growth and diversification. General Electric never manufactured cyclotrons or their specialized components but rather had entered the small market for academic accelerators through the sale of conventional electrical equipment—coils for the electromagnets, rectifiers for the oscillators, and so on—directly to universities. "While we could not attempt to build the cyclotron itself," wrote W. H. Branch, Harnwell's contact in Schenectady, "I believe we would entertain a request to furnish the parts with the understanding that you would make the installation since we have no facilities for testing." Branch visited Philadelphia on December 13 to get more details and then discussed options with the vice president of engineering, research director William Coolidge, and the president of the GE X-Ray Corporation, "who would handle such a device commercially if it ever got into regular production." They declined to proceed. By then, the outbreak of war in Europe and the advancing requirements of national rearmament had redirected corporate priorities. "Our engineering departments and laboratories are carrying such an extremely heavy load of work, most of which is essential to the National Defense Program," Branch wrote in early January 1941, "that we feel we could not undertake the designing and building of this cyclotron without interfering with our Defense work." He suggested that the University of Pennsylvania postpone construction of a cyclotron and reassess options for an accelerator at a later date: "We feel sure that through the natural course of technical developments which are bound to occur, a delay of two or three years may well result in your obtaining a much better or simpler or less costly, or possibly quite different type of device than would result from the duplication of a cyclotron. . . . Under normal conditions, we would be glad to undertake such a project."[28]

With Westinghouse and General Electric out of the running, Thomas Gates, the president of the university, investigated the impact of defense priorities on the acquisition of critical materials needed to proceed with the construction of the cyclotron "ourselves and not by contract." On January 13, one week after GE withdrew from the competition, he conferred directly with Edward Stettinius, the former chairman of the United States Steel Corporation who now served on the Advisory Commission to the Council for National Defense that President Roosevelt had established in the spring of 1940. Writing on Gates's behalf the next day, William DuBarry, the university's vice president, told Donner that "Stettinius said that there were half a dozen cyclotrons now in operation and that the general policy was not to duplicate expensive equipment, unless it was absolutely necessary." Three days later, on January 17, shortly after he canceled the physics department's search for a candidate to operate the cyclotron, DuBarry suspended all work on the project.[29]

The campaign to line up industrial rather than academic consumers of stock cyclotrons yielded similarly disappointing results. In the summer of 1940, the American Cyanamid Company, a manufacturer of agricultural fertilizers and other nitrogen-based commodity chemicals that had recently diversified into the production of dyestuffs and pharmaceuticals, weighed the option of acquiring a cyclotron to produce radioisotopes for in-house tracer studies and also for commercial sale. By then, Condon had abandoned plans to build a cyclotron for radioisotope production along the same lines at East Pittsburgh. Referencing American Cyanamid's plan after a conversation with the director of research at Lederle Laboratories Incorporated, a maker of vaccines, antitoxins, and other biologicals that the company acquired in 1930, DuBridge wrote Lawrence, "They are much further advanced in their thinking about building a cyclotron than the Westinghouse people, exclusive of Condon."[30] American Cyanamid's initial interest in nuclear physics, however, did not evolve into the dedicated in-house research program that Condon hoped to populate with high-voltage equipment from Westinghouse. To the contrary, its embrace of research in the field suffered from the same lack of institutional commitment that had frustrated his R&D initiatives at East Pittsburgh.

Shortly after he succeeded company founder Frank Washburn as president of American Cyanamid in 1922, William Bell—an attorney who had worked for tobacco magnate James Duke, an early investor—launched an ambitious diversification program to offset competitive pressures in established agricultural markets. In 1929, Bell purchased the Calco Chemical Company—a New Jersey–based producer of dye intermediates—and then executed a broader but related strategic expansion into pharmaceuticals, the development of which relied on the products the new dyestuffs subsidiary turned out. By 1940, American Cyanamid had expanded under Bell's stewardship to become the fourth-largest chemical manufacturer in the United States. Now beholden to a large biological and medical research program split among the General Research Laboratories—the company's central R&D organization in Stamford, Connecticut—and the laboratories operated by the Calco and Lederle subsidiaries, Bell prepared to exploit the clinical potential of nuclear physics to expand the pharmaceuticals business.[31]

On May 27, 1940, Bell called on Alfred Loomis, a Wall Street attorney and investment banker turned physicist who maintained an elaborate laboratory at his home in Tuxedo Park, New York, to discuss options for the acquisition of radioisotopes to be used for biomedical research at American Cyanamid. "Bell . . . is so keen on the cyclotron development," Loomis wrote Ernest Lawrence the next day, "that he came to see me with the heads of his three laboratories, to talk over the whole problem of how they could use the various aspects of cyclotron research and products in their laboratories." A well-connected advisor to Lawrence who helped raise funds to support his research, Loomis urged Bell to visit Berkeley. He asked Lawrence to extend a personal invitation, which followed in late June. "They want to discuss whether or not to build a cyclotron themselves," Loomis explained, "or whether by making grants to universities they could get sufficient radio-active material at least for preliminary investigations." Bell arranged a visit to Berkeley to see the new sixty-inch cyclotron first-hand. Ralph Barnes, the head of physics research at the General Research Laboratories in Stamford who had previously worked with Condon on the analysis and interpretation of infrared spectra at Princeton, accompanied Bell to California with the research directors from Calco and Lederle.[32]

After spending a week at the laboratory, a visit that included an intro-duction to the medical research Lawrence's younger brother John super-vised, Bell returned to the East Coast favorably impressed. "The general program, under your direction, of numerous medical applications of radio-active material and of neutrons themselves," he wrote John Lawrence in early August, "seems to me to promise more for humanity than anything else I know." Thanking Ernest Lawrence for "the courtesy extended to us during our visit," Calco's research director also left Berkeley "fascinated by the work and I am sure that it has possibilities in many fields of research."[33] By the end of the month, Bell had settled on the acquisition of a sixty-inch cyclotron, and he began "studying the question of construction costs." In September, Barnes contacted Condon directly to solicit a bid. Negotiations also got under way with the Research Corporation to license Lawrence's cyclotron patents. In the meantime, Condon, who had instructed his staff at East Pittsburgh to proceed with "the development of . . . measuring equipment suitable for commercial manufacture," sent a newly built Gei-ger counter from East Pittsburgh to Stamford, while Barnes waited on Lawrence to dispatch some samples of radioisotopes produced at Berkeley to start a small interim research program.[34]

Details of the negotiations between Westinghouse and American Cy-anamid are not known beyond Condon's recollection, recorded in January 1941, that East Pittsburgh had participated in "several conferences" with the Westinghouse X-Ray Company, presumably on the design of the cy-clotron. By that time, however, Bell's initial enthusiasm had waned, most likely because of the large investment required and the uncertainties that precluded an optimistic assessment of the market for radioisotopes. The prospect of institutional competition from universities that already produced and distributed materials free of charge raised the most pressing doubts. Because the Research Corporation expected to receive royalties in cases where producers sold radioisotopes, Bell's vice president in charge of research insisted that universities, if they chose to do so, abide by the same rule and pay equivalent fees or agree to a fixed sale price at the outset that did not undercut American Cyanamid's incentive for commercialization. Howard Poillon, the Research Corporation's president, rejected the second option—price fixing—on the basis of its questionable legality. He also deemed the collection of royalties from universities unnecessary on the grounds that academic cyclotrons already operated near capacity to satisfy their own research needs. Even if they did decide to sell some excess output to other users, such quantities and the income received, Poillon predicted, would be negligible. Academic producers posed no serious threat to American Cyanamid's commercial interests. Ernest Lawrence, whom Poillon consulted for advice, agreed. "It is both the inclination and the function of university staffs to go forward into new fields," he wrote Poillon in late September 1940, "and I am sure that [American Cyanamid] need not worry about university competition in commercial production." Still, Poillon understood the obstacles that the company and other industrial firms primed to test the waters, such as Westinghouse, faced. "The fears [American Cyanamid] expresses," he explained to Lawrence, "are those which will be existent in the mind of any commercial concern, and will bear greater weight if that concern is considering a cyclotron for the sole purpose of producing radioactive materials."[35]

In October, presumably after discussions with Condon about the acquisition of a cyclotron for commercial radioisotope production had been dropped, Stanley Beard, Lederle's director of research, announced a more modest and less risky plan to obtain materials for in-house studies from external sources. Bell had already proposed this option during initial discussions with Alfred Loomis in the spring of 1940. "There seemed to be a possibility," Beard wrote immediately following a conversation with Bell on October 10, "that part-time service from two of the Eastern cyclotrons might be made available for several months for radioactive phosphorus and possibly other salts." In late February 1941, after his negotiations with Condon had ended, Ralph Barnes set aside funds to hire a nuclear physicist at Stamford to undertake tracer studies with externally produced radioisotopes, and he initiated a search for suitable candidates. "I must admit that our progress is slow," Bell, who had inserted himself into the recruiting process, wrote Lawrence a month later. "It is impossible to make a good start without a program[,] and our people, unfamiliar with the subject, can't plan the projects." Like their managerial counterparts at Westinghouse, Barnes and Bell relied on outside expertise—Lawrence, in this case—for recommendations. In May, Lawrence persuaded Bell to hire one of his students—Richard Raymond—to work alongside Joseph Kennedy, a young instructor in the Berkeley chemistry department who had co-discovered the fissionable element plutonium earlier that year. Both men, now assigned to the new applied nuclear physics group at Stamford, reported directly to Barnes.[36]

An encouraging debut—"Raymond has made an excellent start with us," Bell told Lawrence excitedly in mid-July—quickly foundered on mismatched priorities at Stamford. "One rather odd defect about these [tracer] problems," Bell acknowledged, "is that some of our people, while they realize that there are fantastic possibilities in the use of atom tracers, fear that they may appear foolish in suggesting what they have in mind, and, therefore, hesitate to make suggestions." The shift from commercial to defense R&D further disrupted plans to get the tracer work on a firm footing. In early August, after barely two months on the job, Raymond resigned and moved to MIT to work on microwave radar at the Radiation Laboratory. Bell acknowledged the source of Raymond's dissatisfaction. "The great problem," he confided to Lawrence, "is to line up projects and, on our part, this is difficult to do because we have such amateurish ideas of how to use this new tool in their solution." Meanwhile, Joseph Kennedy, who had obtained radioisotopes for his research from the University of Rochester, triggered the termination clause in his employment contract shortly after Raymond departed for Cambridge. He returned to Berkeley, leaving Barnes with no option but to grow the radioactive tracer program internally. Barnes dispatched a physical chemist from Stamford—who had never seen a cyclotron in operation—and a biochemist from Lederle to Berkeley for three months "to gain first-hand experience in this field." They resumed early in 1942 the work Kennedy and Raymond had barely started at Stamford the previous year and proceeded on a far less ambitious scale than Bell had originally envisioned when Barnes first queried Condon about the acquisition of a cyclotron from Westinghouse.[37]

Bell's enthusiasm for nuclear physics foundered on the hard reality of an uncertain commercial market for radioisotopes and lack of interest among the research staff, even though the breakthroughs coming out of Berkeley initially resonated with the technical requirements of product development in the company's pharmaceuticals business. A scenario similar to the one at American Cyanamid played out decisively at East Pittsburgh. In mid-October 1940, shortly after Lee DuBridge had declared his intent to work out an arrangement with Westinghouse to sell radio-isotopes produced in the Rochester cyclotron, he left the university on a temporary wartime assignment to direct the Radiation Laboratory at MIT. The shift to military R&D at East Pittsburgh—particularly in microwave electronics—put an end to Condon's unrealized ambitions for commercial radioisotope production and assembly-line manufacturing of cyclotrons. "I suppose I started a little too early trying to interest people in this as a commercial development," Condon wrote research director Lewis Chubb in January 1941. "Now everybody is a little tired of the story, and besides they are all so busy with defense work that they do not care to listen." During the summer, Condon shut down the atom smasher, suspended nuclear physics research, and shifted the assigned staff to electronics R&D.[38]

MASS SPECTROMETERS

While nuclear physics gave way to new priorities for military hardware, mass spectrometry research continued apace on the guiding assumption that lucrative commercial applications might be forthcoming. One of Walker Bleakney's most promising graduate students at Princeton, John Hipple, whom Condon had recruited to East Pittsburgh on a postdoctoral fellowship in 1938, parlayed his mentor's oil-prospecting research into new wartime markets for synthetic rubber, high-octane aviation fuel, and other strategic materials slated for national defense purposes. Although this initiative recorded some successes, it foundered on the same insufficient market demand, managerial indifference, and technological uncertainties that undermined Condon's prior plans to mass-produce and sell radioisotopes and cyclotrons. The evolution of the mass spectrometer from an expensive, complex, and often temperamental instrument devoted to unique problems of academic interest into a rugged, low-cost tool broadly suited to routine chemical analysis outside the laboratory proceeded in fits and starts. Any understanding of the events that unfolded on Condon's watch at Westinghouse must begin with a brief explanation of how the field first acquired legitimacy as a potential technological resource for industry and then adapted to changing market forces during the war.

At Princeton, John Hipple had conducted many of the mass spectrometric analyses of gas samples that outside researchers and laboratories requested from Walker Bleakney on a regular basis. By the summer of 1937, a few months before Westinghouse publicly announced the postdoctoral fellowship program, Bleakney had nearly completed the construction of a new mass spectrometer, one that "will be larger than any previous instrument of its type . . . [and] includes all of the best features found in our experience," he wrote on July 13. When the Research Corporation—the equivalent of a modern-day venture capital firm—learned by way of Ernest Lawrence, an early beneficiary of its largesse, about Bleakney's interest in soil gas analysis, its president, Howard Poillon, offered to solicit more stable funding from the petroleum industry and explore patent opportunities based on the design improvements Bleakney had just introduced. Although he welcomed the opportunity to collaborate, Bleakney acknowledged the high cost of his instrument, the specialized skill and training required to operate it, and the lack of portability necessary to conduct the on-site analyses that oil firms desired.[39] These constraints prompted Bleakney to design and build a more compact instrument for field use. The Research Corporation obtained a patent on its essential features in November 1940, but the portable unit never entered the market because of ongoing improvements that precluded a commitment to standardized mass production. The oil representatives to whom Poillon had reached out for support also questioned the usefulness of the technique itself. "Replies indicated a wide diversity of opinion as to the value or importance of such geochemical analysis," Carroll Wilson, Poillon's deputy in charge of the negotiations, wrote in the biannual report he submitted to Bleakney at the end of 1939. A petroleum geologist at Princeton who monitored Bleakney's progress expressed the same ambivalence. "It is a method which is as yet being tested and perfected," he explained in March 1940, "and it has not yet gained general recognition as one of the dependable techniques for oil exploration."[40]

In a bid to obtain more legitimacy, Wilson offered to run tests on Bleakney's instrument for a nominal fee, but the firms he invited to submit samples for analysis declined to proceed. "We are not interested in this at the present time," replied a geologist at the Pure Oil Company on May 17. An exploratory visit to Gulf Oil's new research laboratories outside Pittsburgh at the end of April yielded the same outcome. Favorable action on Bleakney's patent application at the end of the year did not compensate for these setbacks.[41] The patent's narrow scope—it lacked broad fundamental coverage applicable to all mass spectrometers—combined with the pressing requirements of military preparedness to bring the research program to a halt by the middle of 1941. Bleakney had already accepted an appointment to Division A (armor and ordnance) of the newly established National Defense Research Committee (NDRC) and moved on to military projects while still on the Princeton payroll. "Until the burden of work in connection with the defense program relaxes considerably and frees some of Bleakney's time, as well as the time of the Research Corporation staff," Wilson wrote in his last biannual progress report before the United States entered the war, "there seems to be little that can be done in the way of advancing this matter."[42]

National defense priorities also redirected John Hipple's mass spectrometry research at Westinghouse. Shortly after he arrived at East Pittsburgh in the spring of 1938, Hipple built a mass spectrometer that took advantage of a specially designed spherical magnet to study atomic and molecular structure. He also used the new instrument to detect trace impurities in gas mixtures produced during the heat treatment of transformer steels, a line of inquiry similar to the preliminary gas analyses Walker Bleakney carried out on behalf of the oil industry.[43] Early in 1940, he separated small quantities of U-235 and U-238 present in uranium tetrabromide gas to determine the constituent isotope responsible for fission. Before he obtained conclusive evidence, however, Hipple learned that similar experiments conducted by Alfred Nier at the University of Minnesota and John Dunning at Columbia University (and independently at General Electric) had already confirmed the slow-neutron fission of uranium-235.[44]

Condon, disappointed that Westinghouse "got left [out] on the uranium isotope separation job," instructed Hipple to abandon the fission research and concentrate instead on the completion of a portable mass spectrometer for gas analysis. When his fellowship expired three months later, Hipple accepted Condon's offer to join the research staff of the electrophysics department at East Pittsburgh on a permanent basis.[45] By then, mass spectrometry, and Hipple's instrument in particular, stood poised to emerge as a potentially valuable tool for process control in oil refineries, a related but different field of application that revived interest in the industry as soil gas analysis fell out of favor in the oil-prospecting business. The purported speed and efficiency of mass spectrometry compared to the chemical methods of hydrocarbon analysis already in use at the plant level appealed to oil producers as a cost-effective alternative. Like nuclear physics, however, mass spectrometry still lacked an established market, but the seemingly attractive opportunities for growth prompted Condon and

4.2. John Hipple shows Condon the innards of the mass spectrometer he built at East Pittsburgh in 1938. Edward U. Condon Papers, American Philosophical Society

Hipple, as well as a handful of other entrepreneurs who had cut their teeth by way of soil gas analysis, to test the waters and seek commercial outlets. With Bleakney's research sidelined, Westinghouse entered the field of chemical process control virtually unchallenged.[46]

The only serious competition Westinghouse faced at the outset came from the United Geophysical Company, a much smaller firm founded in California in 1935 to provide petroleum exploration services to the oil industry. Two years later, the in-house R&D organization that designed and manufactured the seismic instruments United used in the oil-prospecting business became a separate, wholly owned subsidiary called the Consolidated Engineering Corporation (CEC), which began selling its products to other customers. In the spring of 1940, while Hipple labored to complete his portable mass spectrometer, the president and founder of United Geophysical—Herbert Hoover Jr. (the son of the former US president)—visited East Pittsburgh. "[Hoover] has five physicists working on mass spectrometers for soil gas analysis and for plant control problems in oil refineries," Condon explained to Chubb, "and [he] started a correspondence concerned with avoidance of any patent competition between his group and Westinghouse." Hoover, who had announced CEC's progress at a meeting of the American Institute of Mining and Metallurgical Engineers in New York City prior to the consultation with Condon, also visited Gulf Oil's nearby research laboratories to make a sales pitch before Carroll Wilson arrived there on behalf of the Research Corporation to drum up support for Walker Bleakney's studies at Princeton. "Hoover . . . said that he expected to get broad patent protection and had developed for this purpose a mass spectrograph superior to any others of which he knew," Wilson told Bleakney shortly afterward.[47]

Hoover may have been too optimistic. Although Harold Washburn, CEC's vice president in charge of research, also hoped to obtain promising results, he had lost confidence in the commercial viability of soil gas analysis for oil exploration before Hoover visited Gulf in Pittsburgh. His ambivalence also carried over to other potential applications. In the fall of 1941, now three years into their work, Hoover and Washburn collectively acknowledged in a regional petroleum industry trade journal that "it may be some time before instruments can be made available for routine purposes in refineries and chemical plants generally."[48] By then, however, American Cyanamid, the Shell Oil Company, and the Mellon Institute of Industrial Research—under the aegis of the industrial fellowship Gulf Oil funded—had already started their own in-house development programs.

Condon exploited their interest to deepen Westinghouse's expertise in the field. He started locally, at Gulf Oil, by way of a postdoctoral fellowship recipient who turned Hipple's large mass spectrometer loose on the analysis of butane mixtures. This collaborative arrangement, Condon wrote in January 1941, "will give us a better understanding of just how useful a mass spectrograph can be in a fairly complicated plant control problem." Separately, Condon received inquiries about Hipple's progress on the portable mass spectrometer from US Steel, the Texas Company (a large refiner and distributor of motor oil and other petroleum products), and also American Cyanamid, Shell, and Gulf. He convened at East Pittsburgh a meeting of representatives from these firms to see the completed prototype first-hand—"successfully demonstrated on June 25," he wrote in his quarterly research report.[49]

Descriptions of Hipple's instrument began to appear in the scientific and industry trade literature shortly after the demonstration. "The first portable mass spectrometer ever built," the *New York Times* reported, went on display for "an estimated 40,000 engineers, technicians, purchasing agents, and business men" in attendance at the eighteenth biennial Exposition of the Chemical Industries in New York City in early December.[50] Condon, meanwhile, turned over to the research products department at the radio division in Baltimore the planning for quantity production and commercial sale of a standard instrument based on the prototype. Production got under way in the summer of 1942.[51] "We now have an instrument designed for commercial sale," Condon explained to MIT president Karl Compton in August: "Three of them have been sold and the first mass spectrometers ever to be made in a factory are now being made in our Radio Division. These will be completed within a month or two and after that we will be in a position to make as many as may be needed—provided there are proper priorities and materials allotments." With a working instrument in hand, Condon emphasized the importance of hydrocarbon analysis to the production of synthetic rubber and high-octane aviation fuel. To break into the market, he took advantage of a contract dispute between Consolidated Engineering and the Standard Oil Company of New Jersey, whose research arm—the Standard Oil Development Company—had just placed an order for three mass spectrometers, CEC's first commercial sale. The purchase agreement required Standard (and all buyers) to share improvements to the instruments after their installation, including royalty-free licenses on patents. The company refused to abide by this provision and promptly canceled the order, which opened the door for Westinghouse to sell mass spectrometers to a small but dedicated customer base.

In late July, Condon arranged—in consultation with the associate director of Standard Oil's Esso Laboratories in Elizabeth, New Jersey—a cooperative program to develop techniques for hydrocarbon analysis relevant to synthetic rubber production using stock production models of Hipple's gas analyzer. "We will have three or four instruments set up," Condon told Compton. Condon confirmed the participation, scheduled to begin at East Pittsburgh in September, of representatives from Esso, Humble Oil, Shell, the Hercules Powder Company, and the Standard Oil Company of Louisiana, with more to follow. The details of how they structured the arrangement are unknown, but the available evidence indicates that Esso initially sought Westinghouse's participation to avoid CEC's disclosure rules, which the company later withdrew. "The agreement is that the techniques so developed by this group," Condon explained, "shall be made freely available to the entire industry, so that the only advantages that accrue to the participants in the program is [*sic*] a slightly earlier knowledge of the results and the fact that their men get trained in using the procedures."[52]

Esso's early cooperation with Westinghouse and other oil and chemical firms slowed CEC's progress, but only temporarily. CEC delivered its first commercial mass spectrometer—a larger and more elaborate instrument that lacked the portability Condon and Hipple had prioritized—to the research department at the Atlantic Refining Company in Philadelphia shortly after Westinghouse commenced production in Baltimore. Plans to expand sales and diversify into new markets proceeded apace after the war. "We are . . . interested in developing any type of mass spectrometer for which there is a market," Harold Washburn wrote Ernest Lawrence, who had taught the young upstart at Berkeley and helped CEC break into the field, in December 1946. Anxious to tap into Lawrence's wartime connections to obtain "future business," Washburn asked him for leads "so that on future Government contracts we will at least have a chance to be considered."[53]

Although it weighed options to license Walker Bleakney's original 1940 patent the same year Washburn queried Lawrence about new market possibilities, Westinghouse did nothing to stop CEC's advance. It remained a competitor for a short time, then exited the analytical instrument business, most likely because of insufficient consumer demand. Participation dwindled to the production and supply of spare parts to users who had previously purchased gas analyzers. Defections from the research staff also sapped the incentive to continue a viable program. John Hipple left East Pittsburgh early in 1947 to join Condon in Washington, DC, as head of mass spectrometry research at the National Bureau of Standards.[54] Without competitive pressure from Westinghouse and General Electric, another early entrant into the field, Consolidated Engineering, dominated the market. It purchased the rights to Bleakney's patent from the Research Corporation in 1953 and offered for sale two years later a new mass spectrometer that incorporated his contributions to the art.[55] By then, Westinghouse had gone on to exploit more promising sources of growth—microwave electronics and nuclear power—that also debuted during World War II.

CHAPTER 5

Westinghouse at War

IN THE SUMMER of 1940, after Germany knocked France out of the war and Great Britain stood alone as the only major power against the Luftwaffe and the threat of invasion, Westinghouse prepared for an influx of defense orders, which accounted for more than one-third of the new business booked by year's end.[1] The surge in defense business also restructured R&D at East Pittsburgh. Shorn of the nuclear physics program that he shut down in mid-1941 to make room for research in microwave electronics, Condon fully embraced the synergies between science and engineering to pursue an R&D strategy that remained faithful to the postdoctoral fellowship program but also fulfilled the wartime priorities Lewis Chubb and Marvin Smith now deemed most pressing. Under their guidance, Westinghouse became a leading supplier of microwave radar equipment to the armed forces, while Condon assumed a critical role in the perfection of the electromagnetic separation process Ernest Lawrence had pioneered at Berkeley to produce weapons-grade uranium for the first atomic bombs.

While he helped to put Westinghouse R&D on a wartime footing, Condon never lost sight of the institutional imperatives that structured his justification for fundamental research at East Pittsburgh. Academic-industrial collaboration guided the postwar planning for R&D that he began to lay out in detail in the spring of 1944. Condon called for the establishment of a new corporate laboratory on the West Coast, initially to take over on a permanent basis the uranium separation work Lawrence had started only a few years earlier. Condon also envisioned a more elaborate research program in such technologically cutting-edge fields as atomic power that promised to take advantage of the local expertise in Berkeley's science and engineering departments. When Chubb and Smith demurred in favor of preserving limited corporate resources for more pressing wartime obligations, Condon proposed a less ambitious alternative that embodied the same cooperative vision. He opened discussions with the university administration to obtain a full-time faculty appointment in the engineering school and negotiated a preliminary agreement with Smith to secure funding for graduate research fellowships that he would oversee on behalf of the company's interests. A breakdown in the working relationship with Lawrence early in 1945, however, wrecked the negotiations and also signaled Condon's diminished status within the company. This outcome and a series of run-ins with Smith and other executives in Pittsburgh in the months that followed prompted Condon's resignation from Westinghouse later that year.

WINDING DOWN NUCLEAR PHYSICS

During a visit to the United States in January 1939, the eminent physicist Niels Bohr brought news of the recent discovery of uranium fission in Germany. On January 26, Bohr along with Enrico Fermi, the 1938 Nobel Laureate in physics who had fled fascist Italy only three weeks earlier, opened the Fifth Washington Conference on Theoretical Physics. "Certainly the most exciting and important discussion was that concerning the disintegration of uranium of mass 239 into two particles . . . with the release of 200,000,000 electron-volts of energy," *Science* proclaimed in a conference summary published shortly afterward. While Condon sat in the audience that day, his research group in East Pittsburgh fine-tuned the atom smasher for routine operation. In a little more than a year, by the spring of 1940, they had discovered photofission (the splitting of uranium and thorium atoms by high-energy gamma rays) and also produced a wealth of data on neutron energy thresholds in nuclear reactions. Meanwhile, the White House had already begun to mobilize the nation's scientific resources to investigate the military applications of uranium fission. The president established the Advisory Committee on Uranium, headed by Lyman Briggs, the director of the National Bureau of Standards, in the fall of 1939 to coordinate the research just getting under way. The pace of national preparedness quickened in the months that followed. Activated on June 27, 1940, the National Defense Research Committee absorbed Briggs's committee and put the government's atomic energy program on a firmer institutional footing.[2]

While in New York City on business a few days later, Westinghouse's Lewis Chubb met briefly with Frank Jewett, president of the National Academy of Sciences and also head of the NDRC division responsible for communication and transportation. Jewett, who had just stepped down as president of the Bell Telephone Laboratories, told Chubb about the Advisory Committee on Uranium and urged him to contact Briggs directly. On July 2, the day after the *Physical Review* published the preliminary results of the photofission research Condon's group had just completed at Westinghouse, Chubb offered to share with Briggs, on Jewett's recommendation, the company's newfound expertise in nuclear physics. "Since we are doing considerable work here in the field," Chubb wrote Briggs, "I thought I would . . . find out the best method of obtaining general information of what is going on so far as to preclude any possibility of wasteful duplication. . . . We feel that there is a possibility that we are working somewhat in the dark and would like any possible advice from you as to the best way of correlating our work with other activities which you know of." He assured Briggs that "Condon and his staff will be very glad to cooperate," and he summarized the investigations under way at East Pittsburgh, "which are essential to a complete understanding of the problem." The discovery of photofission topped the list, followed by measurements of energy thresholds for the fast neutron fission of natural uranium and thorium.[3]

Briggs, who acknowledged that "we have all been deeply interested in your discovery of photo-fission," pressed Chubb to continue work along that line. "This seems to be highly desirable," he wrote. Briggs also identified another priority. "The effective separation of U235 and U238," he told Chubb, "is of course an outstanding problem." At Columbia in March 1940, John Dunning and Alfred Nier had already confirmed that the lighter isotope, uranium-235, in natural uranium split when struck by slow neutrons, which indicated neutron absorption by the heavier and far more abundant isotope, uranium-238. Their results also suggested that fast neutron bombardment of uranium-238 produced fission, but not frequently enough to sustain a chain reaction. Consequently, the path to a self-sustaining chain reaction—the prerequisite for power production or a bomb—quickly focused on the problem of how to separate the constituent isotopes in natural uranium. Research on gaseous diffusion and development of the high-speed centrifuge had already begun at Columbia and the University of Virginia, but Briggs asked Chubb to keep him "advised of any attack on this problem that might be made in your laboratory."[4]

Westinghouse recorded no major breakthroughs. On Condon's instructions, John Hipple abandoned his mass spectrometric analyses of uranium to focus on the development of a commercial gas analyzer for the chemical and petroleum industries shortly after Dunning and Nier confirmed the slow neutron fission of uranium-235. Dialysis showed some potential value as a separation method, but Condon, who had followed work along this line at East Pittsburgh, concluded in March 1941 that it "does not seem that this is as useful a method as the centrifuge, even though it works in principle." By then, Westinghouse had already received an NDRC contract to build a centrifuge for isotope research at Columbia, but this technique never transitioned to the production line. The following year, in 1942, Joseph Slepian filed a patent application for an isotope separator, also funded by the NDRC, called the ionic centrifuge, but it too failed to get beyond the experimental stage.[5]

Meanwhile, despite Briggs's initial enthusiasm, research on photofission proceeded no further than the full-length discussion of the results published in the *Physical Review* in January 1941. Condon highlighted in March photofission's "scientific importance," but he conceded that it "apparently doesn't affect the application matter directly." In early July, Briggs, who still served as chairman of the NDRC uranium committee, appointed Condon to the newly established subcommittee on theory of nuclear chain reactions, under the direction of University of Wisconsin physicist Gregory Breit, an old acquaintance from Princeton days and now a consultant to the Naval Ordnance Laboratory in Washington, DC. In October, Breit asked Condon to confirm the reliability of the fission threshold measurements for uranium-238 obtained in the atom smasher "in connection with the proposed use of the fast neutron chain reaction." He specifically referenced the original research results Condon's group had published in the *Physical Review* in July 1940.[6] No new data existed. Condon had already shut down the atom smasher and reassigned his nuclear physics group to work on microwave electronics, a field of study that originated in the pre-war fundamental research program and now consumed an increasingly large share of laboratory space and resources at East Pittsburgh.

RAMPING UP MICROWAVE ELECTRONICS

Unlike studies of uranium fission, which lacked a clear-cut connection to Westinghouse's business interests, microwave research complemented established product lines in vacuum tube electronics. The company had started working on high-frequency tubes in 1930 but abandoned the research four years later because of poor commercial prospects and financial pressures brought on by the Depression.[7] Recent breakthroughs in the field, however, prompted Condon to revisit the subject and start a small program within his research group at East Pittsburgh. In May 1939, he announced the names of five more postdoctoral fellowship recipients, bringing to ten the total number of newly minted PhDs on the payroll. One of them, Alden Ryan, had just obtained a doctorate in physics from Iowa State College. Ryan planned to investigate, Condon wrote in his quarterly progress report, "[the] dielectric properties of materials at ultra-high frequencies . . . including studies of new devices for generating oscillations in this frequency band." Such frequencies corresponded to wavelengths in the ten-centimeter range.[8]

Condon encouraged Ryan to start with the klystron, a new microwave oscillator invented at Stanford University in 1937. The Sperry Gyroscope Company signed an exclusive patent agreement with Stanford the following year to develop the klystron for its aircraft detection and navigation business, and in May 1939 it opened a tube manufacturing facility south of Palo Alto, in nearby San Carlos.[9] By then, Westinghouse had expressed interest in the new technology. Engineers from East Pittsburgh and the lamp division in Bloomfield, New Jersey, visited San Carlos a month after the factory opened in preparation for negotiations to cross-license the klystron patents Stanford had recently assigned to Sperry Gyroscope. In September, Sperry's chief research engineer visited East Pittsburgh and promised to give Condon one of the first klystrons built at San Carlos. Condon, meanwhile, developed a method to calculate klystron efficiencies, and he worked out the overall theory of the device as a primer for the research Ryan prepared to undertake on the dielectric properties of materials at high frequencies.[10] Ryan received the eighth production unit of the model 400 klystron from San Carlos in October and immediately began measuring its wavelength and power output. Condon assigned other postdoctoral appointees to the project in the months that followed. The pace of development quickened after the summer of 1940, when Westinghouse became the first US company to obtain patent rights from Sperry. Ryan's replacement (Condon did not renew his postdoctoral appointment for a second year) began to assemble power supplies for the ten- and forty-centimeter wavelength klystrons (derivatives of the two types of tubes originally manufactured at San Carlos) built at Bloomfield and shipped to East Pittsburgh for testing.[11]

Condon's expectations for microwave research did not stop there. Frederick Seitz collaborated with postdoctoral fellows Sidney Siegel and Thomas Read to incorporate new work in the physics of solids into the company's existing knowledge base in practical metallurgy and materials technology. Condon envisioned the same synergies between the cutting-edge microwave research under way at Stanford and the vacuum tube technology Westinghouse already produced for a mass market. He handpicked William Hansen—a thirty-year-old, Stanford-trained electronics prodigy and one of the klystron's co-inventors—for a temporary research assignment at East Pittsburgh. Condon extended an invitation to Hansen in the fall of 1939, shortly after Seitz completed his first summer appointment at Westinghouse and returned to the University of Pennsylvania. Condon had known Hansen for nearly a decade; they had met at Stanford in the summer of 1930, when Condon guest lectured in the physics department. "It occurred to me," he wrote Hansen, "that you might like to work here on some basic physics problem but under circumstances where you could get familiar with some industrial problems." Confident in a satisfactory resolution of the patent negotiations between Westinghouse and Sperry Gyroscope, thereby precluding a potential conflict of interest, he preselected Hansen for a postdoctoral fellowship to begin in the fall of 1940, during a planned sabbatical from Stanford. Hansen accepted the invitation, and then, in June, Condon offered him a full-time position at East Pittsburgh. While Hansen pondered his future, the quickening pace of klystron development at Sperry intervened to complicate their plans. In December, as the United States moved closer to war, Sperry transferred the San Carlos operation, and also Hansen and other physicists and engineers working on the klystron, to new R&D laboratories on Long Island, New York, closer to the company's manufacturing facilities.[12]

The defense mobilization that got under way in earnest in 1940 limited the discretion Condon had previously wielded to recruit academic experts for temporary research assignments at East Pittsburgh. Military obligations intervened to block Hansen's appointment and preclude Seitz's return for a third summer session in 1941. The transition from peacetime to wartime, however, differed more in degree than in kind. Condon still managed to execute collaborative arrangements modeled on the postdoctoral fellowship program, but the scope of research no longer varied so widely across disparate scientific disciplines. Instead, it supported the narrower, technological requirements of weapons making. Rather than continue to tap a diminishing pool of university recruits, Condon looked for fresh research opportunities within the new institutional apparatus the federal government established to coordinate and fund weapons R&D at the national level. He appealed directly to his old friend and mentor at Princeton, MIT president Karl Compton, who now served as chairman of NDRC Division D, which coordinated research on detection, controls, and instruments. Section D-1, also known as the microwave committee, focused on high-frequency radar applications.

Headed by Alfred Loomis, the Wall Street tycoon turned physicist, the microwave committee at the outset limited its work to the development of transmitters capable of delivering sufficiently high power at wavelengths at or below ten centimeters. By comparison, the radio tubes Westinghouse built for the radar sets used by the US Army Signal Corps operated at much longer wavelengths, in the three-meter range. These systems required large antennas for transmission and detection, a technological limitation that improved performance at the expense of size and weight. Microwave transmission made system portability on aircraft and other mobile platforms—an NDRC objective—a distinct possibility, but the current technology could not meet the new operational requirements. Low power combined with excessive heat buildup to degrade performance.[13]

In early September 1940, before Hansen left Stanford for the East Coast, Condon queried Compton about the status of the NDRC microwave program and how Westinghouse might be of service. "We would be glad to help here in connection with the work in any way we can," he explained. On October 14, NDRC chairman Vannevar Bush, perhaps at Compton's urging, asked Condon to serve as a consultant to Division D. Two weeks later, Condon and William Shoupp, one of the first postdoctoral fellows he had hired to operate the atom smasher at East Pittsburgh, traveled to MIT for a weeklong conference on applied nuclear physics.[14] On the first day of the conference, they attended a closed-door meeting, run by Section D-1 chairman Alfred Loomis, during which he outlined plans for the radar research program at the new Radiation Laboratory that Compton had just established on campus. He also introduced the group to what Shoupp later called "an astonishing development"—the resonant cavity magnetron, a small, lightweight, high-power microwave transmitter invented earlier that year at the University of Birmingham. A British delegation had just brought it to the United States to help launch the US radar program. Unlike the conventional transmitters Westinghouse manufactured, the compact cavity magnetron generated short pulses of high power at a wavelength of only ten centimeters. Shoupp, whom Condon had transitioned to electronics R&D at East Pittsburgh after he shut down the atom smasher in mid-1941, later recalled the significance of this technological breakthrough. "The high power and light weight of the transmitting tube and the fact that at microwave frequencies it should be possible to construct small directional antennae," Shoupp wrote shortly after the war, "indicated that it would be possible to develop small light weight radar sets, possibly small enough for use in airplanes. In fact, a 30 inch diameter parabola using 10 [centimeter] waves would make an antenna as directional as a 50 foot antenna at the longer, conventional 3 meter radar wave lengths."[15]

Before the conference ended, Lee DuBridge, who had just arrived in Cambridge from the University of Rochester to direct the Radiation Laboratory, handpicked Condon to serve as a part-time consultant. He commuted to and from Pittsburgh on alternate weeks. To get started, Condon drafted "a rather full set of notes on the underlying theory of the microwave field" for distribution to new recruits at MIT. He also began to outline a research program at East Pittsburgh to coincide with the expected increase in the production of transmitter and receiver tubes at the lamp division in Bloomfield and the circuitry for radar sets at the radio division in Baltimore.[16] Two weeks into the new year, Marvin Smith, the vice president of engineering, put Condon in charge of a new microwave committee comprising representatives from all three locations to spearhead this new initiative in collaboration with the work just under way at the Radiation Laboratory. He envisioned a bright future. Microwave electronics, Condon explained to Smith, is "a field that is wide open at the present time. . . . [Immediate] applications are preponderantly military . . . with a corresponding civilian use existing in practically every military application." Condon aimed to hire up to thirty-five physicists and engineers—nearly half of the total to be assigned to East Pittsburgh—"based upon the problem at hand, the obtainable personnel, and the amount of effort being expended by other research groups."[17]

Condon moved quickly to exploit the resnatron, a high-power microwave transmitter whose invention at Berkeley in 1938 preceded the debut of the resonant cavity magnetron at MIT. Even though early versions of the new tube—the brainchild of electronics prodigy David Sloan—operated at a wavelength five times longer than the ten-centimeter threshold the NDRC had established for radar applications, Alfred Loomis agitated for an accelerated development program. "If a tube of 25 to 50 kilowatts at 20 to 25 centimeters were available," Loomis wrote Ernest Lawrence, Sloan's graduate advisor, in July 1940, "there are some very pressing problems that could be powerfully attacked." Elaborating further, he predicted that "if funds are not available from other sources, I am sure that the National Defense Research Committee can supply the essentials."[18] Later that year, the NDRC awarded its first radar R&D contract to Berkeley to continue the project, but the resnatron now faced an increasingly crowded field of competitors. "The problem of getting large power for micro waves is being attacked in different laboratories from three or four different points of view, and I think the Sloan tube will have to meet stiff competition," Loomis told Lawrence anxiously in mid-September, less than two weeks before he and other members of the microwave committee got a first glimpse of the new cavity magnetron at his Tuxedo Park estate. Lauriston Marshall, Sloan's collaborator in the electrical engineering department, agreed with Loomis's assessment. Even though the wavelength for peak power had dropped to thirty-five centimeters by the end of the year, Marshall conceded in the spring of 1941 that "plans for the future development of the high-frequency tube have 'been in the air' for some time."[19]

Condon offered Sloan and Marshall a way forward that put the project back on track, given the microwave committee's newfound preference for the cavity magnetron. He petitioned Karl Compton to transfer the NDRC contract from Berkeley to Westinghouse. Condon anticipated the likelihood of both military and commercial outlets, a strategy consistent with his earlier comments to Marvin Smith about the company's immediate and long-term priorities in the field. "We plan to go ahead with development of commercial tubes of this type," he wrote Compton in mid-April 1941, "to make them available for military micro-wave applications, and subsequently for general use whenever this becomes permissible." Three days after Condon wrote Compton, Benedict Cassen, whom he had recently welcomed back to East Pittsburgh, filed a patent application for a high-voltage x-ray tube that relied on similar technology.[20] Just two weeks earlier, Marshall had highlighted the role that the resnatron might play in the continuation of Cassen's research, which Condon now supervised. "This work," Marshall wrote Sloan on April 5, "has been going on at a wavelength of 3 or 4 meters. It is now believed that our development at 30 centimeters may offer considerable promise. The Westinghouse Company would also like to investigate the possibilities of our device in the medium power field." Marshall and Sloan relished the opportunity to renew industrial interest in the resnatron after the collapse of negotiations with Sperry Gyroscope the previous summer. Now paid consultants to Westinghouse, they prepared, Marshall wrote on May 21, "to go ahead with a full experimental program" that coincided with a major facilities expansion at East Pittsburgh to accommodate three new organizational units dedicated exclusively to microwave research: power tubes, receiver tubes and circuits, and systems and applications.[21]

Discretionary research projects—once the public face of Westinghouse research and a primary selling point of the postdoctoral fellowship program—all but disappeared in favor of investigations that focused on short-term military requirements. The defense mobilization exhausted the pool of prospective applicants. Condon recorded a 50 percent drop in the number of applications submitted for the five fellowships scheduled to begin in 1941. That decline bottomed out at 80 percent in 1942, shortly after the atom smasher shut down. Unlike the four previous groups of fellowship recipients, each of which included five appointees, the latest cohort of only three had been chosen on the basis of how closely their proposed research fit into the company's priorities for defense R&D. Westinghouse suspended the fellowship program at the end of 1942.[22]

Condon did not lament this turn of events. To the contrary, the rapid shift from fundamental research to technological development dovetailed nicely with his own sensibilities and management priorities. Condon had effectively demonstrated the practical usefulness of theoretical physics at East Pittsburgh, even though his academic training and experience resided at the periphery of the company's core technological strengths. He broadened the scope of metallurgical knowledge to incorporate new theories of atomic behavior, and he also sought to reduce, by way of statistical analysis, the number of defects in the transformers that Westinghouse manufactured for electric power distribution systems. The postdoctoral appointees who worked for him demonstrated the same versatility. Nuclear physics and microwave electronics relied on shared skills and competencies that Condon exploited for military applications.

John Coltman, a newly minted PhD in nuclear physics from the University of Illinois, came to East Pittsburgh on a postdoctoral fellowship in the spring of 1941. Condon immediately assigned him to the electronics group to work under William Shoupp, one of Coltman's classmates at Illinois who had encouraged him to apply for the fellowship. A short-wave radio enthusiast with a mechanical flair, Coltman cut his teeth on radar receivers then switched to the three-centimeter wavelength magnetron.[23] The transition from high-voltage to microwave research, however, proceeded in fits and starts. "It is now clear," Condon wrote in January 1942, "that the war will push us deeply into [the microwave field] before we can do much research." Coltman confronted a steep learning curve that confirmed Condon's prediction. The superb functional attributes of the resonant cavity magnetron—high power and small size and weight—did not emerge full blown from a well-ordered research program in which the theory of the device had already been thoroughly worked out.[24] "Its mechanism of operation was only a matter of conjecture," Coltman later explained. "A few empirical design principles were known, but an imposing array of unexplained phenomena beset the experimenter." Given the priority assigned to "the development of a better 3 [centimeter] magnetron . . . it was necessary because of the definiteness of the goal to relinquish many interesting opportunities for research of a more fundamental nature." The group Coltman assembled under Shoupp's direction tripled the efficiency of the three-centimeter magnetron and boosted the peak power tenfold, but the work did not proceed from first principles, the strategy that had initially guided the nuclear physics program.[25] It followed more closely the pattern of technological innovation illustrated in Stephen Angello's research on electron conduction in copper-oxide rectifiers, discussed in chapter 3. The copper-oxide rectifier showcased a mechanical simplicity that masked the largely unknown physical basis of its operation, a shortcoming Condon revisited early in 1939 as a potential source for new research findings.

Angello exploited the opportunity Condon had identified, initially for his doctoral dissertation at the University of Pennsylvania. Then, in March 1942, he joined the electronics group at East Pittsburgh full time to pursue a narrower, technologically driven objective that focused on the development of crystal detectors for microwave radar receivers. "The study of high frequency rectification by crystals, meaning, of course, by high micro-wave frequencies, is exactly what we want Angello to be doing for us," Condon advised Frederick Seitz on the eve of his dissertation defense.

"At present it looks like crystals are here to stay . . . and it is very important to give the subject a very thorough modern going-over to see if we can get better materials than the present ones." Angello focused on silicon, already a leading candidate for microwave detection purposes, and the effort paid off. The radio division in Baltimore began to manufacture silicon detectors based on his research in the summer of 1943.[26] By then, the resnatron, the product of another cooperative initiative Condon had spearheaded, had emerged from East Pittsburgh ready to be introduced into combat as a radar countermeasure device. Westinghouse manufactured more than three dozen resnatrons, each weighing in at five hundred pounds. The army used them to jam German radar, most notably during the Allied invasion of northern France in June 1944.[27]

NUCLEAR PHYSICS REDUX

In late March 1943, while the radio division prepared to scale up production of the silicon crystal detectors Stephen Angello had pioneered at East Pittsburgh, Condon dashed off a brief memorandum to his staff. "I have been called away on a special job which will probably require my absence from the Laboratories for several months," he wrote cryptically, "although I will probably be back for brief visits in the meantime." Condon traded the smoke and din of industrial Pittsburgh for the arid climate and quiet of the New Mexico desert. On his way to San Francisco on business in February, Condon ran into J. Robert Oppenheimer, who had boarded the same train in Chicago to return to Berkeley. Condon and Oppenheimer had known one another for nearly twenty years; they first became acquainted in Göttingen in 1926. Now Oppenheimer, who had already recruited much of his scientific staff, asked Condon to serve as his deputy at the secret bomb-making laboratory under construction at Los Alamos.[28]

Early in 1943, Los Alamos, located approximately thirty-five miles northwest of Santa Fe, the state capital, looked more like a ghost town than the bustling community of civilian scientific and military personnel it would become by year's end. Maj. Gen. Leslie Groves, who managed the Manhattan Engineer District for the War Department, had handpicked Oppenheimer to serve as the scientific director of the new laboratory. Given the large-scale expansion of the site—infrastructure, housing, and laboratory space—Groves instructed Oppenheimer to pick an associate director—a second-in-command—to help plan and coordinate the work in collaboration with the local military personnel responsible for administration on behalf of the Army Corps of Engineers. Oppenheimer selected Condon, whose industrial experience at Westinghouse especially appealed to Groves. "I felt that Oppenheimer should have help in handling the many administrative details of his job . . . so I had urged him to select as an associate director a physicist with [an] industrial background," Groves later wrote.[29]

After further discussion with Oppenheimer in Berkeley, Condon agreed to visit Los Alamos on his way back to Pittsburgh. He reached Los Alamos on March 13 and toured the site with Groves and his staff. The next day, Condon arrived in Pittsburgh and immediately sought approval for a leave of absence from Chubb, Marvin Smith, and Westinghouse chairman Andrew Robertson. Smith initially opposed the move; microwave electronics research at East Pittsburgh still required Condon's full attention. Condon persisted—"I wanted keenly to go," he later wrote—until Smith reluctantly agreed. In early April, a few days after Chubb had named William Shoupp acting head of the new electronics division, Condon returned to Los Alamos, finding the site a huge construction zone overrun with scientists and their families and truckloads of laboratory equipment awaiting installation.[30]

Condon quickly got to work. On April 5, Robert Serber, who had accompanied Oppenheimer to Los Alamos from Berkeley, delivered the first of five lectures to introduce the "basic ideas of the problem"—bomb design and assembly—to the scientific staff. Condon, who had arrived a few days earlier, attended Serber's short course—the prelude to a longer, more intensive review of the subject and research program to begin immediately afterward—and he wrote up the lecture notes. "The object of the project," Condon wrote on the opening page, "is to produce a practical military weapon in the form of a bomb in which the energy is released by a fast neutron chain reaction in one or more of the materials known to show nuclear fission." Every new member of the scientific staff who arrived at the laboratory in the months that followed received a copy of Condon's notes, known as the *Los Alamos Primer*.[31]

Condon's absence still angered Marvin Smith. To calm nerves at East Pittsburgh, Condon wrote directly to Chubb and Smith and dismissed the rumors of his impending departure from the company. He reminded Chubb that he had initially agreed to a provisional three-month appointment at Los Alamos on a trial basis. He also highlighted in a separate letter to Smith the potential commercial benefits that might accrue to the company in the long term. "The subject has progressed farther than I thought and will have very far-reaching implications for us in the years to come," he wrote. "It is no longer near[ly] as speculative as a year ago and in fact it is no longer speculative at all that the subject will have some big implications for engineering." Elaborating further, Condon interpreted the work under way at Los Alamos as a necessary adjunct to the now largely defunct fundamental research program. Condon had never formally relinquished his role as manager and facilitator of the latter type of research; he still held the post of associate director. "It is part of my job as developer of fundamental research for the company," he told Chubb, "to keep in intimate touch with anything as important as this [work at Los Alamos]. . . . As soon as it gets out of the woods and looks like a sure thing, I will want to get out of it and get into the next uncertain thing that looks promising." In the interim, to mollify Smith, Condon promised to spend at least one week each month back at East Pittsburgh.[32]

Smith's immediate reaction to Condon's proposed work schedule is not known, but the uncertainty about whether he or Oppenheimer would sign off on it generated some personal anxiety about the future. "The problem is important and fascinating," Condon confided to his wife Emilie. "But I am quite worried about how long it will require. [I]t looks to me like a very long job perhaps requiring more than three years. I doubt very much whether I could keep the Westinghouse connection alive that long and even if I did I would be way behind in time lost in developing my place in the company organization." Moreover, Condon's brief tenure as deputy director at Los Alamos had already been marred by repeated clashes with General Groves over security procedures at the site. On April 20, the day after Groves had visited Los Alamos, Condon wrote Emilie, "It seems quite clear that the government intends to retain a very tight control over this in such a way that my knowledge of the subject cannot be made of much use to Westinghouse. . . . I know enough of the way things go with the Army to know that I wouldn't want to stay with this work."[33]

Condon referenced Groves's policy of compartmentalization—the intentional segregation of research personnel and the projects they managed to minimize the likelihood of an unauthorized release of sensitive technical information. Condon opposed this policy (other scientists at the site did too, though more discreetly), and he did little to implement its provisions, a pattern of behavior that provoked the general's ire. By the end of April, Condon's patience had expired. "The thing that upset me the most," Condon told Oppenheimer in his letter of resignation, "is the extraordinary close security policy. . . . An aspect of this policy for which I am completely at a loss to find justification is the tendency to isolate this group [at Los Alamos] intellectually from the key members of the other units of the whole project." Condon had also come to dislike the administrative functions of his position. "I don't particularly feel that I am good at that," he had explained to Emilie one week earlier, "and secondarily don't like it well enough to pay the price for my curiosity about this subject (strong as it is)." Condon quietly left Los Alamos on April 27.[34]

Three days later, on Friday, April 30, Condon, now back to work at Westinghouse, toured the manufacturing aisles at the main factory complex next to the research laboratories in East Pittsburgh. The switchgear and control division had just begun to manufacture the large electromagnetic separation tanks, ion sources, collectors, and other components soon to be assembled and installed at the Clinton Engineer Works under construction at Oak Ridge, Tennessee, to produce enriched, weapons-grade uranium-235 for the bombs to be built at Los Alamos. "I am back where I belong!" Condon wrote Oppenheimer on May 3. "Friday afternoon I took a walk in the shops and saw the large pumps and tanks that are being made for [Oak Ridge]. They are coming along fine and are quite an impressive sight."[35] Condon went back to work in the electronics division until later that summer, when Ernest Lawrence asked him to help shore up a major gap in the electromagnetic separation work under way at Berkeley.

Late in 1941, Lawrence had removed the accelerator apparatus from the vacuum tank nestled between the magnet poles of his thirty-seven-inch cyclotron and inserted in its place the components of a mass spectrometer to separate uranium-235 from the more naturally abundant uranium-238. Lawrence scaled up the equipment, placing several tanks between the poles of the huge electromagnet—weighing in at nearly five thousand tons—for the yet-to-be-completed 184-inch cyclotron. The rapid push toward full-scale production, however, proceeded in fits and starts. Magnetic irregularities in the materials that framed out the tanks blurred the beams of uranium chloride ions at the collectors. Specially machined iron shims attached to the top and bottom of each tank varied the strength of the magnetic field between the poles, which allowed any stray ions in a beam to be pulled back into focus at the collector. This adjustment promised to give an appreciable rise in the production rate of enriched uranium-235 without degrading the purity of the material. Oppenheimer had worked out the mathematical theory of this ion-focusing technique before he left Berkeley for Los Alamos, but the engineering details still required careful study. Lawrence recruited Condon to pick up where Oppenheimer had left off.[36]

Unlike his brief and troubled stay at Los Alamos, Condon received full backing from Marvin Smith, especially since the work to be done—expected at the outset to last only a few months—supported the manufacturing job already under way in the switchgear and control division at East Pittsburgh. Condon stepped off the train in his native California on September 21, 1943, and immediately took up residence at Berkeley, where he and six physicists and engineers dispatched from Westinghouse began to work on problems of ion optics and focusing methods basic to the development of magnetic shims for the separation tanks to be manufactured in quantity. Condon's group joined a dozen company engineers already at the laboratory who translated the experimental results the physicists obtained into machine designs sent to East Pittsburgh to guide the production of the process equipment. Manufacturing took priority over R&D. "The experimental work was being done under such frantic conditions," Condon later recalled, "that most of the information was transmitted to the design engineers . . . before the experimental results had been fully checked, so there was a great deal of lost [momentum] as conflicting opinions were being incorporated into [the] designs." Commenting on Lawrence's visit to Oak Ridge in late January 1944, Condon, who had remained in Berkeley, told Emilie, "They are having all kinds of trouble and delays at Tennessee and I am rather discouraged about the whole business."[37]

PLANNING POSTWAR RESEARCH

By early spring, Condon had already decided, in consultation with Chubb, Smith, and Lawrence, to stay on the West Coast "for the duration, if necessary." His pessimistic assessment of the project—the haphazard and rushed development of the production equipment installed at Oak Ridge—prompted a reexamination of the company's research priorities. Looking ahead to the postwar period, he bemoaned the emphasis on manufacturing at the expense of a broader program of research and development. Condon understood the urgency of short-term production commitments. "I feel that in the meantime it has become clear, that we have a continuing job to do in the evolution and further development of this field," he wrote Chubb in mid-April. "We are in a fundamentally unsound position in manufacturing so much stuff of which we have so little technical knowledge. . . . We should do more to get into a more sound position based on the assumption that this [work] will be with us for a long time to come."[38]

Condon recommended that Westinghouse hire and train "a first rate engineering group that . . . does not just lean on the laboratory here [at Berkeley]." Directed by "a competent engineering manager close to research [at East Pittsburgh] with no other responsibilities" and drawing initially on Berkeley for experienced personnel, the new group would cooperate with Lawrence's staff "on the basis of scientific and technical quality" rather than simply serve as a conduit for the transfer of manufacturing knowledge back to the switchgear and control division. Condon also urged Chubb to act quickly to prevent Westinghouse from losing its competitive advantage to other firms also working on the project. "If they show initiative," he cautioned, "they will be getting the replacement orders and orders for conversion units to improved models instead of us and we will find ourselves quickly out of [the business] altogether." Already, General Electric and the Eastman Kodak Company, whose subsidiary, the Tennessee Eastman Corporation, operated the Oak Ridge site for the army, had shown interest "in the part of the job that has been ours exclusively up to now." Functionally, Condon wanted to pull the entire project out of the switchgear and control division. "The project is big enough," he wrote, "to justify essentially a special operating division just to foster it."[39]

Chubb shared Condon's proposal with Marvin Smith, who prioritized production over R&D. "While we recognize the merit of your suggestion that we take a more active part in the research and engineering development of this project," Smith wrote Condon, "we think we must give this phase of the matter further consideration and proceed cautiously in view of limitations in manpower and facilities. . . . We are still having difficulty meeting the shipping schedules that are desired even under the present scope of our contract." Smith did, however, show some willingness to be flexible. He invited Condon to "develop the specific arrangements that would be necessary to enlarge our participation along the lines you have suggested." He intended to discuss Condon's proposal with Westinghouse president George Bucher "to consider the advisability of changing our present policy of being a manufacturer only."[40]

Smith promised to follow up with Condon directly, but there is no evidence of further communications between them. The proposal remained on the shelf until October. By then, Condon had learned some of the details of Lawrence's plans for postwar research at Berkeley, which prompted him to restart discussions with Smith and Chubb. Expecting the work under way to continue after the war, Lawrence wanted to transfer the bulk of it elsewhere at that time to focus once again on high-energy physics research. He did, however, intend to keep his hand in the isotope-separation work, pointing out to Condon that the best option should include collaboration with his staff through the establishment of new R&D facilities located nearby.[41] "Since we [Westinghouse] have been the manufacturers of the process equipment," Condon wrote Smith in a letter that outlined their plan for collaboration, "the feeling is that it would be natural for us to organize and operate the facilities on this basis." Elaborating further, he told Smith, "I have felt for some time that it would be a good thing for us to establish a unit of our research organization on the Pacific Coast. I can see that by proper handling of this situation we can do so practically getting the whole move financed by the Government merely by doing aggressively all that they are asking us to do on this job."[42]

Citing overcrowded conditions at East Pittsburgh, comparatively higher building costs in the East, the impending availability of "many good physicists who have been tied up on war projects," and the added benefits of proximity to the scientific and engineering resources at the University of California, Condon outlined a proposal for a new corporate laboratory in Berkeley. Only a few weeks earlier, Isidor Rabi had told Condon that the Radiation Laboratory at MIT "will soon send out notices to all and sundry inviting them to approach . . . Laboratory personnel for jobs to be taken on some time during the next year. . . . If you would like me to watch out for your . . . interests, I will be glad to do so." Condon envisioned a staff of twenty research engineers—including new recruits from the anticipated reactivation of the postdoctoral fellowship program—and thirty shop technicians and support personnel. In a related move, he requested Smith's permission to work closely with the company's Pacific Coast district to broaden the appeal of the proposed laboratory as a regional corporate asset. "This would involve," he told Smith, "our getting some facilities during 1945 and preparing to take over more and more of the research load from Lawrence's laboratory as this situation develops."[43]

Condon prepared to expand the research staff at Berkeley in accordance with a new contract agreement completed on September 30 between Westinghouse and the Stone and Webster Engineering Corporation, the firm the army had selected in 1942 to construct the uranium production facilities at Oak Ridge. He also hoped to bring into this group "two or three young theoretical physicists to work for me . . . on our own expense to do paper work studies of the atomic power question." This line of thinking did not emerge suddenly. Three years earlier, in March 1941, Condon had summarized in a memorandum to Smith "some information about U.S. Government interest in the atomic power from uranium question which may be useful in considering our policy toward specific questions which involve Westinghouse." Although he recommended no further action "at this time," Condon anticipated a bright future. "The main thing to realize," he explained, "is that there is more to this than one or two little development jobs—if we get in on this nicely, there may be a very large special business for the Navy." Now, in the fall of 1944, he reiterated the same argument in the context of a major expansion of corporate research in California. "I think we should right now establish contact with Admiral [Harold] Bowen [director of the Naval Research Laboratory] or other top figures in the Navy," he wrote Smith, "and attempt to get their blessing to our starting studies in the field and perhaps to get agreement as to some special job we can do for them. Even if that should fail, I would like to have some men studying the question thoroughly."[44]

Condon's plans foundered on lackluster support from East Pittsburgh. Chubb, who had visited Berkeley after Condon submitted the laboratory proposal to Smith, apparently felt that it lacked direction. This assessment prompted Condon to acknowledge that a successful laboratory "would depend on its being above a certain minimum size at the outset." Despite Rabi's assurance of ready access to research personnel coming off radar R&D at MIT, Chubb and Smith may have anticipated a more competitive labor market or simply did not want to divert manpower and resources from other high-priority projects. Even Lawrence proceeded cautiously. Military priorities precluded a wholesale transition of his research staff back to peacetime investigations. He too preferred to wait until hostilities ended.[45] Nor did Condon obtain the services of the theoretical physicists he had requested to investigate commercial possibilities in the atomic power field. There is no evidence that Smith or Chubb seriously considered this option.

Unable to establish a permanent corporate presence in California, Condon hatched an alternative plan modeled on the same expectations, only narrower in scope. He tried to obtain a faculty position in the engineering college at Berkeley that also included a provision to remain on the Westinghouse payroll as a consultant. This proposal did not transpire out of desperation to remain at Berkeley, although he had on occasion expressed a desire to return to the West Coast. Like the aborted laboratory plan, it originated by way of a more comprehensive strategy to strengthen

Westinghouse's ties to the academic research community. Early in 1942, more than a year before he left Pittsburgh to assist Lawrence, Condon conveyed to Marvin Smith his long-term aspirations for university collaboration that laid the groundwork for the anticipated move to Berkeley after the war. He drew inspiration from a deeply held conviction about the mutually reinforcing roles of corporate and academic R&D that justified the postdoctoral fellowship program and also the temporary research appointments he personally arranged to bring Frederick Seitz, Stephen Angello, David Sloan, and Lauriston Marshall to Westinghouse to continue their pathbreaking work in the physics of solids and microwave electronics. The same guiding principle structured the cooperative ventures that he had conceived but failed to execute successfully: collaboration between Westinghouse and the University of Pittsburgh to support radioisotope research in the medical school and the far more ambitious proposal to set up a national institute for scientific research in Pittsburgh. Condon expounded on the collective action of the universities to supply the scientific knowledge that industry required to turn out what he called "important engineering developments." He placed special emphasis on this relationship. "[It] is," he wrote in a draft proposal prepared for Smith after their first meeting in February 1942, "the one which I am tempted to elaborate more fully because it lies closest to my present interests."[46]

An overhaul of the company's continuing education program for engineers topped his list of recommended actions. Westinghouse established an in-house training course for newly hired engineers in 1910. That same year, the engineering school at the University of Pittsburgh organized a program of cooperative study with Westinghouse and other local firms. Participating undergraduate students obtained practical shop-floor experience to supplement their work in the classroom. Opportunities for graduate instruction debuted in 1924, when the chairman of the physics department offered—on his own initiative and "in a very small way"—advanced training for interested Westinghouse engineers. Then, in 1927, a formal and far more elaborate program dedicated to graduate instruction and research in the engineering school got under way. The experienced engineers in the manufacturing divisions at East Pittsburgh who taught the new hires recruited each year now enjoyed status as faculty appointees alongside their academic peers in the graduate school, while the university got for its new cadre of degree candidates ready access to fully updated and well-stocked laboratory facilities.[47] Enrollment grew quickly, but the output of original research did not meet expectations. "The difficulty has been that the Westinghouse lecturers have not interested their students early enough in their courses in finding problems for research," explained Lee Sieg, dean of the graduate school, in June 1930. Nor is it likely that the engineering school provided much in the way of added motivation. "I realize . . . that any official group inspecting our school . . . will rate us low in our facilities for and attention to research," Sieg's counterpart in the engineering school wrote five years later. "This point, however, does not trouble me greatly as our professors have opportunity for commercial connections which I consider more valuable to us as a going undergraduate school than too much personal laboratory research."[48]

In the proposal he drafted for Marvin Smith, Condon did not mention the cooperative engineering program by name or the quality of research in the engineering school at the University of Pittsburgh. It is likely, however, that he knew about their status, given the broad reforms he recommended and the input he received from the company's longtime chief engineer. Like Lee Sieg, Condon prioritized research from the outset, but he envisioned a wholly different instructional model. "[I] contemplate in effect operating a graduate university with courses and practical training specifically adapted to our needs," Condon explained. "The faculty would be rather heavily loaded with research men of broad fundamental training and a much larger proportion of student work assignments would be to the Research Laboratories than is the case at present." Such an arrangement—structured by "a personal acquaintance of all the young engineers with the facilities, staff, and activities of the Research Laboratories"—promised to yield "the best kind of interplay of research and engineering." It also presaged a new type of company engineer specifically tuned to Condon's sensibilities. "I would like to see the training program for engineers so worked out," he wrote, "that all our young engineers would feel like 'alumni' of the Research Laboratories." These priorities prompted Condon to call for an increase in the number of research fellowships and grants the company regularly awarded to graduate students and faculty in the universities, and he looked forward—based on his own recent experience—to "the bringing of a few university professors to the Research Laboratories for summer work or while on sabbatical leave of absence [and] perhaps the loaning of some of our research men for special short-term teaching jobs in the better universities." All of these proposed initiatives owed their legitimacy to the ethos of cooperative research embodied in the postdoctoral fellowship program. "Of course our present system of Westinghouse Research Fellowships," Condon wrote, "would continue as a basis for a continuously fresh development of our fundamental research and a means of recruiting high-type research personnel."[49]

It is not known if Smith read or even received the proposal Condon prepared. Only a handwritten draft has survived. The ideas and recommendations in it, however, implicitly guided the employment negotiations he opened with the University of California in the fall of 1943 and then pursued more vigorously in 1944 after plans for the West Coast laboratory fell apart. Shortly after he arrived in Berkeley to assist Lawrence, Condon received an offer from university president Robert Sproul to head the new engineering college just established at the university's Los Angeles campus. Condon declined the invitation because of the more pressing demands of war work that had prompted Lawrence's request for assistance, but he left open for future consideration the option of permanent employment in California. Informal discussions continued with Sproul and Morrough O'Brien, the university's dean of engineering, into 1944. "I am more in love with the place than ever," Condon wrote Emilie from Berkeley in early April. "Not only is the climate fine, but the whole region is beautiful and the intellectual atmosphere of the university is wonderful." On October 13, just four days after he submitted the West Coast laboratory proposal to Smith, Condon shared preliminary plans for a return to Berkeley on an academic appointment with John Hutcheson, a close friend and manager of the radio division in Baltimore whom Chubb had promoted to associate director of research in Condon's place the previous year.[50]

Hutcheson urged Condon to put in writing to Smith and Chubb an explanation of "how you can keep in touch with the most recent advances being made in physics and how you can advise Westinghouse Management regarding these and the possible effect they might have on Westinghouse business." On Hutcheson's advice, Condon presented to Smith in late October "an attractive alternative which would get for the company a good many of the advantages which I saw in the western laboratory plan." As a paid consultant and full-time member of the Berkeley faculty, Condon expected to "keep . . . management advised of new developments in fundamental physics which affect our business" and train graduate engineering students in fields of direct interest to the company. Consulting work for other regional universities and technical assistance to the Pacific Coast district office promised to broaden the appeal of his position outside East Pittsburgh. At its core, however, the proposal embodied the same collaborative instincts that Condon had recently communicated to Smith to reform—by way of the research laboratories—the company's in-house training program for engineers. "I think it would be good to set up several Westinghouse fellowships to get graduate students [in the engineering school] to work on projects at the University here," he wrote. "I expect that by correlating with East Pittsburgh some really worthwhile jobs could be handled in this way, and that it would be a means of doing some good recruiting and training of young men for the laboratories at East Pittsburgh and Bloomfield."[51]

Smith reacted cautiously, but his willingness to negotiate pointed to Condon's diminished status at East Pittsburgh. "Although I am reluctant to adopt any plan that takes you away from the Company to the extent that we are talking about," he told Condon, "we will be glad to give further consideration to any plan that meets your personal desires and still makes the maximum use of your services to Westinghouse." Initially groomed to succeed Lewis Chubb—"I once indicated to you a desire to do the job," he reminded Smith—Condon now relinquished that role to Hutcheson, whom Smith had already put in line to become director of research in advance of Chubb's impending retirement. "I think the need for a good manager for the laboratories has now been adequately met by the appointment of Hutcheson," Condon told Smith, "so there is no need for me to offer to take that on. . . . I can be more useful if I have my full energies free to work on fundamental science."[52] Behind these self-styled justifications, however, stood a headstrong personality that sometimes alienated those with whom Condon worked. Smith's assessment of Condon's future at Westinghouse late in 1944 must be understood within this context. He disapproved when Condon agreed to serve as Oppenheimer's deputy at Los Alamos without prior notification. Condon returned to East Pittsburgh after less than one month on the job, but only because he clashed with General Groves and disliked the administrative responsibilities of his position. Tensions erupted again after a run-in with Lawrence at Berkeley as the war entered its final year.

On January 10, 1945, following further negotiations with Dean O'Brien in which he elaborated on "plans for strengthening the work in fundamental sciences in the engineering college," Condon submitted a formal proposal for employment to President Sproul. Westinghouse, through its recently established educational foundation, prepared to set aside more than $60,000 spread over five years to pay Condon's salary in the electrical engineering department, plus fund at least two graduate research fellowships, after which the university would be expected to cover the requisite costs if the company withdrew its support. "I could hardly believe my ears," Condon told Leonard Loeb, recounting the details of the offer that Smith—chairman of the educational foundation's board of trustees—had put forward during a brief visit to East Pittsburgh in December. Separate from the faculty position, Westinghouse planned to keep Condon on retainer as a company consultant and expected him to spend two months of every year at East Pittsburgh. "Inasmuch as my work in this connection would be also in fundamental science," Condon wrote Sproul, "it seems to me that this is a contact which could only operate to increase the effectiveness of the professorship." At a meeting in the president's office on January 17, Sproul verbally approved the proposal. Condon sold the family home in Pittsburgh and, in anticipation of the move to the West Coast, purchased a new house in Berkeley, where Emilie and their three children had been living in temporary quarters since the summer of 1944.

After a promising start, the transition to Berkeley stalled. Failure to secure the additional personnel required by the revised Stone and Webster contract angered Lawrence, who complained about Condon's handling of the arrangements directly to East Pittsburgh without his knowledge. Hutcheson intervened personally to mollify Lawrence and get the project back on track, but the damage had already been done. Lawrence's behavior prompted Condon to leave Berkeley on January 31. "I thought I wasn't of any more use out there," he wrote Leonard Loeb from Pittsburgh. Marvin Smith did not immediately withdraw support, but he used the incident to try to convince Condon to "settle down to my old job here." Still, Condon remained hopeful and prepared to move forward with the negotiations, but only if the university appealed directly to Smith and renewed its commitment through a formal offer from President Sproul. Condon expressed concern that, without some official response from the university administration, Smith might back out.[53]

Condon asked Loeb to approach Sproul on his behalf. "I believe it would be very propitious at this time," Loeb wrote Sproul in late February 1945, "to let [Condon] know personally that you still desire him." Sproul, who assumed "the arrangements were merely taking a bit of time," told Loeb, "I have already gone 'all out' for Dr. Condon as a member of our faculty, and I can hardly do more. Perhaps, you will be able . . . to inform him of this reaffirmation of the thoughts I expressed to him personally on January 17."[54] Despite Loeb's request that he reach out to Sproul again, Condon dropped the matter entirely, still smarting from Lawrence's seemingly deliberate behavior to undermine the appointment, what he considered to be a lack of formal commitment from the university, and news from

Hutcheson, after a brief stopover with Smith in Berkeley in late March, that O'Brien no longer wished to continue the negotiations. "After . . . such clear indications that [the] people in control don't want me I have rather soured on the idea," he wrote Loeb. "I don't think now that anything can be gained by pushing it." Hutcheson dispatched a replacement from East Pittsburgh to pick up where Condon left off, and he remained in Berkeley until the end of July, when the project closed down.[55]

Condon prepared to go back to work at Westinghouse. "There is an awful lot of work to do to get the science of physics rolling again after five years of war gadget making and I am going to devote myself to that," he wrote Emilie in mid-April.[56] The resumption of prewar responsibilities, however, could not mask the personal frustration that had so badly damaged his standing at East Pittsburgh. Marvin Smith expected no further disruptions, but they flared up again in June when Condon ignored the objections General Groves had raised on behalf of the War Department and prepared to travel to Moscow with an American delegation to mark the 220th anniversary of the founding of the Russian Academy of Sciences. Condon did not attend the celebration, but the political fallout and seemingly blatant disregard for corporate protocols thwarted any remaining ambitions he may have had for an industrial career at Westinghouse. Troubled relations with upper management combined with an emerging interest in postwar atomic energy policy to hasten his exodus from Pittsburgh. Condon resigned from Westinghouse in October and moved to Washington, DC, to head the National Bureau of Standards.

CHAPTER 6

Cold War in Washington

EDWARD CONDON'S transition from industry to government initially had less to do with the content of R&D than the political implications of the wartime weapons it helped produce. Like many of the scientists who had worked on the atomic bomb, Condon took an active interest in domestic politics as the war drew to a close. Shortly after the bombings of Hiroshima and Nagasaki, he began to pay close attention to the political machinations just under way in Congress to sort out the disposition of the Manhattan Engineer District and establish new institutional mechanisms to manage the peacetime control of atomic energy. Through his participation in a progressive-left coalition opposed to the War Department's position, Condon met Henry Wallace. One of the movement's leading luminaries and Franklin Roosevelt's former vice president and agriculture secretary, Wallace now headed the Department of Commerce. Wallace, whom Roosevelt had appointed commerce secretary in March 1945, handpicked Condon to be the new director of the department's National Bureau of Standards. Condon prepared to draw on prior experience at Westinghouse to manage scientific research on behalf of the small firms Wallace prioritized as being most in need of government assistance after the war.

6.1. Henry Wallace. US Department of Agriculture History Collection, National Agricultural Library

The first director not selected from within the bureau's ranks, Condon transitioned to a large federal bureaucracy that bore little resemblance to the more intimate R&D organization he had known at Westinghouse. No longer did he personally select young PhD scientists to conduct research on topics of their own choosing. Civil service requirements structured hiring practices at the bureau. They imposed salary caps for new employees and often assigned preference to in-house job candidates who had already accumulated many years of service. The style of management differed too. At Westinghouse, Condon answered to a single chain of command that approved the liberal use of corporate funds for research projects he deemed worthwhile. The Bureau of Standards did not confer the same flexibility. The execution of Henry Wallace's programmatic initiative for technical aid to small businesses required financial support from a constituency outside the Department of Commerce—lawmakers in Congress who appropriated funds to federal scientific research agencies that competed against each other for limited resources. The academic-industrial collaboration Condon had pioneered at Westinghouse and now planned to extend to the bureau did not easily accommodate these procedural requirements of public administration. Moreover, the onset of the Cold War and the corresponding shift to permanent military preparedness rendered a seamless transition untenable. Alternatively, Condon forged cooperative relationships with the army and the navy to tap the universities for new recruits, shore up the bureau's finances, and provide technical assistance to the small business community.

FROM PITTSBURGH TO WASHINGTON

In May 1945, shortly after he had returned to Pittsburgh from Berkeley, Condon received an invitation from Edwin Smith, executive director of the National Council of American-Soviet Friendship (NCASF), to join a delegation of American scientists scheduled to attend a celebration in Moscow in mid-June to mark the 220th anniversary of the founding of the Russian Academy of Sciences. Condon already served on the council's science committee, established in 1943 to facilitate the exchange of technical information between the United States and the Soviet Union. Especially active on behalf of Westinghouse, he obtained Russian publications on topics of company interest, and he also helped coordinate the collection and shipment of publications written by the research staff at East Pittsburgh to Russian libraries. Condon agreed to make the trip, initially without any objection from upper management. "Westinghouse is doing a lot of business with Russia and this trip is regarded as a very important event by the company," Condon wrote his wife, Emilie, on June 5, two days before the scheduled departure of the American delegation from New York City.[1]

This optimism quickly faded when the War Department intervened. Maj. Gen. Leslie Groves, whose security measures at Los Alamos Condon had sharply criticized two years earlier, took direct action to prevent his participation. On June 6, after he had arranged to delay for another day the flight from LaGuardia Airport, Groves telephoned Westinghouse president George Bucher and demanded that Condon remain behind. Groves lacked confidence in Condon's judgment, especially his handling of classified information. The other invited Americans, who, like Condon, possessed an intimate knowledge of restricted data from wartime research, had consented to Groves's request not to attend the celebration. On June 7, Bucher told Condon to return to Pittsburgh. Condon quietly ignored Bucher's instructions and appealed his case directly to President Truman. Although the White House did not intervene, Condon's persistence angered Groves. The irate general complained to Bucher, who, now deeply embarrassed, prepared to dismiss Condon for insubordination. "It now looks like the end at Westinghouse," Condon wrote Emilie two days later.[2]

Bucher backed off after Condon apologized and promised Marvin Smith "to be careful not to do things on the outside without carefully checking with him." He pledged to "stay close to the job" and settled back down to work at East Pittsburgh. The reconciliation did not last. Already hamstrung by Smith's refusal to support a permanent R&D presence on the West Coast and the failed bid for a consulting job that left room for an academic appointment at Berkeley, Condon began to investigate other career opportunities. NCASF president Edwin Smith had helped him prepare for the trip to Moscow. At some point, they discussed the National Bureau of Standards, and while it is not known if Smith initially informed Condon about the impending retirement of longtime director Lyman Briggs or if Condon had obtained that information elsewhere, he urged the new secretary of commerce to arrange a meeting. "Condon . . . who is a friend of mine," Smith wrote Henry Wallace on June 11, "has some interesting ideas in respect to scientific research and your department. I think you would be interested to talk with him."[3] A similar recommendation came independently from Harlow Shapley, a progressive-left astronomer at Harvard who also advised Wallace. "The directorship [of the Bureau of

Standards] is a position of dignity and importance, especially if in the future it tends in the direction of original research in special fields and is not wholly a service station to science and industry," he reminded Wallace on August 1. Shapley believed that Condon—"a brilliant theoretical physicist . . . with micro-wave radio engineering as one his specialties"—fit the bill. Moreover, Shapley, who had participated in the anniversary celebration at the Russian Academy of Sciences, most likely knew about Condon's run-in with Groves and the status of his job at Westinghouse. "Two months ago," he explained to Wallace, "[Condon] told me that he was not satisfied with his industrial research position and hoped that I would keep him in mind if a position especially suitable to his talents came to my attention."[4]

Wallace, meanwhile, followed customary procedure and relied on the bureau's visiting committee—an advisory body to the secretary of commerce comprising eminent scientists and engineers who served on a rotating basis—to recommend a suitable replacement for outgoing director Lyman Briggs, now more than one year past the mandatory retirement age of seventy. Unaware that Shapley had shared Condon's name, as well as those of two other academic scientists, with Wallace, the committee solicited suggestions directly from Briggs.[5] Staying in-house, he submitted the names of two bureau employees at the end of August: Eugene Crittenden, a thirty-six-year veteran and chief of the electricity division who had served as assistant director under Briggs since 1933, and forty-seven-year-old Hugh Dryden, chief of the mechanics and sound division and the bureau's wartime head of guided missile research for the navy. The bureau's ranks favored the youthful Dryden, but Wallace changed course and decided to ignore the committee's deliberations.[6] Instead, he acted on the advice from Shapley and NCASF president Edwin Smith.

Shapley highlighted Condon's appeal in a direct reference to unpleasant working conditions at Westinghouse. Smith, however, identified a strategic rather than a tactical motive; he pointed out to Wallace that Condon "has some interesting ideas in respect to scientific research in your department." Smith's observation merits some discussion. On June 9, two days before Smith wrote to Wallace and encouraged him to arrange a meeting, Condon—who sensed his impending discharge from Westinghouse—alluded to the possibilities that a move to the Bureau of Standards might entail. "I . . . may try to meet Henry Wallace to see if I can fit into one of the bureaus in his department," he wrote Emilie. "In my own opinion I could be valuable to him as director of the Bureau of Standards."[7] These statements suggest an interest in the bureau's top job for reasons that may have had nothing to do with a desire to leave Westinghouse, despite his recent troubles there. It is possible, and perhaps likely, that Condon saw an opportunity to realize the expectations for the national institute of scientific research he had first proposed in Pittsburgh before the war.

A direct extension of the postdoctoral fellowship program he established at Westinghouse, the national institute would cultivate industrially relevant R&D by way of a regularly replenished staff of newly minted PhD scientists recruited to conduct research on topics of their own choosing in a university atmosphere. This guiding principle of the institute's mission now reemerged after the war in an R&D environment strikingly different from the one Condon had known in Pittsburgh. The federal government rather than industry stood poised to provide the bulk of funding for academic research in the physical sciences and engineering disciplines. Condon adjusted his initial conception of the institute to fit this new political reality. Some of the details got a hearing during his first appearance before the Committee on Appropriations in the House of Representatives in January 1946. Rep. Louis Rabaut (R-Michigan) asked Condon, "What are your proposals for changes in the Bureau? Do you have some new ideas?" Condon singled out the importance of "research in fundamental science of a long-range and basic character that is known to be of general value." He also identified the Office of Research and Inventions (ORI), which the secretary of the navy had established the previous May to fund extramural research in the universities, as the most suitable institutional template from which to develop the bureau's new programmatic functions. "I would hope to see us allowed to do for peacetime fundamental science," he told Rabaut, "something of the sort that has recently been announced as part of the Navy's research plans that involve a high degree of collaboration and intimate cooperation at the working scientists' level with universities throughout the country. . . . I think the Navy's plan sounds very good, especially because they seem to have worked out a scheme that involves putting this in the hands of the working scientists themselves."[8]

Condon did not mention his proposed national institute of scientific research by name in his congressional testimony, but the similarities to the bureau by way of the navy's R&D ambitions are not merely coincidental. Wallace got a sense of Condon's thinking only a few months before Representative Rabaut solicited the statement about the bureau's future direction. In late September 1945, Wallace and a select group of scientists attended—by invitation only—a conference that University of Chicago president Robert Hutchins organized to discuss the political, social, and military implications of atomic weapons and to formulate plans for future action on matters of policy. Condon also attended the conference. The gathering provoked much enthusiasm among the participants, and it profoundly influenced Condon's own thinking. Late each night, Condon composed a letter to his wife that described in detail the daily proceedings. "I feel that I am making acquaintances that will at last give you and me an avenue toward effective political action with the people we admire from a distance," he wrote Emilie on September 21. Bent on making "every use of the flexibility inherent in my Westinghouse job," he looked forward to "rigorous action toward constructive use of the atomic bomb situation for [the] establishment of a world government. . . . It is a long chance but the only one that matters . . . right now."[9]

Condon met Wallace for the first time on September 22, the last day of the conference. "He comes up in every way to my expectations of him," Condon wrote Emilie that evening. "We talked about the [Department of] Agriculture regional research laboratories and about [the] possibility of doing something like that for small business. He says he is very anxious to put tremendous vitality into the Department of Commerce[,] and the Bureau of Standards he wants to be a first-rate center of pure research and an outstanding source of technical research of value to small business." Wallace did not hire Condon on the spot; instead, he used the occasion to obtain a first impression. Condon, meanwhile, looked forward to a continuation of the dialogue on atomic energy policy that the Chicago conference had stimulated: "I told [Wallace] if I could be of temporary immediate assistance on technical points relative to bomb policy or other matters that I could give him a good deal of my time."[10] Before he left Chicago, Wallace invited Condon to Washington for another meeting.

Condon called on Wallace on October 3, and the next day, Lyman Briggs learned the outcome of their conversation. "Condon combines recognized ability as a physicist with an interest in and firsthand knowledge of [the] application of research to industry," Briggs wrote Gano Dunn, president of the J. G. White Engineering Corporation and chairman of the bureau's visiting committee. "I think the latter weighed heavily in [Wallace's] decision [to choose Condon] combined with the belief that it is desirable to introduce some new blood." On October 5, Wallace forwarded Condon's name to President Truman for approval, and he highlighted one additional qualification that Briggs had not mentioned to Dunn. "[C]ondon is . . . interested," Wallace explained to the president, "in the practical service functions of the Bureau—of helping business and industry, especially small business, to put scientific and technological developments to the most effective use in promoting employment and production."[11]

Returning to Pittsburgh only once, Condon remained in Washington to monitor—with other like-minded scientists, some of whom he had met in Chicago—the legislative maneuvering just under way in Congress to sort out postwar atomic energy policy. Although he, like Wallace, opposed the legislation that the War Department drafted and that Rep. Andrew May (D-Kentucky) and Sen. Edwin Johnson (D-Colorado) introduced in Congress on October 3 on the belief that its provisions gave the military departments a controlling stake in the peacetime management of atomic energy, Condon did not criticize their position publicly. Wallace insisted on this course of action. Condon's nomination to run the Bureau of Standards still required the president's consent and congressional approval. "I am being discreet and only advocating careful study and thorough consideration [of the legislation] in order to get time for a better bill to be drawn up," he wrote Emilie on October 16.[12]

The absence of public pronouncements that might embarrass the White House did not shield Condon from criticism among the opposing factions as the debate intensified. At the Pentagon two days later—on October 18—Condon met with advisors to Secretary of War Robert Patterson and some members of the interim committee he had established to draft what became the May-Johnson bill. Although conducted "on a very nice and polite basis," the meeting had a clear objective. "I was . . . asked to lay off and help get the bill through [Congress] fast," he explained to Emilie. "I refused and said it had to be discussed and understood by Congress, that I didn't want to stop such a movement, and anyway I couldn't even if I wanted to." Whatever it did to generate negative perceptions of Condon and the professional associations he cultivated, this behind-the-scenes maneuvering apparently did not upset Wallace. On October 23, Wallace wrote in his diary, "The more I see of Condon the better I like him."[13] The path to Condon's nomination, however, did not sit well with some members of the bureau's visiting committee. MIT president Karl Compton, who served on the committee with Gano Dunn, later recalled, "We had the impression that Secretary Wallace had acted precipitately and cavalierly in proceeding so promptly with his action, so soon after he had asked our help." The committee's displeasure had no discernible impact. Condon resigned from Westinghouse on October 24. Truman, meanwhile, raised no objections; he officially nominated Condon to be director of the Bu-

6.2. Secretary of Commerce Henry Wallace presides as Condon recites the oath of office to become the fourth director of the National Bureau of Standards, November 5, 1945. Courtesy Harry S. Truman Presidential Library and Museum

reau of Standards five days later. Wallace administered the oath of office to Condon on November 5.[14]

THE ONCE AND FUTURE BUREAU OF STANDARDS

In the fall of 1945, the National Bureau of Standards comprised thirteen divisions devoted to research in the physical sciences and engineering disciplines: electricity, weights and measures, heat and power, optics, chemistry, mechanics and sound, organic and fibrous materials, metallurgy, clay and silicate products, simplified practice, trade standards, codes and specifications, and ordnance development. The bureau's early growth coincided with the rapid expansion of the nation's science-based industries, all of which required quantitative standards to mass-produce goods for a consuming public. Representative examples include standards for electrical quantities—such as voltage, current, and resistance—and standards for heat, power, radiant energy, density, and pressure. On behalf of medical practitioners, the bureau operated an elaborate x-ray laboratory that turned out dosage and protective installation standards for high-voltage diagnostic equipment used in hospitals. The bureau also tested for quality and uniformity the myriad materials and products the federal government purchased.[15] Military priorities overshadowed these civilian functions in wartime. During World War II, the Bureau of Standards managed the early development of the atomic bomb, established an elaborate radio propagation research program, and introduced the navy's radar-guided air-to-surface missile—or glide bomb—called the Bat. On the eve of Condon's appointment, the War and Navy Departments (which merged with the newly independent US Air Force to become the National Military Establishment in 1947, renamed the Department of Defense two years later) funded nearly three-quarters of the bureau's operating budget.[16]

Condon inherited these wartime obligations, but he did not simply pick up where Lyman Briggs left off. He managed the Bureau of Standards according to principles that originated in Henry Wallace's state-centered prescription for postwar economic growth. This break with the past signaled a sweeping reassessment of the bureau's mission that merits some explanation. Also new to the Department of Commerce, Wallace envisioned the exploitation of scientific and technological breakthroughs introduced during the war to revitalize the bureau's civilian functions on behalf of the small business community. Wallace cut a familiar figure in this role. A progressive agrarian reformer from Iowa, he had nurtured from an early age an abiding interest in science—especially plant breeding—and had worked with the Department of Agriculture's extension service to develop new strains of hybrid corn. This experimental research culminated in a business partnership with two local entrepreneurs. In 1926, they formed the Hi-Bred Corn Company to serve regional markets.[17] Appointed secretary of agriculture in 1933, Wallace put this personal experience to work on behalf of struggling farmers who faced sharp declines in crop prices. He presided over an expansion of scientific research in the Department of Agriculture to help farmers lower fixed costs and increase productivity. "The

6.3. Aerial view of the grounds of the National Bureau of Standards in northwest Washington, DC, 1952. National Institute of Standards and Technology

research job, far from being done, is only well begun," Wallace declared during a national radio broadcast shortly after he took office. "We shall need new varieties of cereals and grasses to resist diseases better than those we now have. We shall have to keep cutting costs of production by increasing yields per acre. Methods of cultivation, like methods of feeding and managing livestock, must be subject to continuing investigation if we are to keep abreast of the continually changing economic world about us."[18]

By the time Roosevelt handpicked him for the vice presidency in 1940, Wallace had become one of the leading luminaries of the New Deal's liberal wing. No longer focused solely on agricultural reform, he embraced and spoke out in favor of deficit spending and other fiscal policies designed to stimulate the economy and maximize employment. Such thinking informed the policies he developed while serving as chairman of the Bureau of Economic Warfare (BEW). Established to purchase strategic materials from overseas sources for the war effort, the BEW—through its contract provisions—became the vehicle by which Wallace intended to promote abroad the social and economic progress he advocated at home. The existence of the BEW and Wallace's liberal vision of how the agency should operate drew sharp criticism from fiscal conservatives in the administration, especially Jesse Jones, the former chairman of the Reconstruction Finance Corporation (RFC) whom Roosevelt had appointed secretary of commerce in 1940. Jones had resigned from the RFC in 1939 to become the federal loan administrator, a more powerful position from which he still directed the RFC and a host of other federal lending agencies. As the federal loan administrator, a post that he held while commerce secretary, Jones controlled access to the funds Wallace needed to execute the BEW's procurement contracts. Unable to agree on a suitable division of labor or resolve differences about their respective roles and missions, Jones and Wallace began to feud publicly. The acrimony between them exasperated Roosevelt, who intervened in the summer of 1943. He abolished the Bureau of Economic Warfare and stripped Jones of his lending authority. His influence diminished, Wallace paid an especially high price in the run-up to the 1944 presidential election. Roosevelt dropped him from the ticket and selected a new running mate—Missouri senator Harry Truman. Wallace's still strong appeal among New Deal liberals, however, precluded further punitive action. He requested and Roosevelt approved his transfer to the top job in the Department of Commerce, a decision that galled Jones and prompted his resignation early in 1945.[19]

At the Commerce Department, Wallace prepared to carry forward the fiscal policies that structured late New Deal economic thought. Rather than regulate production through the enforcement of antitrust laws to break up large firms—a hallmark of the New Deal's early years—Wallace and other like-minded reformers rejected the restructuring of capitalist institutions in favor of government spending to stimulate consumption. They deemed such measures necessary after the war to spur economic growth and boost employment, lest the cancellation of military contracts trigger another depression. The same priorities guided Wallace's thinking about the value of public investment in R&D on behalf of small businesses. They required government assistance to lower the barriers to entry that large firms had erected in the marketplace through comprehensive patent protection. In this case too, Wallace did not endorse the wholesale dismantling of large corporations, which now stood poised to resume prewar patterns of competition and growth.[20] He acknowledged the advantages of mass production. "Our technological civilization often requires, in the public interest, the pooling of patents and know-how by large corporations to build a better product for mass distribution," he proclaimed in *Sixty Million Jobs*. In this manifesto for postwar prosperity that he wrote and then published in September 1945, Wallace called for a more technologically robust small business community to stimulate consumption across all industries in an expanding national economy. "Unless the little man also has access to the bounties of technology," he argued, "free enterprise will suffer to the detriment of the full employment of labor and our resources."[21]

On the face of it, Wallace did not propose anything radically new. Politicians had long derided monopoly power as the scourge of the small business owner, and they seized on that distinction to secure the passage of protective legislation in Congress.[22] Similarly, at the end of World War I, Samuel Stratton, the first director of the Bureau of Standards, had staked the agency's peacetime future in part on the production of knowledge and the development of new technologies for industries that lacked in-house resources for R&D. Wallace's programmatic initiative on behalf of the small business community, however, far exceeded in scale and scope anything Stratton or his successors conceived. It drew inspiration and institutional legitimacy from the Department of Agriculture's long tradition of scientific and technological aid to farmers, which Wallace himself had exploited in Iowa to establish his foothold in the hybrid corn business. "The field of agriculture offers a lesson for industrial research," he declared in *Sixty Million Jobs*. "For more than two generations, the government has taken an active part in conducting, sponsoring, and coordinating research. . . . [The] ability [of farmers] to achieve . . . notable results is due largely to the vast number of improvements in crop production, handling, and processing which the years have brought. And here, the Department of Agriculture, the various experiment stations financed by the Federal government and the states, and the four great regional agricultural laboratories furnish abundant and conclusive proof of the benefits of government research."[23]

Connecting the historical experience of the farmer to the anticipated needs of small business structured Wallace's postwar vision for research at the Bureau of Standards. To get started, he solicited advice from outgoing director Lyman Briggs. A veteran of the Department of Agriculture who had transferred to the bureau in 1920 and then advanced to the director's office in 1932, Briggs favored a steady course.[24] "I do not think that any basic change in the postwar activities of the National Bureau of Standards is advisable," he wrote Philip Hauser, Wallace's assistant, five days after Roosevelt died in Warm Springs, Georgia. "Its activities . . . can be expanded to the advantage of industry and commerce, but this would in fact constitute only an enlargement of what the Bureau was doing before the war," he explained. The bureau did not distinguish between the large and small firms that routinely requested technical information—according to one survey, the bureau received more than one hundred personal inquiries for technical advice every day before the war—nor did it publicize the customary service of answering inquiries, given the demands on scarce institutional resources. Briggs remained cautiously optimistic that Wallace could change course without "assurance that the basic research of the Bureau, which provides the very foundation for this technical information service, is substantially increased. Otherwise, we will smother the source of authoritative information under a flood of inquiries."[25]

Wallace prepared to recast in the coming months the largely reactive growth strategy Briggs suggested—the addition of more research personnel to field external queries—into a more proactive variant specifically tuned to the needs of small businesses and akin to the functional role of the experiment stations run by the Department of Agriculture. "This will involve," Philip Hauser wrote, "the coordination of the technological activities of the Department [of Commerce], the strengthening of laboratory aid and technical service functions, increased stimulation and utilization of inventions through modernization of Patent Office procedures and practices and a positive program of stimulation to inventors, and the development of a strong technical advisory service to business operating through an adequate organization of field offices."[26] Wallace publicly announced the reorganization of the Department of Commerce along these lines during a press conference in Washington, DC, on September 20, two days before he met Condon in Chicago. "He put strong emphasis on technical aids to small business," the *New York Times* reported. The budget that President Truman sent to Congress in January 1946 captured the full extent of Wallace's ambitions. It requested for fiscal year 1947 a 78 percent increase over the sum the Commerce Department received in 1946. The Bureau of Standards received an added boost in March, when the president forwarded to Congress a supplemental request for $1 million—equivalent to one-third of the agency's current appropriation—to upgrade physical facilities and equipment.[27]

Despite the president's support, Wallace and Condon got off to a slow start. The Bureau of Standards had earned a reputation for useful but unimaginative research that Condon found difficult to reverse. Commenting on its scientific output in 1942, Frank Jewett, formerly in charge of the

Bell Telephone Laboratories and now president of the National Academy of Sciences and a member of the visiting committee since 1935, described the bureau as "a Government organization which does its assigned job excellently but tends to develop a lack of imagination which results in its being always a follower rather than a leader in its field." Leonard Loeb, who had worked briefly at the bureau during World War I, painted an unflattering portrait of Briggs. "He was not [a] very progressive individual," Loeb told Condon after only three weeks on the job. Nor did options for the future bode well given the feedback Harlow Shapley received, prior to Condon's appointment, from colleagues whose advice he had solicited on Wallace's behalf. One of them, Harvard astronomer Donald Menzel, had participated in the bureau's wartime research on radio propagation. "I have had considerable contact with many departments of the Bureau of Standards over a period of years," he wrote Shapley. "Doubtless there are many good men on the staff. There are, however, very few persons of top rank. I can think of none whom we should consider for appointment to permanent rank at Harvard." Menzel also criticized the civil service requirements that structured hiring practices: "The trouble . . . with most government research institutions, under Civil Service, is that they offer neither the high salaries of industry nor the freedom of the university. . . . The [Commerce] department must insure that the opportunity and new positions thus created go to really good men and not, in general, used to advance men already there on the mere excuse of seniority. Appointments should be made primarily on the basis of ability." Shapley forwarded Menzel's comments to Wallace, adding that they "can be taken as reflecting both his rather expert viewpoint, and my own."[28]

Deserved or not, the bureau's reputation as a second-rate research institution limited Condon's ability to recruit fresh talent for the new R&D programs he and Wallace envisioned. Other obstacles included, as Menzel had observed, civil service salaries lower than those universities and industrial firms paid for equivalent work. Condon's annual salary dropped from $15,000 to $10,000 when he moved from Westinghouse to the bureau, and only Congress had the authority to raise it.[29] The promotion of bureau personnel based on seniority and rank rather than demonstrated ability also inhibited progress. Shortly after Menzel had sent his sobering critique of the bureau's research staff to Shapley, W. Edwards Deming, a consultant in the Bureau of the Budget, sent a cautiously optimistic assessment of prospects for the future to Wallace's assistant, Philip Hauser. Deming pointed out that, based on their ages and years of service, eight of the bureau's thirteen division chiefs planned to retire either right away or within the next few years. He welcomed this mass exodus but, like Menzel, admonished Hauser that "a new director will be helpless without division chiefs of stature. The way the [bureau] is organized, a division chief is extremely powerful. Of late years the policy has been to move up the next man to division chief, regardless of qualifications. Promotion from within is commendable provided you have good men . . . but when practiced ruthlessly is disastrous." Deming anticipated that the judicious selection of replacements outside established protocols "could give a fresh start and eventually rebuild the place scientifically." So did Leonard Loeb. "Get rid of the duds and get live young men for the future," he advised Condon, who had personally solicited his input.[30]

Condon took their advice to heart. The impending retirement, on January 31, 1947, of Frederick Bates, a forty-three-year bureau veteran who headed the optics division, prompted Condon to rethink the merits of staff promotion along the lines Deming, Menzel, and Loeb had suggested. Eugene Crittenden, associate director of the bureau and Briggs's handpicked successor, now emerged as the leading candidate to fill the vacancy until Condon learned that the chief of the division's spectroscopy section, a younger and less senior member of the technical staff, garnered more internal support for the job. Gregory Breit, an old friend with whom Condon had collaborated at Princeton a decade earlier, passed along "through my grapevine and a source unknown to you" an unflattering appraisal of Crittenden's candidacy. "There is . . . some dissatisfaction concerning rumors of Crittenden's being in line for appointment as Chief of the Optics Division," Breit wrote Condon. Although Crittenden had a longer employment record and more seniority, neither he nor the chief of the spectroscopy section got the job. In October, Condon reorganized the optics division and gave it a new and more up-to-date title—the atomic physics division. Rather than promote from within, Condon appointed himself division chief.[31]

Even military research programs started during the war and now poised for peacetime expansion tested the limits of Condon's ability to restructure hiring practices. In 1942, the Joint Communications Board, acting on behalf of the Joint Chiefs of Staff, established the Interservice Radio Propagation Laboratory (IRPL) in the radio section of the electricity division under the direction of John Dellinger, a thirty-five-year bureau veteran. The IRPL operated a worldwide network of field stations that collected and analyzed ionospheric data to assure the optimal selection and efficient allocation of radio frequencies used by the army and the navy. It also provided a service that forecast ionospheric conditions for military and commercial radio operators. Rather than continue to fund the IRPL through the War and Navy Departments, the Joint Communications Board drew up plans in the spring of 1945 to support the R&D already under way at the bureau by way of direct appropriations from Congress.[32] In mid-April 1946, with Henry Wallace's approval, Condon abolished the IRPL and transferred its civilian and military functions to the new Central Radio Propagation Laboratory (CRPL), which received official designation as a separate division of the bureau, effective May 1. Early that summer, still fresh off electronics research at Westinghouse, Condon prepared his first budget estimates for the bureau's operations in fiscal year 1948, and it included a request for funds to acquire new laboratory facilities for the CRPL and also recruit qualified personnel. "This program," Condon wrote Wallace, "requires further extension into the ultra-high frequency and micro-wave region in order to render comparable service to the television industry and to military and commercial radar."[33]

Initially, Condon exploited personal contacts to fill out the research staff. He started with Gregory Breit, who later raised concerns about the leadership of the optics division. Enthusiastic about "the prospects of stimulating an active group," Breit responded favorably to Condon's query in late January 1946 about running the CRPL but on the condition "that I could deal with you directly when really necessary without going through [associate director Eugene] Crittenden. I would not undertake the job as [John] Dellinger's subordinate." Breit lacked respect for the professional standing of these two bureau veterans. "I do not imply by this the presence of negative qualities in either of these men, but only the absence of sufficiently marked [positive] qualities and talents," Breit explained. Negotiations proceeded no further, presumably on the assumption that the management incentives Breit requested could not be arranged. Without an alternative candidate, Condon appointed the aging Dellinger chief of the CRPL. Two months later, in March, Condon tried to recruit Donald Menzel "to take over the section on basic research at ionospheric frequencies," a field of study that closely matched the work he directed at Harvard. Menzel balked at the comparatively low $7,175 salary. "As interested as I am in the type of work presented," he wrote Condon, "the starting salary is not attractive to me. . . . My present Harvard salary is slightly in excess of $10,000, and rumor ha[s] it that all salaries in the higher brackets will shortly be increased."[34]

In April 1948, Condon approached Menzel again, this time to replace Dellinger, who planned to retire at the end of the month. Condon sweetened the offer with news about the likelihood of a substantially larger appropriation from Congress in fiscal year 1949—almost twice the amount received in fiscal year 1948—to pay for an expanded research program. Bent on appealing to Menzel's scientific interests, Condon also highlighted the broad scope of the research already under way at the bureau—"It is maintained on a liberal enough basis to permit fundamental research on radio noise from cosmic sources, study of basic astrophysical correlations with solar activity, and so on," he wrote—and through cooperative programs set up on contract with universities. Other incentives included progress on the acquisition of research facilities, namely the Bureau of the Budget's recent authorization of Condon's request for a new radio laboratory to be built in Washington. Citing recent congressional approval of a onetime salary increase to $15,000 for forty-five R&D positions, split between the army and the navy, Condon remained confident that he could—through a similar request from the Commerce Department—obtain authorization to offer Menzel a compensation package in excess of the current $10,000 ceiling. Congress did not act on Condon's behalf.[35] Menzel chose to remain at Harvard, a decision that left Condon with the less satisfying alternative of selecting him to serve in an advisory role on the bureau's visiting committee for a five-year term, beginning in 1949. Condon, meanwhile, reverted to the established policy and promoted from within; he appointed a Dellinger protégé, who had joined the bureau in 1935, to be chief of the Central Radio Propagation Laboratory.[36]

BUDGET BLUES AND THE COMPETITION FOR RESOURCES

The civil service system and the bureau's own bureaucratic traditions explain some, but not all, of the difficulties Condon experienced. Early in 1947, Condon sharply criticized the army and the navy for what he considered to be their monopolization of the nation's scientific and technological resources. "Although we try to keep up a good front and work for improved conditions," he wrote Under Secretary of Commerce William Foster in February, "the Bureau is in a very run down and sorry state and it is going to be a long, hard pull to tidy it up and make it be a first rate institution. This is especially difficult because of the extraordinary competition which we are getting not merely from the Army and Navy research laboratories operated under Civil Service but the even more severe competition which we get from the laboratories operated under private contract from the Army and Navy." The reallocation of capital and labor during the war had restructured the political economy of the federal scientific research establishment. "I am not arguing against adequate research for . . . defense," Condon explained further, "but am arguing against wasteful and inefficient practices in connection with the operations which have mushroomed so greatly in the past year or two."[37]

Lee DuBridge, wartime director of the Radiation Laboratory at MIT and the soon-to-be installed president of the California Institute of Technology, had already captured the mood of the times to which Condon referred with such displeasure. DuBridge believed the wartime practice of outsourcing R&D offered better prospects for the advancement of science and technology than the stifling bureaucracy that structured the conduct of research in government-owned laboratories. The Radiation Laboratory stood out as an exemplar of efficient contracting between the government and the industrial suppliers who developed and manufactured in quantity the microwave radar equipment the armed services required. "The contract gives the Government full authority and responsibility in regard to policy," DuBridge wrote Condon in January 1946, "but leaves the matters of personnel, working conditions, [and] management . . . in the hands of a private contractor, free from Government red tape and particularly from Civil Service regulations." Condon got another taste of the times when, in April, Vice Adm. George Hussey, chief of the Navy's Bureau of Ordnance, solicited his input to find a civilian technical director for the Naval Ordnance Laboratory. "The rebuilding of the staff to carry out the post-war mission of the Laboratory is now in progress," he explained to Condon. Hussey already had an advantage—construction of a new $15-million R&D campus in suburban Maryland, which Leonard Loeb, as a naval reserve officer during the war, had helped plan and organize. "They are having a hard time in getting good men," Loeb advised a former student in search of a job. "I think perhaps you might have a better chance there than at the Bureau of Standards which tends to be stuffy, whereas [the Naval Ordnance Laboratory] is young and dynamic."[38]

These patterns of behavior guided Condon's thinking about the bureau's budget, which he prepared for the first time for fiscal year 1948. "You're going to hear a lot more about the National Bureau of Standards if . . . Condon . . . has his way," reported the *Washington Post* in August

1946, shortly after the Commerce Department released some details of the budget—still in draft form—to the public. "The hard-hitting director is making plans for a year hence, and they call for, in part: An annual budget that would be five times this year's six-million-dollar appropriation, at least twice the number of employees, additional laboratories, and better salaries for his able scientists and technicians." Condon saw federal agencies engaged in the equivalent of a land grab; they proceeded in an uncoordinated frenzy to obtain larger budgets for research and development. "Especially is this true of the War and Navy Departments," he wrote in the preface to the 1948 budget estimate Henry Wallace reviewed two months before the *Washington Post* printed the highlights. "The different military bureaus are actually engaged in an inflationary competition for the services of too few available trained scientific workers," he explained. Condon believed that perceptions about the allocation of resources for R&D mattered as well. A natural lag in industrial reconversion "placed an undue emphasis on military developments," but the large reductions in defense spending that accompanied the end of hostilities proceeded apace regardless of how much money the military departments devoted to research and development. Civilian agencies, such as the Bureau of Standards, lacked the same cover. Once stripped of their wartime contracts, "a proper expansion" of peacetime R&D "appears as a very great percentage expansion of that work. . . . Against such a background," Condon told Wallace in conclusion, "an attempt has been made to develop a policy for the National Bureau of Standards."[39]

Condon acted quickly to gain a foothold in atomic energy research, which the War Department still controlled through the Manhattan Engineer District. Transfer to civilian oversight in a new agency—what became the Atomic Energy Commission (AEC)—still lay in the future. In January 1946, Condon and Rolla Dyer—the director of the National Institute of Health (NIH)—proposed to build, preferably on the new NIH campus just outside Washington, an atomic reactor to produce radioisotopes for cooperative research in physics, biology, and medicine. Who first suggested the idea is not known, but the proposal captured Condon's long-standing interest in radioisotope production and research, a field he had tried but failed to exploit on a commercial scale at Westinghouse before the war. The plan also gave expression to the type of university-industry collaboration that Condon first envisioned in Pittsburgh and now prepared to foster through an expansion of in-house R&D at the Bureau of Standards and the National Institute of Health.

Dyer and Condon petitioned Henry Wallace to act on their behalf. On January 22, Condon drafted a letter that outlined the research program, and he asked Wallace to forward it over his signature to the president for approval. "The recent developments in atomic energy research," Condon wrote, "have opened up new possibilities for research in medicine, physics, and many other fields whose importance cannot be overestimated. . . . Access to radioactive materials and the radiations themselves, specialized personnel and equipment with a high degree of teamwork and cooperation among several disciplines, is necessary." He requested $2 million annually for operating expenses and a onetime appropriation of $6 million from Congress for the reactor and a new laboratory to accommodate research personnel from the National Institute of Health, the Bureau of Standards, and interested universities. Expecting the president's assistance to obtain the requisite funds, Condon added in closing, "Your approval might take the form of a favorable indication to the Bureau of the Budget to proceed with specific plans for a joint project of the two agencies." Truman declined to intervene. "No steps in this direction can be taken until the Congress acts on the control of . . . Atomic Energy," he wrote Wallace two days later.[40]

Condon abruptly switched tactics to preserve the collaborative structure of the research program he and Dyer envisioned. Also in January, Isidor Rabi, Condon's old friend at Columbia University, spearheaded negotiations with the Manhattan Engineer District to build an atomic reactor and associated laboratories for nuclear physics research somewhere in the Northeast. In the spring, Columbia and eight other regional universities formed a nonprofit corporation—Associated Universities Incorporated (AUI)—to manage on behalf of the War Department what became Brookhaven National Laboratory, the first peacetime high-energy physics facility dedicated to unclassified academic and industrial research in the United States. After Rabi briefed him on the details, Condon saw an opportunity to bypass the legislative logjam in Washington. "I would like very much to make suitable arrangements to have the Bureau [of Standards] represented in the planning of the project," he wrote Lee DuBridge at the University of Rochester, one of the AUI's founding members. "If the interests of the Bureau of Standards and of the [National] Institute of Health can be provided for by participation in the regional laboratory project, I would prefer to do that instead of requesting a separate project of our own."[41]

Condon requested $25.5 million from Congress for fiscal year 1948, which came in at more than four times the amount the Bureau of Standards had received the previous year. Wallace imposed no restrictions on the choice of research topics. "The distribution of funds . . . for Research and Development," his budget director wrote Condon in August 1946, "is left to your discretion to select those projects and the amounts of each which best conform to existing policy directives." Atomic energy research received high priority. Condon wanted ample funds to support neutron radiation research, the production of radioisotopes, and "other researches essential to [the] peace-time applications of atomic energy." The Commerce Department's Office of Budget and Management (OBM) did not approve his initial request without modification. It cut nearly $7 million Condon had set aside for new construction on the bureau grounds in northwest Washington, DC. The Bureau of the Budget in the Executive Office of the President took additional reductions, which dropped the total from $17.1 million to $10.6 million. Congress approved a final budget of $7.9 million, a 35 percent increase—equivalent to a little more than $2 million—over the amount received in 1947, but still only a fraction of what Condon had hoped to get.[42]

Consistent with the reasoning behind the president's decision to reject the request for a nuclear reactor on the NIH campus earlier in the year, the Bureau of the Budget struck the line item Condon had inserted for atomic energy research. Work in this and other strategic industries—such as synthetic rubber—whose postwar status remained in flux, "represented questions of policy which are still open," he recalled in March 1947, less than three months after the War Department transferred the Manhattan Engineer District to the newly established Atomic Energy Commission. "That was a policy question that should be resolved after the Commission is fully organized," he explained. The negotiations with Rabi and DuBridge to incorporate the research priorities of the Bureau of Standards and the National Institute of Health into the programmatic functions of the AEC's new regional laboratory in upstate New York fared no better. Construction of the research reactor commenced during the summer, but three more years passed before it became fully operational. Sidelined because of the slow progress, Condon had to settle for a secondary role on the periphery of these developments. Two months before the groundbreaking, he agreed to serve on Brookhaven's scientific advisory committee. The appointment fell far short of the ambitions Condon and Rolla Dyer had originally envisioned for cooperative atomic energy research in Washington.[43]

Condon regrouped and inserted a fresh request for nearly $3 million—

which also included a provision for biological and medical applications—in the bureau's budget for fiscal year 1949. This time, however, he had to rely on the support of W. Averell Harriman, the former US ambassador to the Soviet Union whom Truman had appointed secretary of commerce to succeed Henry Wallace in September 1946. Unlike Harriman and other anticommunist hardliners in the president's cabinet, Wallace had remained an outspoken and increasingly isolated advocate of less hostile relations with the Soviet Union. Truman dismissed Wallace from the Commerce Department on September 20, one week after he gave a controversial speech in New York City that not only embarrassed the president but also undermined the White House's hawkish foreign policy, of which Harriman had been one of the principal architects.[44]

Harriman did not support Condon's funding request. Postwar demobilization prompted sharp cuts in the federal budget, while a resurgent Republican Party swept the 1946 midterm elections and gained control of both houses of Congress. The new cadre of fiscally conservative lawmakers in Washington combined with Truman's preference for a balanced budget to tighten the purse strings even further. Acting on instructions from the White House to stick "strictly to the President's budget directive of vigorous economy," Harriman pruned Condon's budget—he cut nearly half of the allocation for research and testing—to fall in line with the reduced funding levels the bureau had been forced to accept in fiscal year 1948. "These objectives have been achieved," Harriman wrote James Webb, Truman's director of the Bureau of the Budget, in September 1947, "only by means of drastic reductions in the [B]ureau [of Standards] estimates." In line with Harriman's actions, the Commerce Department's Office of Budget and Management struck the $3 million Condon set aside for atomic energy research and instructed him instead to solicit funds directly through a cash transfer from the Atomic Energy Commission.[45]

Plans to take over on a permanent basis the management of the government's wartime synthetic rubber research program foundered on the same institutional uncertainties of postwar reconversion that obstructed the expansion of atomic energy R&D. The Bureau of Standards lacked the political clout and legal authority to achieve Condon's stated objectives. In June 1946, Condon had solicited Wallace's support to continue the research started on funds transferred to the bureau from the Rubber Reserve Company during the war. He singled out for acquisition the government's synthetic rubber pilot production plant and evaluation laboratory in Akron, Ohio, and the extramural research already under way on contract to universities and industrial firms. Condon estimated that the government spent through the Reconstruction Finance Corporation's Office of Rubber Reserve, successor to the Rubber Reserve Company, $3 million annually on rubber R&D. To expedite the transfer, he reached out to the interagency policy committee on rubber, which the Office of War Mobilization and Reconversion (OWMR) had recently established to determine a course of action for the maintenance of a peacetime synthetic rubber industry.[46]

In September, after the interagency policy committee received the bureau's proposal and the corresponding cost estimate for fiscal year 1948 made its way to the Bureau of the Budget for consideration, Condon learned that the navy had emerged as a potential competitor. "I have a grape-vine report that the Naval Research Laboratory hopes to take over this work and may soon try to get Secretary [of the Navy James] Forrestal to recommend that the transfer be ordered by the President," he wrote Wallace on September 18, two days before Truman dismissed him from the cabinet. "It is not my desire to rush the transfer but simply to get into the situation at this time in such a way as to prevent the transfer to the Navy by default." Condon discussed this last-minute maneuvering with James Newman, special counsel to the president. On Newman's advice, Condon drafted a letter requesting that "no action be recommended to the President until the case for assigning this work to the National Bureau of Standards has been carefully examined," and he asked Wallace to send it over his signature to OWMR director John Steelman, who also served as a special assistant to Truman. "You may prefer," Condon also urged in closing, "to recommend to the President directly that such a transfer be made," but Wallace's departure from the Commerce Department precluded that option. On October 11, Steelman promised Averell Harriman, just four days into his new job as commerce secretary, that "no decision on this matter will be reached without giving full weight to the interests of the National Bureau of Standards."[47]

Steelman took no action on Condon's behalf. The funding request Condon submitted for rubber research in fiscal year 1948 did not get past the Bureau of the Budget. He tried again for fiscal year 1949, but in the summer of 1947 the Commerce Department's Office of Budget and Management acted first to block the allocation of funds. "The Rubber Program was omitted because we are without authority to take it over," the OBM's director explained to Condon. The approved budget for fiscal year 1949 that Condon presented to the House Appropriations Committee on January 20, 1948, captured the sharply diminished scope of the remaining option. It included a request for $50,000 to pay for a less ambitious rubber research program in the bureau's organic and fibrous materials division. Then, on March 31, Congress passed the Rubber Act, which ended further negotiations. This legislation required the Reconstruction Finance Corporation to turn over to private industry the government's surplus synthetic rubber plants and R&D installations.[48]

THE MILITARY ORIGINS OF CIVILIAN RESEARCH

The unresolved legal status of atomic energy and synthetic rubber research restricted the scope of Condon's ambitions for R&D at the Bureau of Standards. It also undermined the justification for technical assistance to industry that Henry Wallace had proclaimed on behalf of the small business community. The legal ambiguities built into the bureau's founding legislation exacerbated these symptoms. Failure to clarify them in a timely manner reinforced dependence on the military departments for financial support to strengthen the bureau's civilian functions. Cash transfers to the Bureau of Standards from the Department of Defense and the Atomic Energy Commission—the civilian agency wholly responsible for the production of fissile material for nuclear weapons—increased sharply from $9.4 million in 1949 to $23.8 million in 1951. During the same three-year period, direct appropriations from Congress rose only 7 percent, from $8.5 million to $9.1 million. This lopsided growth in the budget prompted Condon to observe in August 1951, "The Bureau is in better condition than it was six years ago. . . . [but] all the funds . . . go to the military these days so that I had to get it [*sic*] from them in order to do anything at all."[49]

The bureau's founding legislation, or organic act, did not prohibit the transfer of funds from other government agencies. To the contrary, transfers gave the director added flexibility to pursue new lines of research. Rather than amend the organic act, which required congressional approval, Condon's predecessors—especially founding director Samuel Wesley Stratton—took advantage of transferred funds and special, onetime appropriations from Congress to execute research projects that gradually became established in-house R&D programs. This pattern of behavior, which Condon fully exploited, drew sharp criticism from some lawmakers. "I do not like this transfer business too well," Sen. Joseph Ball (R-Minnesota) told Condon in April 1948 during a hearing of the Senate Appropriations

Committee to review the bureau's budget request for fiscal year 1949. "It seems to me that there is an institution out there and the committee that has jurisdiction over it sees one-third of the picture."[50]

Henry Wallace had tried to update the bureau's enabling legislation to capture the scope of current operations that Senator Ball criticized, but opposition to his stated goals and the slow pace of progress prompted Condon to take advantage of transferred funds as an alternative strategy for institutional growth. Late in the summer of 1945, just before Condon replaced him, Lyman Briggs drafted, on Wallace's instructions, an amendment to the organic act. After he spoke to Briggs about it, visiting committee chairman Gano Dunn confirmed that "the principal part of the amendment is for the purpose of transferring to the Organic Act certain authorizations; definitions of scope and activities of the Bureau that in the past have been covered by supplementary legislation, executive orders, and customary procedure." Wallace, however, also envisioned a more ambitious objective. The amendment included a special provision that gave the bureau, for the first time, broad discretion in "the prosecution of basic research in physics, chemistry, and engineering to promote the development of science, industry, and commerce."[51] Briggs had seen this type of programmatic initiative before, and he may have discussed it with Wallace. Ten years earlier, Briggs had petitioned—albeit unsuccessfully—the president and Congress to expand research at the bureau along the same lines, partially to restore funds lost to deep budget cuts in his own agency but, more broadly, to help revive sick industries felled by the Depression.[52] The extent to which Wallace relied on Briggs's thinking on this matter, if at all, is not known. Either way, the brief clause Briggs wrote in section 2 of the amendment fully captured Wallace's intent to transform the Bureau of Standards into a nationally recognized technical resource for the small business community.

Not everyone agreed with Wallace's interpretation of the bureau's functions. Vannevar Bush, who also served with Gano Dunn on the visiting committee, opposed the unrestricted expansion of R&D. "While I believe that it is important to the effective operation of the bureau and to its ability to attract scientists of high caliber that it be free to conduct basic research in science," Bush told Dunn, "I think it should be unmistakably clear that the major emphasis should remain on its unique assignment in the field of standards." Bush's objection originated in a contrasting vision for the federal support of science embodied in the yet-to-be-established National Science Foundation. "Publicly and privately supported colleges and universities and the endowed research institutes must furnish both the new scientific knowledge and the trained research workers," he wrote in the NSF's founding document, *Science, the Endless Frontier*. "It is chiefly in these institutions that scientists may work in an atmosphere which is relatively free from the adverse pressure of convention, prejudice, or commercial necessity."[53] Bush also rejected Wallace's claim that small businesses required the helping hand of government. "I think the direct governmental aid to small business, either in the form of aid for their research, or indeed otherwise, is likely to be rather futile," he wrote a few months before Briggs began to draft the amendment to the bureau's organic act. "I have a strong conviction that the small industrial units which are really worthwhile are decidedly able to take care of themselves. What they need is not government aid, but lack of government interference."[54]

The draft amendment in its original format did not survive Bush's objections or Wallace's abrupt departure from the Commerce Department. Early in 1947, Averell Harriman reassured congressional leaders that "the proposed legislation is in effect only an amplification of the outline of the functions and activities as contained in the original Organic Act of the Bureau. . . . No new fields of work beyond the present scope of the Bureau are included in the amplified language." When the bill that contained the amendment made its way to Congress two years later, in 1949, Wallace's original provision for basic research had been completely stricken from the text. The amendment to the organic act became law on July 22, 1950, and with it, the bureau's dependence on transferred funds for institutional growth continued.[55] The Department of Defense, more than any other federal agency, shored up the bureau's finances. Military largesse helped offset shortfalls that Wallace's original version of the amendment had been designed to rectify through a broader legal mandate to conduct R&D.

Condon established with military support an elaborate research program in digital electronic computing that showcased the type of industrial service Henry Wallace envisioned on behalf of the small business community but failed to make permanent through a revision of the bureau's organic act. Interest in this field originated at the Bureau of Standards before the war in response to a request from the Works Progress Administration (WPA) to put unemployed scientists and technicians back to work. In January 1938, the bureau established the Mathematical Tables Project (MTP) in New York City to provide computation and consulting services to the scientific and engineering communities. Military applications for the Office of Scientific Research and Development and the navy's Bureau of Ordnance dominated the MTP's research agenda during the war. The newly established Office of Research and Inventions carried this military relationship into the postwar period. The ORI provided the bulk of the Mathematical Tables Project's operating budget in fiscal year 1946.[56]

Separately, in one of his first official acts, Condon established a mathematical analysis function at the Bureau of Standards that complemented—and drew into closer association—the work already under way in the Mathematical Tables Project. He intended to apply mathematical methods, especially statistical analysis, to the bureau's routine functions to maximize operating efficiencies and improve the robustness of research results. This objective did not emerge suddenly. Condon had maintained an abiding interest in the subject for nearly twenty years, first at the Bell Telephone Laboratories and then at Westinghouse. For the new position of mathematical assistant to the director, Condon handpicked John Curtiss, a thirty-six-year-old Harvard-trained mathematician. Now on the faculty at Cornell University, he had just completed wartime service as a naval reserve officer assigned to the research and standards branch in the Bureau of Ships. Curtiss's pioneering studies of fuel consumption rates in naval vessels translated into more efficient planning of fleet operations. He also introduced statistical sampling techniques into the navy's purchase specifications to improve the quality and reliability of shipbuilding materials. Condon did not hire Curtiss by way of the civil service system. Instead, he tapped into the overhead fees charged against navy R&D funds transferred to the Bureau of Standards to pay Curtiss's salary.[57]

On January 12, 1946, shortly after he decided to hire Curtiss, Condon met MTP director Arnold Lowan in New York City to discuss the project's future. "Condon contemplates making the MTP a division of the National Bureau of Standards," Lowan wrote retired director Lyman Briggs two days later, "provided this may be justified not solely on the basis of services of the type we have been rendering for the Army and Navy during the war, but also on the basis of peacetime needs." Only a few months earlier, the navy's Bureau of Ships—Curtiss's former employer—and the Office of Research and Inventions had jointly drafted a proposal to establish a computation center. It also included a broader mandate for government-wide service that appealed to Condon. On January 30, the day after he had publicly proclaimed before the House Appropriations Committee the navy's enlightened support of academic research as a model for the bureau's own programmatic initiatives, Condon sat down with the ORI chief, Adm. Harold Bowen, to discuss options for collaboration. They agreed to establish at the Bureau of Standards a new applied mathematics division built around the Mathematical Tables Project but also tailored to support the navy's ambitions for a computation center.[58]

In line with these developments, Condon restructured the bureau's electronics R&D to support the new computing initiative. Electronics research, much of which the bureau carried out on behalf of the Army Ordnance Department, had advanced rapidly during the war. In 1942, Lyman Briggs consolidated this work in a separate ordnance development division, and early in 1945, he solicited $500,000 from the army's chief of ordnance to build a new dedicated laboratory on the bureau grounds. When the laboratory opened later that year, the ordnance development division maintained a staff of almost four hundred employees, equivalent to nearly one-fifth of the bureau's entire workforce.[59] On January 30, 1946, the same day he met with Admiral Bowen, Condon expanded the division's scope of work to include civilian applications. He established a separate electronics section that consolidated in one location all of the bureau's R&D on electron tubes and printed circuits. "It is expected that new projects will accrue to this Section under the sponsorship of civilian agencies of the Government," Condon wrote. "Thus, [it] will become the Bureau's center for general development work in applied electronics." Military priorities, however, still dominated this process of accommodation. "Our agreement with the Army," wrote Harry Diamond, chief of the ordnance development division, in June, "is that we are to serve as their electronic ordnance laboratory." The Ordnance Department's Frankford Arsenal in Philadelphia continued to turn out pilot production models of new fuze designs that Diamond's staff developed at the bureau.[60]

Plentiful army resources allowed Condon to assemble a first-rate research staff to populate the one-stop shop for potential civilian clients he envisioned. New employees enjoyed excellent working conditions. One of them, Willard Bennett, had just come off a wartime assignment in the Army Signal Corps. Rather than resume a university career, Bennett joined the new electronics section in March 1946, and he began canvassing university physics departments to round out the research staff. "[Condon] has already set in motion some changes which I believe are likely to be very significant improvements in time," he wrote Berkeley's Leonard Loeb two months later, while inquiring about potential recruits. "There are excellent opportunities here, both for physicists interested in atomic physics and for men interested in various types of so-called electronic development, including magnetrons, velocity modulation tubes, and all the similar appurtenances of modern radar and radio development."[61]

Consumers of electronics technology, especially the small firms that Henry Wallace had collectively singled out as the type most in need of government assistance, also extolled the research spirit to which Bennett referred. Already out of the Commerce Department, Wallace likely would have been pleased to hear the praise a representative of the Orthon Corporation, a small electronics firm based in New York, heaped on the ordnance development division following a visit to Washington in December 1946 to exploit the bureau's latest work on printed circuits and the miniaturization of electronic devices. "As a small business," he explained, "our facilities for research and for proper consultation on electronic problems are limited. . . . We wish to commend your Bureau on this type of assistance which you are rendering to small businesses such as our own." The bureau, at Condon's behest, also cosponsored conferences and symposia on printed circuit technology that drew broad participation from industry and other government agencies.[62]

Orthon's representative proclaimed a public benefit that owed its institutional legitimacy not to Wallace's state-centered prescription for postwar economic growth but rather to the short-term technological requirements of the armed services. Earlier in the year, the Office of Research and Inventions had authorized the transfer of $425,000 to the Bureau of Standards—equivalent to nearly 15 percent of the congressional appropriation it received in fiscal year 1946—to continue the Mathematical Tables Project and construct a digital electronic computer, with the understanding that the equipment would be located in the proposed computation center and used primarily for navy projects. Following Condon's lead, John Curtiss envisioned technical aid to business and industry—the primary mission of the Bureau of Standards—as a logical extension of the military priorities that had guided the project from the outset. "It should be emphasized most strongly," he wrote, "that there is no intention to try to make this Computing Center the unique central facility of its type in the Government. Rather, it is hoped that its existence will give impetus and guidance to other computing laboratories, both Government and non-Government, through the development of techniques and methods which are of interest to all applied mathematicians, through creating a pool of highly skilled consultants available to all other computing facilities, and through the solution of certain types of problems for which other laboratories are less suitably equipped." Condon agreed in September to select the center's location, set up the physical plant, recruit the requisite technical personnel, and obtain additional financial resources from Congress and other interested government agencies.[63]

The Commerce Department's Office of Budget and Management approved the request for $250,000—slightly more than half of the ORI's commitment—to support the new computing initiative that Condon had inserted in the fiscal year 1948 budget, the first one he submitted to Congress. Meanwhile, the direct solicitation of financial support from other federal agencies foundered on tepid interest. The Coast and Geodetic Survey expressed interest in the bureau's program, but it favored a more conservative approach. "This agency," Curtiss reported in October 1946, "would prefer to wait to do business with a going concern rather than to fight for an appropriation for support of a hypothetical laboratory." He received a similar response from the Weather Bureau: "The mathematical aspects of the weather problem should be further developed from a theoretical point of view before large scale computation is attempted."[64] Only the Bureau of the Census responded proactively on a scale that rivaled the ORI's investment, but its interest had no connection to the navy's plans for a separate computation facility or Condon's expectations for a new division of applied mathematics. Once under way, however, the Census Bureau project, which the army supported, laid the groundwork for his broader computing ambitions.

Like the navy, the Census Bureau had paid close attention to advances in computer technology, with an eye toward more efficient data collection, processing, and analysis. "We have been impressed by some of the developments of computing equipment that have taken place during the war," census director James Capt wrote Condon in March 1946, "with the possibility of adapting such equipment to serve Census requirements." Capt mentioned the electronic numerical integrator and computer (ENIAC)—the first general-purpose electronic computer, built at the University of Pennsylvania during the war—and he requested Condon's assistance "in investigating the possibilities of developing this or other equipment to serve the objectives we have in mind." John Mauchly, one of the inventors of the ENIAC, had already visited the Census Bureau and other government agencies while on the lookout for customers to establish a market for the new technology. Capt asked Condon to prepare a budget to cover the costs of design and construction of automatic sequencing computer equipment at the Bureau of Standards or contract administration through an outside vendor based on operational specifications the Census Bureau provided. Condon proposed a working budget of $300,000, but he questioned Capt's seemingly optimistic expectation of a two-year time frame for completion. "It should . . . be recognized that this project is of a pioneering nature," Condon wrote. "High speed sequence-controlled computing devices are all in the experimental stage. . . . Complete success within two years in meeting the specifications of the Bureau of the Census, especially as to reliability, cannot be guaranteed at this date."[65]

The Army Ordnance Department, which had paid for the ENIAC, enlisted the aid of the ordnance development division at the Bureau of Standards to undertake what Harry Diamond called "grass-roots research and development work on basic items for electronic digital computing machines." The Ordnance Department also set aside for Capt's use the $300,000 that Condon requested. In early September, just before Capt released these funds, and one week after the Bureau of Standards acquired the money the Office of Research and Inventions had previously promised for the computation center, Diamond, on Condon's instructions, worked out the management guidelines for R&D and production. He tasked his own division and John Curtiss's office with joint administration of all funds through contracts to outside vendors, and he limited in-house technical studies to a supporting role focused on broad programmatic oversight. Responsibility for the reliability of the electronic components and the engineering necessary for system integration resided in the ordnance development division. Diamond assigned Curtiss, who also served as liaison to the contractors the Bureau of Standards hired, technical compliance of the computing equipment with the stated operational requirements that the sponsoring agencies established. Later that year, the bureau awarded a research and study contract for the Census Bureau computer to the Electronic Control Company, the small start-up firm John Mauchly and his partner, ENIAC coinventor J. Presper Eckert, had founded in Philadelphia.[66]

In the summer of 1947, Condon elevated the bureau's budding computer initiative to division status. On July 1, he established the National Applied Mathematics Laboratories (NAML), comprising four sections, and he appointed Curtiss chief. In addition to a new statistical engineering laboratory modeled on the mathematical functions Condon had hired Curtiss to perform, the old Mathematical Tables Project, now scheduled to receive a larger professional staff and electronic computing equipment, acquired renewed status as the computation laboratory, still located in New

York City.[67] The navy's original plan for a computation center attracted the interest of the air force and culminated in the founding of the institute for numerical analysis located on the Los Angeles campus of the University of California. At this new facility, Condon and Curtiss expected mathematical research to proceed alongside the training of specialists in the field and the provision of computation services to the sponsoring agencies as well as to the defense industries concentrated in Southern California. The machine development laboratory, the last section in the new division, consolidated on behalf of future clients the consulting and contract management services Harry Diamond had initially established to manage the acquisition of computers for the Census Bureau and the navy.

In late June 1948, as James Capt's original deadline for completion approached, the Bureau of Standards awarded the follow-on design contract for the Census Bureau computer to the Electronic Control Company. This joint effort culminated in the introduction of the universal automatic computer (UNIVAC), the first general-purpose digital electronic computer designed for commercial use. The execution of the UNIVAC contract through the new National Applied Mathematics Laboratories placed the Bureau of Standards at the forefront of a nascent computer industry, but small firms did not always reap the benefits. Much larger, technologically diversified enterprises, such as General Electric and the Raytheon Manufacturing Company, also obtained contracts from the bureau to design and build computers, often for military clients. Even the Electronic Control Company, the small business Eckert and Mauchly had founded after the war to exploit their pioneering work on the ENIAC, survived only briefly as an independent concern. The Remington-Rand Corporation, an established tabulating and office equipment manufacturer, acquired the financially troubled firm early in 1950, more than one year before the Census Bureau took delivery of the UNIVAC.[68]

The preponderance of these and other large firms in the national economy—which New Deal liberals decried but the onset of the Cold War helped to perpetuate—blunted the impact of government aid to industry that Henry Wallace had sought on behalf of the small business community. Defense priorities also upended the delicate balance between academic and industrial research that Condon had carefully nurtured at Westinghouse before the war. Onerous civil service requirements and other bureaucratic constraints on growth prompted him to take advantage of military patronage to advance the bureau's mission of industrial service. He revised the founding principles of the postdoctoral fellowship program at Westinghouse to reflect the organizational imperatives of the new postwar political economy of science. When he testified for the first time before the House Appropriations Committee in early 1946 to outline a vision for the bureau's future, Condon had told Representative Louis Rabaut, "I would hope to see us allowed to do for peacetime fundamental science something of the sort that has recently been announced as part of the Navy's research plans that involve a high degree of collaboration and intimate cooperation at the working scientists' level with universities throughout the country." Two years later, John Curtiss described the most salient functional attributes of the new institute for numerical analysis at the University of California in terms strikingly similar to those Condon had proclaimed in 1939 to justify the establishment of a national institute of scientific research in Pittsburgh. "The intention is to maintain a small, permanent staff at the top research levels and to carry out most of the work through temporary appointments of strong research men on leave of absence from their regular places of employment," Curtiss wrote in *Science* in March 1948. "The Institute will provide facilities for visiting scholars to investigate and develop in their own special fields of research the mathematical techniques studied by the [National Applied Mathematics] Laboratories."[69]

The founding of the National Applied Mathematics Laboratories signaled a commitment to military R&D that continued for the remainder of Condon's career at the Bureau of Standards, even in technical fields that received funds from Congress rather than the Department of Defense. "Of the directly appropriated funds [to the Central Radio Propagation Laboratory], I think it is fair to say that about half of the work . . . is directly related to military needs," he told the Senate Subcommittee of the Committee on Appropriations on May 18, 1950. That same day, Condon announced a major reorganization of the recently renamed electronics and ordnance development division to accommodate the proliferation of research projects, especially guided missile R&D for the navy. Research on naval weapons, such as the Bat glide bomb, originated in the bureau's mechanics and sound division during the war, then spread to the ordnance development division in 1946. Condon, recognizing the need for space and room to grow by the end of the decade, transferred the work to a new missile development division located on surplus property the navy owned in Southern California, not far from the NAML's institute for numerical analysis in Los Angeles.[70]

The electronics research that Condon had consolidated at the section level early in 1946 to serve military and civilian clients followed a similar trajectory. It too acquired separate divisional status under the same name and continued to support the R&D already under way in the National Applied Mathematics Laboratories. Shorn of these responsibilities, what remained of the electronics and ordnance development division—now rechristened the ordnance development division—focused more narrowly on research priorities for the army, its original sponsor. "In the five years that I have been here," Condon wrote Ernest Lawrence on the day after Christmas, "the [R&D] program has been about 60 [percent] military and the proportion has been very steadily increasing in recent months. One of our weapon projects alone now has a budget 50 [percent] greater than the entire direct appropriation of Congress for civilian functions of the Bureau."[71]

Department of Defense largesse accounted for an increasingly large share of the budget of the Bureau of Standards after June 1950, when war broke out on the Korean Peninsula. "This town is full of war talk, war plans and war preparations and there is a lot to do in adapting the Bureau's program to the changed conditions," Condon wrote Emilie one month after the North Korean army crossed the thirty-eighth parallel. "Whatever else it may do it is making it a lot easier to get quicker actions and decisions on a lot of things." Condon resigned from the bureau one year later, but his departure did not slow the proliferation of military projects. Allen Astin, Condon's associate director since 1950 and handpicked successor, had filled a prominent role in the bureau's wartime electronics research and served as chief of the ordnance development division.[72] Astin presided over further growth in defense R&D. When hostilities in Korea ended in a ceasefire in the summer of 1953, two years after Condon had stepped down, the Department of Defense provided 75 percent of the bureau's swollen $48 million budget. Adding in funds transferred from the Atomic Energy Commission, the figure rose to nearly 80 percent. The army and the navy contributed half of that dollar amount just for ordnance and guided missile research.[73]

CHAPTER 7

Recessional

ON MARCH 1, 1948, a subcommittee of the Committee on Un-American Activities in the House of Representatives publicly censured Edward Condon. The Special Subcommittee on National Security released a report that questioned his loyalty and branded him a security risk.[1] Condon had become politically active as World War II drew to a close. His interests, like those of other like-minded scientists just off wartime research assignments, focused on the handling of nuclear weapons and the future of atomic energy policy, which also touched on larger questions of security, secrecy, and international cooperation in science. Given the perceived threat of domestic communism and the government's increasingly strong stand against it, however, outspoken behavior on such sensitive topics, no matter how innocuous it might have seemed, proved damaging.[2] He made for an easy target. MIT president Karl Compton, a close friend who had known Condon for more than twenty years, summed up the cumulative effect of his behavior one month after the special subcommittee accused him of disloyalty and the scientific community weighed options for a response. "Condon has not always been discreet," Compton wrote Frank Jewett, the recently retired president of the National Academy of Sciences. "He is inclined to be headstrong and to put his own independent judgment ahead of the judgment of his colleagues or superiors."[3] The anticommunist fervor that gripped Washington put Condon on the defensive and ultimately drove him from the National Bureau of Standards.

FROM CENTER TO PERIPHERY

Prior acts confirmed Karl Compton's appraisal of Condon's temperament. Condon had ignored instructions from Maj. Gen. Leslie Groves and Westinghouse president George Bucher to remove himself from the American delegation that traveled to Moscow in June 1945 to mark the 220th anniversary of the founding of the Russian Academy of Sciences. Later that year, he refused the War Department's request to back away from criticism of the May-Johnson bill and then proceeded to antagonize supporters of the legislation. On November 6, after only one day on the job at the Bureau of Standards, Condon agreed to serve as science advisor to the legislative study group that Connecticut senator Brien McMahon had established a few weeks earlier to buy time to prepare an alternative to the May-Johnson bill. His participation opened old wounds in the War Department over the disclosure of classified information that McMahon deemed necessary to execute the group's functions. General Groves especially detested Condon's appointment.[4]

Clashes with Groves and the War Department played into the hands of the House Un-American Activities Committee, but there is no evidence that they compromised Condon's ability to manage classified military programs at the Bureau of Standards. To the contrary, defense R&D consumed an increasingly large share of the bureau's budget on his watch. Two weeks after the HUAC subcommittee called him "one of the weakest links in our atomic security," associate director Eugene Crittenden and Harry Diamond, chief of the ordnance development division, met personally with Secretary of Commerce Averell Harriman to dispel any doubts about Condon's trustworthiness. "Speaking for the Division Chiefs," Harriman wrote in a memorandum for the record after the meeting, "they had unanimously wanted to convey to me their confidence in . . . Condon. . . . Diamond said that he could particularly state, since he himself was engaged in classified work for the armed forces, that . . . Condon had been scrupulously careful in connection with classified material." The Commerce Department's internal loyalty board also cleared Condon of any wrongdoing.[5]

Harriman had been less sanguine about Condon's adherence to security protocols than Crittenden and Diamond. A few months earlier, in the fall of 1947, after HUAC chairman J. Parnell Thomas (R-New Jersey) publicly condemned Condon's affiliation with alleged communist-front organizations, including the American-Soviet Science Society, Harriman quietly tried to ease his controversial subordinate out of the director's office. On the day before Thanksgiving, Condon met personally with Under Secretary of Commerce William Foster, who explained Harriman's position. In the notes he wrote afterward, Condon recounted the details of their conversation. "Harriman had concluded that I am not a 'good security risk,'" Condon recalled, "and therefore if an offer came along that was attractive to me he would 'encourage' me to accept it." Harriman asked Karl Compton, who served on the bureau's visiting committee, to identify employment opportunities on Condon's behalf.[6] Compton passed Condon's name to University of Minnesota president James Morrill, who had inquired earlier in the year about potential candidates to head the university's institute of technology. Morrill extended an offer, but Condon did not accept it. The meeting with Foster had left a bad impression, and Condon—unaware of Compton's involvement—saw no advantage to be gained by swapping administrative responsibilities.[7]

While Condon determined to stay put, Harriman applied pressure elsewhere. On March 26, 1948, shortly after he met with Crittenden and Diamond, Harriman instructed Foster to obtain "reports on security within the Bureau" directly from the security officer rather than through Condon's office. "This is not an indication of my lack of confidence in the Director of the Bureau or in the present practices," Harriman explained to Foster, "but I believe it is sound procedure that I should have current knowledge of this important aspect of the Bureau's operations from the security officer direct." Frank Jewett assessed Condon's status at the Bureau of Standards in similar but broader terms. "Under present conditions and with Congress and the public generally in the mood they are [in]," Jewett wrote Karl Compton three days later, "it is not at all beyond the bounds of possibility that unless Condon is completely cleared in the minds of the public his days as Director of the Bureau are numbered. The public is in no mood to brook any taint of disloyalty or incapacity in that important position."[8] Jewett's sobering prediction about the future did not transpire. Neither did Harriman's attempt at reassignment. Harriman left the Commerce Department at the end of April; Condon remained on the job for another three years.

Condon, however, began to clamor for a change in November 1950, when he telephoned chemist Henry Eyring, a colleague from Princeton days and now dean of the graduate school at the University of Utah, to inquire about employment opportunities in Salt Lake City. The meteoric rise of Wisconsin's firebrand senator Joseph McCarthy had alarmed him. In a letter he wrote to Eyring the following month, Condon recalled his justification for the initial inquiry—the expectation of a "renewed outbreak of the kind of dishonest red-baiting attacks that I went through two years ago." Now, four days before Christmas, he backpedaled as the Bureau of Standards prepared for a windfall of new military R&D programs in the wake of battlefield reversals on the Korean Peninsula that put US forces on the defensive. "I cannot help but feel that the changed war conditions have so greatly modified the situation that I ought not to think in terms of leaving here," he explained to Eyring. "In the last few weeks there has been a mounting tempo of things which the military departments are urgently requesting us to do for them. . . . Maybe with the country in such serious difficulties there will be less trouble caused by the likes of McCarthy, that even though such people themselves may not behave, they will be less listened to."[9]

The change of heart did not last; McCarthy's popularity soared. Six months later, in mid-July 1951, the retiring director of research at the Corning Glass Works, a well-known maker of glass and other specialty materials located in upstate New York, invited Condon to take the job at twice his annual government salary. This solicitation offered a respite from the unsavory politics that Condon had initially hoped to escape when he first reached out to Eyring. He accepted Corning's offer on August 3. Pleased with the large increase in pay, Condon also relished the opportunity to get back into research. "The laboratory organization is quite small," he wrote Emilie after a tour of the cramped buildings and grounds, "which appeals to me in that it will be possible for me to regain close working contact with all that is going on instead of just being an administrative manager as things are here [at the Bureau of Standards]." During a meeting with the president at the White House three days later, Condon announced his resignation, effective September 30.[10]

Although he knew next to nothing about glass technology and the craft traditions at Corning that had produced so many innovations in the field, Condon jumped at the opportunity to learn the business and diversify the company's technological strengths through an expansion of fundamental research. He spearheaded plans to build a new research laboratory stocked with state-of-the-art equipment. He also justified the anticipated growth of the R&D staff on the basis of the same principles of university-industry collaboration that had structured the fundamental research program at Westinghouse before the war. Early in 1953, Condon proposed to upper management that Corning support ten fellowships—five to be awarded annually at the graduate and postdoctoral levels for a maximum of two years—for undirected research on glass, metals, and crystals—broadly construed as "the physics and chemistry of the solid state"—at select universities throughout the United States. "Research proposals are not expected to be of a kind bearing directly on improvement of glass technology," Condon wrote. "Rather it is hoped that through the application of the Corning Fellowships a lasting contribution to fundamental science will be made, through the stimulation of interest in modern physics and chemistry as applied to glass."[11]

Condon's proposal did not duplicate the most radical element of the program at Westinghouse—in-house support of the research itself—but it clearly pointed in that direction. The fellowships and the liberal publication policy that they collectively supported promised to burnish Corning's R&D reputation within the scientific community. Just as he had done at Westinghouse, Condon structured the program to attract for full-time research positions newly minted PhD scientists already familiar with the company's technological competencies. He stipulated that the fellowship recipients and their professors should be invited to Corning periodically during the summer months to collaborate with the local research staff. "In this way," he explained, "all of the Fellows and various professors would get to know our organization and our people, and more of us would know them, and they would carry back into the universities a deeper insight into basic problems than they would get by staying entirely on their own campuses. And by getting acquainted with us they would be subtly influenced, without pressure, to guide their research toward supplying answers to problems of interest to us." Corning activated the program Condon envisioned in the fall of 1953.[12]

The research fellowships signaled what may have been the high point of Condon's brief career at Corning. Throughout this period, questions about his loyalty persisted. The House Un-American Activities Committee had summoned Condon to testify in September 1952, and the Defense Department did not renew the security clearance, which had expired shortly after his departure from the Bureau of Standards, until the summer of 1954. Later that year, Secretary of the Navy Charles Thomas abruptly revoked his clearance pending further review. Although classified military research comprised a small share of Corning's R&D portfolio and senior executives in the company stood ready to support an appeal of Thomas's decision, Condon decided not to proceed. Exhausted and unwilling to endure another round of government investigations that would be necessary for reinstatement, he resigned in December, just as Corning prepared to award the contract for a spacious new laboratory to house the research and development division.[13]

Corning kept Condon on the payroll as a part-time consultant. He received generous compensation equivalent to two-thirds of his annual salary and relocated to Berkeley. Condon visited Corning periodically to work with the R&D staff on problems of mutual interest, and he cultivated on the company's behalf relevant research under way in the universities. This new arrangement also afforded Condon the flexibility to pursue other employment opportunities to supplement his income. "I do not intend to take a university job that makes it necessary to drop Corning for the simple reason that none of them would pay as much," he wrote Emilie in January 1955. Later that month, Henry Heald, the president of New York University, tapped Condon to spearhead a major expansion of the physics department. He welcomed Condon's consulting agreement at Corning and envisioned it as a necessary precursor to a more elaborate program of academic-industrial collaboration. Condon expressed the same enthusiasm. When Heald solicited input from the university's governing council, however, some members reacted negatively on the basis of Condon's prior security clearance problems at Corning and the likelihood that government contracts might be compromised if the appointment proceeded. "I did not suppose that the persecution would be organized to follow me into details of my purely private life," he wrote Heald on March 1. Condon also called attention to a broader national trend that helped explain the governing council's response: "Today the Department of Defense has such an economic strangle-hold on our universities that it is hard to see how things can be worked out in such a way as to have it be really in the best interests of New York University to have me join its faculty."[14] Heald lamented this turn of events—"I am tremendously concerned about the principle involved in this whole business," he wrote Condon—but the die had already been cast. Universities elsewhere readily obtained defense contracts to diversify and expand as the Cold War restructured the economics of academic research.[15]

Out of the running for a faculty appointment at New York University, Condon turned to a more manageable alternative; he had previously agreed to lecture part-time on nuclear magnetic resonance (NMR) and "its applications to the study of solids" at the University of Pennsylvania during the spring semester. The pioneering work in this new field that grew out of wartime microwave electronics research at Stanford University had recently moved off campus to a start-up company called Varian Associates, some of whose founding members—such as William Hansen—Condon knew personally. Still at an early stage of development, NMR showed promise as a tool for the chemical and structural analysis of glass and other materials.[16] Corning had already obtained through a contract agreement with Varian some encouraging results, which Condon reviewed during a visit to the company's Palo Alto headquarters before the semester in Philadelphia commenced. "There is a wide variety of interesting . . . results . . . for different glasses," he concluded, "but the interpretation and meaning of the results will require a lot more work before it is finally clear."[17]

Additional NMR research that Corning supported at Princeton and Washington University helped crystallize Condon's plans for a long-term commitment in the field. So did the diminished prospect of an academic career at the University of Pennsylvania. The assistant secretary of the navy, who happened to be the son of the university's previous president, intervened to block the appointment that the physics department had recommended. This outcome disappointed but did not surprise Condon, given the failed job negotiations at New York University earlier in the year. On June 6, just before the news broke in Philadelphia, he sounded out Corning's new director of research. "Enough of an exploratory sort has been accomplished," Condon wrote, "that we ought to make provision for having the necessary equipment and one or two full time men working on it to make the fullest use of this new [NMR] technique for our problems."[18]

Condon proposed to establish—with himself at its head—a new laboratory "in or near Berkeley" as an extension of the Corning research and development division to study, by way of electron microscopy and nuclear and paramagnetic resonance, the structure and properties of glass. Management approval came quickly, and Condon began to canvass universities for qualified young PhD physicists to fill out the small research staff he envisioned. When options to obtain laboratory space through local contract research organizations failed to pan out, he reached out to the engineering college at the University of California. Corning, at Condon's behest, had already established a graduate research fellowship in the division of mineral technology.[19] "They have plenty of available space . . . on the campus," he wrote from Berkeley, "and the people . . . are very enthusiastic about the possibility of having such a group here." Condon expected to have the laboratory up and running on the university campus in the fall of 1955.[20]

Condon's plan for fundamental glass research bore a striking resemblance to the scheme he had pitched to Marvin Smith and Lewis Chubb at the end of World War II to set up on the West Coast an R&D arm of the Westinghouse Research Laboratories. He proposed to continue the cutting-edge atomic weapons research that Westinghouse had helped advance at Berkeley. Now, ten years later, he revived the same logic to take advantage of local expertise in microwave physics and diversify Corning's knowledge base in materials technology. Condon made this point explicit during the contract negotiations between Corning and the University of California. "We would confine ourselves essentially to fundamental and basic studies from which patents are not likely to arise," he wrote in October. "If we saw something coming out of the work that was likely to lead directly to a commercial development we would take it away from here [Berkeley] and put the real commercial development of it back in Corning."[21] Condon's explanation satisfied neither party. The university reserved the right to grant licenses and receive royalties on patents derived from the sponsored research. Corning refused to abide by these requirements. "It is difficult," the company's lead negotiator wrote in November, "for us to conceive of additional obstacles which could be thrown up to discourage industrial sponsorship of research at the University. . . . They more than offset any advantages which we might hope to derive by carrying on the proposed program of research at the University rather than within our own organization." Early in 1956, Corning withdrew its support and terminated the project. "The company does not feel we [*sic*] can carry on research at such a distance and beyond our control," Corning's new director of research and Condon's successor wrote on January 19.[22]

The demise of the NMR laboratory at Berkeley marked an end to the type of academic-industrial collaboration that Condon had first cultivated at Westinghouse nearly twenty years earlier and expected to continue into the postwar period. Without a project to manage, Condon, who remained on the Corning payroll, decided to return to academia permanently. He had received an initial inquiry from Washington University in January 1955 but nothing concrete transpired, except for a standing invitation to "bring the matter up if I ever felt interested in coming." The political liabilities that had so badly damaged his relations with Corning, New York University, and the University of Pennsylvania had since abated. When he learned shortly after Corning canceled the Berkeley laboratory project that the chairman of the physics department at Washington University had stepped down, Condon restarted the negotiations.[23] This time, his controversial past did not undermine the appointment. The university administration backed his candidacy. He formally accepted the offer to become chairman of the physics department in May and quietly settled into the twilight of his career as an elder statesman of American science. In 1963, Condon relocated for the last time, to the University of Colorado, where he served on the faculty of the newly established Joint Institute for Laboratory Astrophysics. He directed an air force study of unidentified flying objects from 1966 to 1968 and retired from the university two years later. Condon died in Boulder on March 26, 1974.[24]

CONCLUSION

Early in 1970, on the eve of his retirement from the University of Colorado, Edward Condon received a congratulatory letter from Isidor Rabi, a close friend at Columbia University whom he had first met in Munich in 1927. "Hardly anyone has had a more varied career than you," Rabi recalled. "No one else can match it. Always the thread was physics."[25] These retrospective comments point to the most salient and original qualities of Condon's professional life: the diversity of his research interests and the extent to which he unified them in a coherent vision of scientific and technological progress in industry. Condon did not introduce quantum theoretical physics into industry nor did employment at Princeton, Westinghouse, and the National Bureau of Standards signal a radical departure from the types of careers that other physicists of his generation pursued. A conceptually novel approach to academic-industrial cooperation set Condon apart from his contemporaries.

Armed with a strong physical intuition attuned to the synergies between science and engineering, Condon institutionalized at Westinghouse in the 1930s a new R&D culture that embodied the cooperative instincts he had begun to articulate at the Bell Telephone Laboratories a decade earlier. The postdoctoral fellowship program he set up assigned corporate value to academic-style research on a scale previously unknown among Pittsburgh's manufacturing industries. The electrical industry, which Westinghouse and archrival General Electric dominated, had never seen anything like it. Such discretionary freedom, however, laid bare the limits of how far the new knowledge Condon cultivated could be put to practical use. Condon's ability to translate the results of research—a combination of science and engineering—into marketable products succeeded to a point. He attributed to upper management an unwillingness to appreciate the potential commercial value of radioisotopes and cyclotrons. That perceived intransigence rested not on outright ignorance or an aversion to risk but on legitimate concerns about market stability and the prospects for long-term growth.

World War II rendered moot these mismatched corporate priorities as Westinghouse tuned its productive capacities to defense mobilization. Nuclear physics gave way to microwave electronics and other military technologies. The atom smasher regained institutional status as a source of new knowledge after the war, but its original programmatic justification—that unrestricted research in the field would produce commercially valuable technologies—no longer drove corporate policy.[26] Broader forces external to the company also shaped decision-making. The development of nuclear technology transitioned from the wartime Manhattan Engineer District to the newly established Atomic Energy Commission in 1946. The founding of the AEC, not the discovery of photofission at East Pittsburgh in 1940, opened the door to a commercial market for civilian nuclear power that Westinghouse exploited. Similarly, in the case of military applications for the navy, Westinghouse took cues from Adm. Hyman Rickover, who had developed close ties to the company and the electrical industry as a whole through his wartime duties in the Bureau of Ships. Rickover tapped Westinghouse's engineering and manufacturing expertise to develop nuclear-powered propulsion systems for the navy's surface and submarine fleets.[27] Even in the case of quantity production of radioisotopes for industrial and medical research, the postwar market owed its growth not to the university cyclotrons that Condon had singled out for commercial development in the late 1930s but to the far more prolific atomic reactors the AEC operated at the newly established national laboratories.[28]

Despite his absence from East Pittsburgh after the war, Condon—by way of the postdoctoral fellowship program—left a lasting imprint on Westinghouse R&D that complemented Admiral Rickover's ambitions. Under the admiral's guidance, the navy awarded a contract to Westinghouse in the summer of 1949 to build and operate a reactor R&D facility—the Bettis Atomic Power Laboratory—located thirteen miles southeast of downtown Pittsburgh. Earlier—in October 1948—the company had established a separate atomic power division to manage its interests in the field, including the work just under way at Bettis. William Shoupp, whom Condon had handpicked in 1938 to work on the atom smasher, rose through the corporate ranks to become the division's first director of research. Sidney Siegel, another one of Condon's postdoctoral appointees just off a temporary assignment to get research on reactor materials up and running at the newly designated Oak Ridge National Laboratory, also transitioned to management. On his return from Tennessee, Siegel moved to Bettis and worked under Shoupp to manage physics research. Metallurgical and materials research in the new division proceeded at Bettis under the guidance of William Johnson, a local PhD from the Carnegie Institute of Technology who had followed Shoupp and Siegel to East Pittsburgh on a postdoctoral fellowship in 1939.[29]

Westinghouse capitalized on Condon's intellectual heirs at East Pittsburgh to develop nuclear power through a government-subsidized market that had not existed before the war. At the same time, military priorities played an increasingly important role in the allocation of corporate resources to stimulate technological innovation. Secretary of Commerce Henry Wallace, one of the last high-ranking New Deal liberals left in Washington, envisioned a different model of economic growth in this emerging institutional ecology of the Cold War. He sought to unleash the productive capacities of small businesses without the in-house technical resources necessary to compete against their larger science-based counterparts. Wallace restructured the Commerce Department to that end, and he tapped Condon's expertise to transform the National Bureau of Standards into a principal source of new knowledge for the small business community. Wallace, however, misjudged the oncoming calculus of the Cold War, and he also misread the priorities and performance of small firms. They had prospered during World War II, remained sharply divided on the question of government intervention in the economy, and often supported policies that big business championed.[30]

Condon lacked the flexibility to execute the mandate Wallace had bequeathed to him. He could not simply apply the lessons of fundamental research at Westinghouse to the revitalization of the Bureau of Standards. Bureaucratic inertia and the procedural requirements of public administration stood in his way; so did the broader restructuring of the federal scientific research establishment that accompanied demobilization. Unable to remake the bureau in Westinghouse's image, Condon forged an alternative strategy for the bureau's postwar growth. He deliberately blurred the boundaries between civilian and military R&D to overcome a persistent image problem, alleviate bureaucratic constraints on personnel staffing and promotion, and create new opportunities for institutional expansion and diversification. Condon owed his programmatic success in the cutting-edge field of digital electronic computing to the Department of Defense, not to the fading influence of the progressive-left politics that culminated in Henry Wallace's dismissal from the Department of Commerce in 1946.

The political economy of science that the United States inherited from World War II shifted to the military departments the financial responsibility for an increasingly large share of academic research in the physical sciences. Condon did not fundamentally oppose the use of defense resources to underwrite the advancement of science after the war. Commenting on the delayed establishment of the National Science Foundation during an address to the National Academy of Sciences five days before he left the Bureau of Standards, Condon praised the navy's support of university research. "Fortunately the vacuum thus left was well filled by the enlightened scientific research program of the Office of Naval Research," he declared. "This was conducted as liberally and as intelligently as any purely civilian program could possibly have been conducted and has made a wonderful contribution to the development of basic science in America during the post-war period."[31]

Access to new sources of patronage, however, also upset the delicate balance Condon had struck between academic and corporate interests. Industry lost, and the military departments acquired the cooperative ally he had initially cultivated at Westinghouse and then proclaimed as an equal partner in the far more ambitious proposal to produce new knowledge on behalf of the business community via a national institute of scientific research.[32] The scope of his cooperative vision narrowed even further in the mid-1950s as academic suitors, some of whom had found the prospect of industrial collaboration attractive, withdrew their offers of employment to preempt accusations of subversive activity from anticommunist hardliners in Washington. For all of its support of science as a bulwark against communism, the Cold War shattered Condon's worldview and undermined his faith in a political order realigned to accommodate the institutional and technological imperatives of permanent military preparedness.

Condon's cooperative vision retreated to the sidelines, but it did not disappear entirely. Parts of it survived and adapted to the new postwar research environment. That process of accommodation merits further historical study, given the sheer scale of corporate R&D in the United States and the diversity of firms that populated the manufacturing sector. The historiography is still focused on a limited number of large corporations.[33] Additional case studies of firms—both large and small—that exploited science for commercial gain are needed to assess more fully how the type of strategic thinking Condon and his peers espoused guided the management, organization, and execution of technological innovation in industry as the Cold War intensified. What the historical literature already makes clear is that the stakes could not have been higher for industrial R&D managers who, like Condon, had come of age during World War II. While the military departments leveraged academic science on behalf of national security objectives, technologically diversified companies—buoyed by the experience of wartime R&D and the prestige and status conferred on the scientific community—extolled the virtues of research of a fundamental or more speculative nature as the basis for new product innovations.[34] Universities responded in kind; they sharply increased the production of young PhDs, especially physicists, for a robust job market. Firms that had already acquired military contracts to expand their R&D portfolios also obtained from the Department of Defense generous financial incentives to help offset the cost of new laboratory facilities built to house these new recruits.[35] Westinghouse took advantage of defense priorities to build an elaborate R&D campus that opened in suburban Pittsburgh, not far from the atom smasher, in 1956.[36]

Other science-based firms followed suit, but they soon faced some of the same obstacles that had frustrated Condon on the eve of World War II. Although he assembled a first-rate scientific staff at East Pittsburgh and improved Westinghouse's image as a source of new knowledge, Condon met stiff resistance from corporate managers who did not always share his optimistic predictions for market growth based on fundamental research. Mismatched priorities of this type foreshadowed the difficulties that hobbled the corporate R&D community during the later years of the Cold War. Companies that had supported cutting-edge research to generate radical technological breakthroughs began to scale back their programmatic expectations to remain technologically competitive in global markets. Shorn of their ivory towers, which even in the heyday of fundamental research typically consumed only a small share of total R&D expenditures, firms turned to the universities with renewed vigor to tap into new fields of study ripe for commercial exploitation. Meanwhile, government policies and initiatives designed to improve national competitiveness, such as the relaxation of antitrust regulations and the establishment of R&D consortia, eased the transition to this new environment by way of technology transfer and research collaborations across industries and between firms.[37] This institutional realignment, which gathered strength in the 1980s, bore an outward resemblance to Condon's vision of cooperative research, but it undermined the legitimacy of the industrial laboratory as a strategic asset tuned to his sensibilities and pointed to the collapse of a Cold War consensus that no longer privileged the status of science in the service of corporate interests.

NOTES

INTRODUCTION

1. Edward U. Condon to Raymond T. Birge, November 30, 1927, 3, Folder: "Condon, Edward Uhler, 1902–1974," Box 6, Raymond T. Birge Papers, University Archives, University of California, Berkeley, CA (hereafter cited as Birge Papers).

2. Birge to Condon, December 10, 1927, 1, Folder: "Letters Written by Birge, September–December 1927," Box 32, Birge Papers.

3. Ronald R. Kline, "Construing 'Technology' as 'Applied Science': Public Rhetoric of Scientists and Engineers in the United States, 1880–1945," *Isis* 86 (June 1995): 216–17. For an introduction to the historical literature on industrial R&D, see David A. Hounshell, "The Evolution of Industrial Research in the United States," in *Engines of Innovation: U.S. Industrial Research at the End of an Era*, ed. Richard S. Rosenbloom and William J. Spencer (Boston: Harvard Business School Press, 1996).

4. See Ronald R. Kline and Thomas C. Lassman, "Competing Research Traditions in American Industry: Uncertain Alliances between Engineering and Science at Westinghouse Electric, 1886–1935," *Enterprise and Society* 6 (December 2005): 601–45. On the history of Westinghouse and the company's role in the establishment and growth of the electrical industry, see Harold C. Passer, *The Electrical Manufacturers, 1875–1900: A Study of Competition, Entrepreneurship, Technical Change, and Economic Growth* (Cambridge, MA: Harvard University Press, 1953); Thomas P.

Hughes, *Networks of Power: Electrification in Western Society, 1880–1930* (Baltimore: Johns Hopkins University Press, 1983); and Steven W. Usselman, "From Novelty to Utility: George Westinghouse and the Business of Innovation during the Age of Edison," *Business History Review* 66 (Summer 1992): 251–304.

5. See Leonard S. Reich, *The Making of American Industrial Research: Science and Business at GE and Bell, 1876–1926* (Cambridge: Cambridge University Press, 1985); George Wise, *Willis R. Whitney, General Electric, and the Origins of U.S. Industrial Research* (New York: Columbia University Press, 1985); W. Bernard Carlson, *Innovation as a Social Process: Elihu Thomson and the Rise of General Electric, 1870–1900* (Cambridge: Cambridge University Press, 1991); and David A. Hounshell and John Kenly Smith Jr., *Science and Corporate Strategy: DuPont R&D, 1902–1980* (Cambridge: Cambridge University Press, 1988).

6. See Joseph D. Martin and Michel Janssen, "Beyond the Crystal Maze: Twentieth-Century Physics from the Vantage Point of Solid State Physics," *Historical Studies in the Natural Sciences* 45 (November 2015): 631–40, esp. 631–36.

7. This explanation of Condon's role as a translator is adapted from Hugh G. J. Aitken, *Syntony and Spark: The Origins of Radio* (New York: John Wiley, 1976).

8. "A Conference on Problems of Modern Physics: Opening of the Charles Benedict Stuart Laboratory of Applied Physics, June 19 and 20, 1942," 2, copy of conference program preserved at the Indiana State Library, Indianapolis, IN. *Science* published Condon's address after the dedication: E. U. Condon, "Physics in Industry," *Science* 96 (August 21, 1942): 172–74 (quote, 174).

9. See, for example, Christophe Lécuyer, "MIT, Progressive Reform, and 'Industrial Service,' 1890–1920," *Historical Studies in the Physical and Biological Sciences* 26, no. 1 (1995): 35–88; John W. Servos, "The Industrial Relations of Science: Chemical Engineering at MIT, 1900–1939," *Isis* 71 (December 1980): 531–49; Servos, "Engineers, Businessmen, and the Academy: The Beginnings of Sponsored Research at the University of Michigan," *Technology and Culture* 37 (October 1996): 721–62; W. Bernard Carlson, "Academic Entrepreneurship and Engineering Education: Dugald C. Jackson and the MIT-GE Cooperative Engineering Course, 1907–1932," *Technology and Culture* 29 (July 1988): 536–67; Robert H. Kargon and Scott G. Knowles, "Knowledge for Use: Science, Higher Education, and America's New Industrial Heartland, 1880–1915," *Annals of Science* 59, no. 1 (2002): 1–20; Bruce Seely, "Research, Engineering, and Science in American Engineering Colleges, 1900–1960," *Technology and Culture* 34 (April 1993): 346–67; and Susan W. Morris, "Resource Networks: Industrial Research in Small Enterprises, 1860–1930" (PhD diss., Johns Hopkins University, 2003).

10. See Roger L. Geiger, *To Advance Knowledge: The Growth of American Research Universities, 1900–1940* (New York: Oxford University Press, 1986); Robert E. Kohler, *Partners in Science: Foundations and Natural Scientists, 1900–1945* (Chicago: Uni-

versity of Chicago Press, 1991); and Robert H. Kargon, "The New Era: Science and American Individualism in the 1920s," in *The Maturing of American Science: A Portrait of Science in Public Life Drawn from the Presidential Addresses of the American Association for the Advancement of Science*, ed. Robert H. Kargon (Washington, DC: American Association for the Advancement of Science, 1974).

11. David C. Mowery and Nathan Rosenberg, *Technology and the Pursuit of Economic Growth* (Cambridge: Cambridge University Press, 1989), 123–37, 150–56, 258–59; Roger L. Geiger, "Science, the Universities, and National Defense, 1945–1970," *Osiris* 7 (1992): 26–48, esp. 27; Seely, "Research, Engineering, and Science in American Engineering Colleges," 367–79. Although industry, in terms of dedicated financial resources, occupied a less prominent position, it still exerted substantial influence on the content and scope of academic R&D after World War II. See, for example, Christophe Lécuyer, "What Do Universities Really Owe Industry? The Case of Solid State Electronics at Stanford," *Minerva* 43 (March 2005): 51–71.

12. See Thomas C. Lassman, "Putting the Military Back into the History of the Military-Industrial Complex: The Management of Technological Innovation in the U.S. Army, 1945–1960," *Isis* 106 (March 2015): 94–120.

13. See Nathan Reingold, "Choosing the Future: The U.S. Research Community, 1944–1946," *Historical Studies in the Physical and Biological Sciences* 25, no. 2 (1995): 301–28; and David M. Hart, *Forged Consensus: Science, Technology, and Economic Policy in the United States, 1921–1953* (Princeton: Princeton University Press, 1998), esp. chaps. 5–8.

14. See, for example, J. Merton England, *A Patron for Pure Science: The National Science Foundation's Formative Years, 1945–1957* (Washington, DC: National Science Foundation, 1982); Daniel J. Kevles, "The National Science Foundation and the Debate over Postwar Research Policy, 1942–1945: A Political Interpretation of *Science, the Endless Frontier*," *Isis* 68 (March 1977): 5–26; Jessica Wang, "Liberals, the Progressive Left, and the Political Economy of Postwar American Science: The National Science Foundation Debate Revisited," *Historical Studies in the Physical and Biological Sciences* 26, no. 1 (1995): 139–66; Daniel Lee Kleinman, "Layers of Interests, Layers of Influence: Business and the Genesis of the National Science Foundation," *Science, Technology, and Human Values* 19 (Summer 1994): 259–82; Nathan Reingold, "Vannevar Bush's New Deal for Research: Or the Triumph of the Old Order," *Historical Studies in the Physical and Biological Sciences* 17, no. 2 (1987): 299–344; Larry Owens, "The Counterproductive Management of Science in the Second World War: Vannevar Bush and the Office of Scientific Research and Development," *Business History Review* 68 (Winter 1994): 515–76; and Michael Aaron Dennis, "Reconstructing Sociotechnical Order: Vannevar Bush and U.S. Science Policy," in *States of Knowledge: The Co-Production of Science and Social Order*, ed. Sheila Jasanoff (London: Routledge, 2004).

15. See Henry A. Wallace, "Why S. 1850 Is the Better Bill," *Science* 103 (June 21, 1946): 724–25, 729–30.

16. See Theodore Rosenof, "The Economic Ideas of Henry A. Wallace, 1933–1948," *Agricultural History* 41 (April 1967): 143–53; Edward L. Schapsmeier and Frederick H. Schapsmeier, *Prophet in Politics: Henry A. Wallace and the War Years, 1940–1965* (Ames: Iowa State University Press, 1970), chap. 9; Alonzo L. Hamby, "Sixty Million Jobs and the People's Revolution: The Liberals, the New Deal, and World War II," *Historian* 30 (August 1968): 578–98; and Alan Brinkley, *The End of Reform: New Deal Liberalism in Recession and War* (New York: Vintage, 1995), esp. chaps. 7–8, 10.

17. See Daniel J. Kevles, *The Physicists: The History of a Scientific Community in Modern America* (New York: Knopf, 1978); Stanley Coben, "The Scientific Establishment and the Transmission of Quantum Mechanics to the United States, 1919–32," *American Historical Review* 76 (April 1971): 442–66; John W. Servos, "Mathematics and the Physical Sciences in America, 1880–1930," *Isis* 77 (December 1986): 611–29; Katherine Russell Sopka, *Quantum Physics in America, 1920–1935* (New York: Arno Press, 1980); Charles Weiner, "A New Site for the Seminar: The Refugees and American Physics in the Thirties," in *The Intellectual Migration: Europe and America, 1930–1960*, ed. Donald Fleming and Bernard Bailyn (Cambridge, MA: Belknap Press of Harvard University Press, 1969); S. S. Schweber, "The Empiricist Temper Regnant: Theoretical Physics in the United States, 1920–1950," *Historical Studies in the Physical and Biological Sciences* 17, no. 1 (1986): 55–98; and Robin E. Rider, "Alarm and Opportunity: Emigration of Mathematicians and Physicists to Britain and the United States, 1933–1945," *Historical Studies in the Physical Sciences* 15, no. 1 (1984): 107–76. For a broad overview of the discipline, see Helge Kragh, *Quantum Generations: A History of Physics in the Twentieth Century* (Princeton: Princeton University Press, 1999).

18. See, for example, Joseph Frazier Wall, *Andrew Carnegie* (New York: Oxford University Press, 1970); David Cannadine, *Mellon: An American Life* (New York: Knopf, 2006); Kenneth Warren, *Triumphant Capitalism: Henry Clay Frick and the Industrial Transformation of America* (Pittsburgh: University of Pittsburgh Press, 1996); Warren, *Big Steel: The First Century of the United States Steel Corporation, 1901–2001* (Pittsburgh: University of Pittsburgh Press, 2001); George David Smith, *From Monopoly to Competition: The Transformations of Alcoa, 1888–1986* (Cambridge: Cambridge University Press, 1988); Ronald W. Schatz, *The Electrical Workers: A History of Labor at General Electric and Westinghouse, 1923–1960* (Urbana: University of Illinois Press, 1983); Paul Krause, *The Battle for Homestead, 1880–1892: Politics, Culture, and Steel* (Pittsburgh: University of Pittsburgh Press, 1992); S. J. Kleinberg, *The Shadow of the Mills: Working Class Families in Pittsburgh, 1870–1907* (Pittsburgh: University of Pittsburgh Press, 1989); Francis G. Couvares, *The Remaking of Pittsburgh: Class and Culture in an Industrializing City, 1877–1919* (Albany: State University of New York

Press, 1984); John P. Hoerr, *And the Wolf Finally Came: The Decline of the American Steel Industry* (Pittsburgh: University of Pittsburgh Press, 1988); Michael Weber, *Don't Call Me Boss: David L. Lawrence, Pittsburgh's Renaissance Mayor* (Pittsburgh: University of Pittsburgh Press, 1988); Kirk Savage, "Monuments of a Lost Cause: The Postindustrial Campaign to Commemorate Steel," in *Beyond the Ruins: The Meanings of Deindustrialization,* ed. Jefferson Cowie and Joseph Heathcott (Ithaca: Cornell University Press, 2003); Joel A. Tarr, ed., *Devastation and Renewal: The Environmental History of Pittsburgh and Its Region* (Pittsburgh: University of Pittsburgh Press, 2003); Allen Dieterich-Ward, *Beyond Rust: Metropolitan Pittsburgh and the Fate of Industrial America* (Philadelphia: University of Pennsylvania Press, 2015); and Tracy Neumann, *Remaking the Rust Belt: The Postindustrial Transformation of North America* (Philadelphia: University of Pennsylvania Press, 2016).

19. A. A. Bright, "Delivery of Radar Equipments to the United States Army, Navy, and NDRC—to July 31, 1945," October 30, 1945, 3, Folder 131.7, Box 131, Lee A. DuBridge Papers, Archives and Special Collections, California Institute of Technology, Pasadena, CA (hereafter cited as DuBridge Papers).

CHAPTER 1. RISE OF A THEORETICAL PHYSICIST

1. Edward U. Condon, "What Next? The Story of One American Physicist" (unpublished manuscript), n.d., 1–7, 10–15, 21–22, 26–28, 36–42, 46, 50–53, 79, 86–87, 96–97, 102, Niels Bohr Library and Archives, American Institute of Physics, College Park, MD.

2. Condon, "What Next?," 102, 121–22; E. U. Condon, "Reminiscences of a Life in and out of Quantum Mechanics," in *Proceedings of the International Symposium on Atomic, Molecular, and Solid-State Theory and Quantum Biology,* ed. Per-Olav Löwdin (New York: John Wiley, 1973), 9–11; Alexi Assmus, "The Americanization of Molecular Physics," *Historical Studies in the Physical and Biological Sciences* 23, no. 1 (1992): 1–8; E. U. Condon, "A Theory of Intensity Distribution in Band Systems," *Physical Review* 28 (December 1926): 1182–201.

3. Scientific Notes and News, *Science* 63 (February 5, 1926): 161–62; Sopka, *Quantum Physics in America,* A20; Birge to the NRC Committee on Fellowships, March 25, 1926, Folder: "Letters Written by Birge, January–May 1926," Box 32, Birge Papers; Condon, "What Next?," 82–83, 94–96, 125.

4. Condon, "What Next?," 125; Condon to Caroline Uhler, September 10, 1926, 1, Folder: "Condon, Emilie H., #1," Box 45, Edward U. Condon Papers, American Philosophical Society, Philadelphia, PA (hereafter cited as Condon Papers); Condon to Birge, September 10, 1926, Folder: "Condon, Edward Uhler, 1902–1974," Box 6, Birge Papers.

5. Condon to Leonard B. Loeb, October 5, 1926, 1, Leonard B. Loeb Papers, University Archives, University of California, Berkeley, CA (hereafter cited as Loeb Papers).

6. Condon to Birge, January 14, 1927, 3, Condon to Loeb, January 23, 1927, 1–2, July 5, 1927, 1, all in Folder: "Condon, Edward Uhler, 1902–1974, 20 Letters, 1926–1933," Box 4, Loeb Papers; Condon to Birge, July 28, 1927, 3, Folder: "Condon, Edward Uhler, 1902–1974," Box 6, Birge Papers.

7. Loeb to Condon, February 10, 1927, 1–2, April 5, 1927, 1, Folder: "Letters Written by Loeb, 1927," Box 16, Loeb Papers; Condon to Loeb, July 5, 1927, 2, Folder: "Condon, Edward Uhler, 1902–1974, 20 Letters, 1926–1933," Box 4, Loeb Papers; Condon to Birge, July 28, 1927, 1–2, Folder: "Condon, Edward Uhler, 1902–1974," Box 6, Birge Papers; Birge to Elmer E. Hall, April 26, 1927, 2–3, Folder: "1926–1928," Box 1, Edwin M. McMillan Historical Research Papers, Series 11, Subseries A, Record Group 434, Records of the United States Department of Energy, National Archives and Records Administration, San Bruno, CA (hereafter cited as McMillan Papers).

8. Western Electric jointly owned the laboratories with AT&T, the telephone system's parent company and the provider of long-distance service. Reich, *Making of American Industrial Research*, 139, 182–84. On the early history of R&D in the Bell Telephone System and the origins of the Bell Telephone Laboratories, see Lillian Hoddeson, "The Emergence of Basic Research in the Bell Telephone System, 1875–1915," *Technology and Culture* 22 (July 1981): 512–44; and Reich, *Making of American Industrial Research*, chaps. 6–7. For a detailed technical history, see M. D. Fagen, ed., *A History of Engineering and Science in the Bell System: The Early Years, 1875–1925* (Murray Hill, NJ: Bell Telephone Laboratories, 1975).

9. Condon to Loeb, August 28, 1927, 1–2, October 28, 1927, 2–3, Folder: "Condon, Edward Uhler, 1902–1974, 20 Letters, 1926–1933," Box 4, Loeb Papers; Loeb to Condon, November 4, 1927, 1–2, Folder: "Letters Written by Loeb, 1927," Box 16, Loeb Papers; Condon to Uhler, August 13, 1927, 1–3, Folder: "Condon, Emilie H., #2," Box 45, Condon Papers. See also Arturo Russo, "Fundamental Research at Bell Laboratories: The Discovery of Electron Diffraction," *Historical Studies in the Physical Sciences* 12, no. 1 (1981): 117–60.

10. Loeb to Condon, September 15, 1927, 1, 3, Folder: "Letters Written by Loeb, 1927," Box 16, Loeb Papers; Birge to Condon, December 10, 1927, 1, Folder: "Letters Written by Birge, September–December 1927," Box 32, Birge Papers.

11. Condon to Birge, November 30, 1927, 3, Folder: "Condon, Edward Uhler, 1902–1974," Box 6, Birge Papers; Suman Seth, *Crafting the Quantum: Arnold Sommerfeld and the Practice of Theory, 1890–1926* (Cambridge, MA: MIT Press, 2010), esp. chaps. 1–3.

12. See Louis Galambos, "Theodore N. Vail and the Role of Innovation in the Modern Bell System," *Business History Review* 66 (Spring 1992): 95–126.

13. Condon to Loeb, October 28, 1927, 3–5, Folder: "Condon, Edward Uhler, 1902–1974, 20 Letters, 1926–1933," Box 4, Loeb Papers; Condon to Birge, November 30, 1927, 3–4, Folder: "Condon, Edward Uhler, 1902–1974," Box 6, Birge Papers.

14. Condon to Isidor I. Rabi, December 7, 1927, 2, Folder: "Condon, Edward U., 1927–1970," Box 2, Isidor I. Rabi Papers, Manuscript Division, Library of Congress, Washington, DC (hereafter cited as Rabi Papers). For an introduction to the company's work in this field, see Fagen, *History of Engineering and Science in the Bell System: The Early Years*, chap. 9; and S. Millman, ed., *A History of Engineering and Science in the Bell System: Communications Sciences, 1925–1980* (Murray Hill, NJ: AT&T Bell Laboratories, 1984), chap. 1. See also Paul J. Miranti, "Corporate Learning and Quality Control at the Bell System, 1877–1929," *Business History Review* 79 (Spring 2005): 39–72.

15. J. L. Heilbron and Robert W. Seidel, *Lawrence and His Laboratory*, vol. 1 of *A History of the Lawrence Berkeley Laboratory* (Berkeley: University of California Press, 1989), 21; "The Bartol Research Foundation," *Science* 65 (June 17, 1927): 589; William F. G. Swann to Henry A. Erikson, December 7, 1928, Folder: "Edward U. Condon, Correspondence, 1928–1929," Box 1, Henry A. Erikson Papers, Archives and Special Collections, University of Minnesota, Minneapolis, MN (hereafter cited as Erikson Papers).

16. Condon to Birge, November 30, 1927, 2, Folder: "Condon, Edward Uhler, 1902–1974," Box 6, Birge Papers; Condon to Rabi, December 7, 1927, 1, 3–4, 7, Folder: "Condon, Edward U., 1927–1970," Box 2, Rabi Papers.

17. Condon to Loeb, January 17, 1928, 1–4, Folder: "Condon, Edward Uhler, 1902–1974, 20 Letters, 1926–1933," Box 4, Loeb Papers; Condon to Rabi, December 7, 1927, 2, February 4, 1928, 2–3, Folder: "Condon, Edward U., 1927–1970," Box 2, Rabi Papers.

18. Loeb to Condon, January 25, 1928, 2–6, Folder: "Letters Written by Loeb, 1928," Box 16, Loeb Papers; Hall to Birge, June 13, 1927, 2, Folder: "Hall, Elmer Edgar, 1870–1932," Box 13, Birge Papers; Condon to Erikson, January 20, 1928, 1, March 20, 1928, Folder: "Edward U. Condon, Correspondence, 1928–1929," Box 1, Erikson Papers.

19. Kevles, *Physicists*, 93, 253; Condon to Loeb, January 28, 1927, 2, Folder: "Condon, Edward Uhler, 1902–1974, 20 Letters, 1926–1933," Box 4, Loeb Papers; "Memorandum of Action Taken by the Department of Physics," February 10, 1925, Folder 2, Box 17, Records of the Department of Physics, University Archives, Rare Books and Special Collections, Princeton University, Princeton, NJ (hereafter cited as Princeton Physics Records); Henry D. Smyth to Abraham Flexner, April 15, 1935, 1, Folder 5, Box 5, Princeton Physics Records.

20. Birge to Hall, April 26, 1927, 2, Folder: "1926–1928," Box 1, McMillan Papers; A. G. Shenstone, "E. P. Adams, Princeton Physicist," *Science* 125 (February 22, 1957): 339; Karl T. Compton to John H. Van Vleck, February 14, 1927, 1–2, Van Vleck to Compton, February 18, 1927, and Compton to Van Vleck, March 1, 1927, all in Folder C, Box 45, John H. Van Vleck Papers, Niels Bohr Library and Archives, American Institute of Physics, College Park, MD.

21. "Princeton to Have Two New Chairs," *New York Times*, May 31, 1927, 20; "Makes $200,000 Gift for Princeton Chair," *New York Times*, October 2, 1927, 16; Kohler, *Partners in Science*, 209–12; "$500,000 Added to Princeton Science Fund," *New York Times*, October 7, 1928, 1; Francis Bitter to Rabi, January 2, 1927, Folder: "Bitter, Francis, 1928–1965," Box 1, Rabi Papers; Charles T. Zahn to Condon, October 27, 1927, Condon to Zahn, November 2, 1927, both in Folder: "Zahn, Charles T.," Box 128, Condon Papers; Condon to Loeb, October 28, 1927, 1, Folder: "Condon, Edward Uhler, 1902–1974, 20 Letters, 1926–1933," Box 4, Loeb Papers.

22. Condon to Birge, November 19, 1927, 1–2, Folder: "Condon, Edward Uhler, 1902–1974," Box 6, Birge Papers; Condon to Rabi, December 7, 1927, 4, February 4, 1928, 1–2, Folder: "Condon, Edward U., 1927–1970," Box 2, Rabi Papers; Compton to Condon, February 3, 1928, Folder: "Compton, Karl Taylor, #1," Box 19, Condon Papers; V. L. Collins to Condon, April 13, 1928, Folder: "Princeton University, #1," Box 104, Condon Papers; Condon to Hall, February 16, 1928, Folder: "Condon, Edward Uhler, 1902–1974," Box 6, Birge Papers.

23. Loeb to Condon, February 27, 1928, 1–4, Hall quoted in Loeb to Condon, March 12, 1928, 1, both in Folder: "Letters Written by Loeb, 1928," Box 16, Loeb Papers; Birge to Edwin C. Kemble, February 18, 1928, Folder: "Letters Written by Birge, January–September 1928," Box 32, Birge Papers; Condon to Loeb, March 7, 1928, Folder: "Condon, Edward Uhler, 1902–1974, 20 Letters, 1926–1933," Box 4, Loeb Papers; Loeb to Condon, March 12, 1928, 1, Folder: "Letters Written by Loeb, 1928," Box 16, Loeb Papers.

24. Condon to Loeb, February 21, 1928, 1–2, December 24, 1933, 1, Folder: "Condon, Edward Uhler, 1902–1974, 20 Letters, 1926–1933," Box 4, Loeb Papers; Loeb to Condon, December 28, 1933, 1–2, Folder: "Letters Written by Loeb, June–December 1933," Box 17, Loeb Papers; Edward U. Condon, interview by Charles Weiner, October 17, 1967, session 1, transcript, 49, Niels Bohr Library and Archives, American Institute of Physics, College Park, MD.

25. Compton to Condon, February 3, 1928, Folder: "Compton, Karl Taylor, #1," Box 19, Condon Papers; University and Educational Notes, *Science* 67 (May 25, 1928): 530; *The Catalogue of Princeton University, 1928–1929*, 349, University Archives, Rare Books and Special Collections, Princeton University, Princeton, NJ (hereafter cited as Princeton Catalogs).

26. Condon to William F. Magie, May 8, 1928, Folder 2, Box 20, Princeton Physics Records; "Minutes of First Meeting of Committee on Mathematical Physics," April 15, 1931, Folder 9, Box 15, Princeton Physics Records; *The Catalogue of Princeton University, 1931–1932*, 116, 351–52, Princeton Catalogs; Notes and News, *American Mathematical Monthly* 38 (November 1931): 545.

27. Condon, "Reminiscences of a Life in and out of Quantum Mechanics," 14;

E. U. Condon, "Nuclear Motions Associated with Electron Transitions in Diatomic Molecules," *Physical Review* 32 (December 1928): 858–72; "Ronald W. Gurney, Biographical Data," April 25, 1953, 1, Folder: "Gurney, Ronald W.," Box 75, Condon Papers; Condon to Rabi, March 6, 1928, 1, Folder: "Condon, Edward U., 1927–1970," Box 2, Rabi Papers; Kragh, *Quantum Generations*, 175–78, 187–88.

28. Compton to Erikson, December 6, 1928, Folder: "Edward U. Condon, Correspondence, 1928–1929," Box 1, Erikson Papers; Birge to Condon, December 23, 1927, 2, Folder: "Letters Written by Birge, September–December 1927," Box 32, Birge Papers; Condon to Loeb, January 17, 1928, 1, Folder: "Condon, Edward Uhler, 1902–1973, 20 Letters, 1926–1933," Box 4, Loeb Papers; Scientific Notes and News, *Science* 69 (April 19, 1929): 421.

29. James S. Thompson to Condon, January 16, 1928, Compton to Condon, January 19, 1928, Thompson to Condon, March 2, 1928, all in Folder: "McGraw-Hill Book Company, #1," Box 84, Condon Papers; Compton to Condon, June 2, 1928, Folder: "Compton, Karl T., #2," Box 19, Condon Papers; Philip M. Morse, *In at the Beginnings: A Physicist's Life* (Cambridge, MA: MIT Press, 1977), 86; Condon, "Reminiscences of a Life in and out of Quantum Mechanics," 14; Condon to Thompson, December 3, 1929, 2, Folder: "McGraw-Hill Book Company, #2," Box 84, Condon Papers; "The Sixth Annual Summer Symposium in Theoretical Physics at the University of Michigan," *Science* 77 (March 31, 1933): 320; "Symposia on Theoretical Physics and Chemical Kinetics," *Science* 69 (April 19, 1929): 420; E. U. Condon and P. M. Morse, *Quantum Mechanics* (New York: McGraw-Hill, 1929).

30. Condon to Compton, June 19, 1928, Folder: "Compton, Karl T., #2," Box 19, Condon Papers; Compton to Condon, July 4, 1928, 1–2, July 20, 1928, 1, Folder: "Compton, Karl Taylor, #1," Box 19, Condon Papers; John G. Hibben to the Faculty, January 6, 1928, Folder: "Condon, Edward Uhler," Faculty Files, University Archives, Rare Books and Special Collections, Princeton University, Princeton, NJ (hereafter cited as Princeton Faculty Files).

31. Erikson to J. B. Johnston, December 8, 1928, Erikson to Condon, December 15, 1928, Condon to Erikson, December 22, 1928, 1–2, January 18, 1929, Compton to Erikson, March 18, 1929, all in Folder: "Edward U. Condon, Correspondence, 1928–1929," Box 1, Erikson Papers.

32. Condon to Erikson, March 18, 1929, March 19, 1929, Compton to Erikson, March 19, 1929, Condon to Erikson, March 20, 1929, Erikson to Condon, March 23, 1929, Condon to Erikson, March 25, 1929, 3, Lotus D. Coffman to Condon, May 2, 1929, all in Folder: "Edward U. Condon, Correspondence, 1928–1929," Box 1, Erikson Papers.

33. Condon to Compton, October 24, 1929, 1, Folder: "Compton, Karl T., #2," Box 19, Condon Papers; Compton to Condon, November 5, 1929, November 6, 1929, 1, 3, Condon to Compton, November 8, 1929, 1, all in Folder: "Compton, Karl T., #3," Box

19, Condon Papers; Condon to Erikson, January 17, 1930, Folder: "Edward U. Condon, Correspondence, 1928–1929," Box 1, Erikson Papers; "Compton Chosen M.I.T. President," *New York Times*, March 13, 1930, 3.

34. Smyth to Edwin P. Adams, September 21, 1933, 1, Folder 4, Box 25, Princeton Physics Records; Condon to Compton, October 24, 1929, 2–5, Folder: "Compton, Karl T., #2," Box 19, Condon Papers; Condon to Luther P. Eisenhart, October 14, 1933, 6–7, Folder: "Dodds, H. W., #1," Box 61, Condon Papers. On the origins and growth of mass spectrometry up to World War II, see F. W. Aston, *Mass Spectra and Isotopes*, 2nd ed. (New York: Longmans, Green, 1942).

35. Smyth to Roswell C. Gibbs, February 4, 1935, 1, Folder 13, Box 11, Princeton Physics Records; Adams to Harold W. Dodds, February 15, 1935, 5–6, Folder 4, Box 134, Records of the Office of the President, Series 15 (Harold W. Dodds), University Archives, Rare Books and Special Collections, Princeton University, Princeton, NJ (hereafter cited as Dodds Papers); Condon to Dodds, May 23, 1934, 9–11, Folder: "Dodds, H. W., #1," Box 61, Condon Papers; "New Hydrogen Atom Is Found to Be Natural," *Washington Post*, April 29, 1934, M1.

36. Condon to Compton, October 24, 1929, 2, Folder: "Compton, Karl T., #2," Box 19, Condon Papers; Condon to Compton, November 26, 1929, Compton to Condon, December 5, 1929, both in Folder: "Compton, Karl T., #3," Box 19, Condon Papers; Condon to Compton, March 31, 1930, 1, Folder: "Compton, Karl T., #4," Box 19, Condon Papers; Condon to Gentlemen, n.d., Folder 3, Box 17, Princeton Physics Records; Condon to Loeb, April 7, 1933, 1, Folder: "Condon, Edward Uhler, 1902–1974, 20 Letters, 1926–1933," Box 4, Loeb Papers.

37. Condon to George H. Shortley, May 15, 1934, Folder: "Shortley, George H., #2," Box 111, Condon Papers; E. U. Condon and G. H. Shortley, *The Theory of Atomic Spectra* (New York: Macmillan, 1935).

38. See Kevles, *Physicists*, chap. 15; and Charles Weiner, "1932—Moving into the New Physics," *Physics Today* 25 (May 1972): 40–49.

39. M. Stanley Livingston and John P. Blewett, *Particle Accelerators* (New York: McGraw-Hill, 1962), 31; Compton to Edward H. R. Green, October 6, 1931, Folder: "Van de Graaff, R. J., Sept.–Dec. 1931," Box 228, Records of the Office of the President, 1930–1959 (Karl T. Compton and James R. Killian), Institute Archives and Special Collections, Massachusetts Institute of Technology, Cambridge, MA (hereafter cited as Compton Papers); "Silk Belt Gathers Huge Electric Charge," *Science News Letter* 20 (September 19, 1931): 184; "Ten Million Volts," *Princeton Alumni Weekly* 32 (January 8, 1932): 291; "Compton Chosen M.I.T. President," 3. The Round Hill research station is discussed in Alex Soojung-Kim Pang, "Edward Bowles and Radio Engineering at MIT, 1920–1940," *Historical Studies in the Physical and Biological Sciences* 20, no. 2 (1990): 313–37.

40. Chairman of the Research Committee to the President, November 18, 1930,

Folder 2, Box 17, Princeton Physics Records; "The Cyrus Fogg Brackett Professorship of Physics at Princeton University," *Science* 76 (July 29, 1932): 96; Rudolf W. Ladenburg to Adams, May 21, 1932, Ladenburg to William D. Coolidge, January 27, 1933, Coolidge to Ladenburg, January 31, 1933, all in Folder 4, Box 19, Princeton Physics Records; "Report of R. Ladenburg on His Activities during the Year 1934–35," 1–2, Folder 15, Box 12, Princeton Physics Records; Ladenburg to Allen G. Shenstone, March 30, 1935, 2, Folder 3, Box 12, Princeton Physics Records.

41. "Report of R. Ladenburg on His Activities during the Year 1934–35," 2, Folder 15, Box 12, Princeton Physics Records. Smyth replaced Edwin Adams, who had stepped down temporarily earlier in the year because of illness. Dodds to Adams, February 23, 1935, Folder 5, Box 134, Dodds Papers; Dodds to Smyth, March 5, 1935, Folder 2, Box 3, Princeton Physics Records; Smyth to Dodds, March 6, 1935, Folder 4, Box 134, Dodds Papers; Heilbron and Seidel, *Lawrence and His Laboratory*, 89–100, 232; Livingston and Blewett, *Particle Accelerators*, 135, 137.

42. Milton G. White, interview by Charles Weiner, February 27, 1973, session 2, transcript, 57–61, Niels Bohr Library and Archives, American Institute of Physics, College Park, MD (hereafter cited as White interview); "List of Persons Who Might Appear as Character Witnesses for Condon," n.d., 4, Folder: "Security Investigations, 1948, #2," Box 108, Condon Papers.

43. Condon to Loeb, December 21, 1933, 3, Folder: "Condon, Edward Uhler, 1902–1974, 20 Letters, 1926–1933," Box 4, Loeb Papers; Condon to Dodds, May 23, 1934, 8–9, Folder: "Dodds, H. W., #1," Box 61, Condon Papers; "Abstracts of Papers Presented at the Washington Meeting of the National Academy of Sciences," *Science* 81 (May 3, 1935): 421–22; Herbert Childs, *An American Genius: The Life of Ernest Orlando Lawrence* (New York: E. P. Dutton, 1968), 225. In an oral history interview he gave in February 1973, White recalled this narrative of events. See White interview, 61–62. The only written evidence that corroborates White's account is in a letter Condon wrote to Princeton president Harold Dodds in September 1937. "It was entirely due to my initiative and personal contacts with . . . E. O. Lawrence," Condon wrote, "that our cyclotron project was initiated and given a good start." Condon to Dodds, September 27, 1937, 2, Folder 5, Box 134, Dodds Papers.

44. Adams to Dodds, February 15, 1935, 7, Smyth to Dodds, April 3, 1935, 2, both in Folder 4, Box 134, Dodds Papers; Faculty Meeting Minutes, May 1, 1935, 91, May 3, 1935, 92, in "Department of Physics, Princeton University, Minute Book, 1928," Folder 9, Box 32, Princeton Physics Records; Ernest O. Lawrence to Malcolm C. Henderson, May 3, 1935, Frames 027501–2, Henderson to Lawrence, May 6, 1935, Frame 027503, both in Reel 13, Ernest O. Lawrence Papers (Microfilm), University Archives, University of California, Berkeley, CA (hereafter cited as Lawrence Papers); Smyth to Henderson, May 4, 1935, Henderson to Smyth, May 7, 1935, both in Folder 1, Box 2, Princeton Physics Records.

45. Milton G. White to Smyth, May 9, 1935, Folder 1, Box 2, Princeton Physics Records; Lawrence to Smyth, May 13, 1935, May 17, 1935, 1, Folder 13, Box 10, Princeton Physics Records; Gaylord P. Harnwell to Louis N. Ridenour, July 23, 1935, Smyth to White, May 13, 1935, both in Folder 1, Box 2, Princeton Physics Records.

46. Smyth to Henderson, June 3, 1935, 1, Folder 13, Box 10, Princeton Physics Records; Condon to Lawrence, March 10, 1936, Frame 011609, Reel 6, Lawrence Papers; White to Lawrence, October 21, 1935, Frame 059096, Reel 27, Lawrence Papers; Henderson to John Johnston, October 9, 1935, Folder 3, Box 5, Princeton Physics Records; Research Committee Minutes, December 9, 1935, Folder 1, Box 12, Princeton Physics Records; Smyth to Dodds, March 3, 1936, 1–2, Folder 2, Box 3, Princeton Physics Records; Condon to Shortley, June 9, 1936, 2, Folder: "Cambridge University Press, #4," Box 14, Condon Papers; "The Cyclotron at Princeton," *Science*, Supplement 83 (March 13, 1936): 8; Heilbron and Seidel, *Lawrence and His Laboratory*, 269, 302–3.

47. Condon to Shortley, June 9, 1936, 1, Folder: "Cambridge University Press, #4," Box 14, Condon Papers; "E. U. Condon," n.d., 1, Smyth, "Annual Report of the Department of Physics for the Academic Year 1935–1936," September 11, 1936, 5, both in Folder 15, Box 12, Princeton Physics Records; George W. Gray, "What Holds the World Together," *Harper's Monthly*, September 1937, 434–38; "The Atomic-Physics Observatory of the Carnegie Institution of Washington," *Science* 86 (July 23, 1937): 74; "Measure the Forces between Cores of Hydrogen Atoms," *Science News Letter* 29 (March 14, 1936): 167; G. Breit and E. U. Condon, "Interaction between Protons as Indicated by Scattering Experiments," abstract, *Physical Review* 49 (June 1, 1936): 866. A full-length write-up of their conference paper appeared in print in November: G. Breit, E. U. Condon, and R. D. Present, "Theory of Scattering of Protons by Protons," *Physical Review* 50 (November 1, 1936): 825–45.

48. J. McKeen Cattell and Jaques Cattell, eds., *American Men of Science: A Biographical Directory*, 6th ed. (New York: Science Press, 1938), *s.v.* "Cassen, Benedict"; "Princeton Seniors in Class Day Events," *New York Times*, June 17, 1930, 20; B. Cassen and E. U. Condon, "On Nuclear Forces," *Physical Review* 50 (November 1, 1936): 846, 849.

49. Heilbron and Seidel, *Lawrence and His Laboratory*, 301, 305; "The Summer Symposium on Theoretical Physics at the University of Michigan," *Science* 83 (June 5, 1936): 544; Lawrence to Howard A. Poillon, August 12, 1936, Frame 049110, Reel 22, Lawrence Papers; Condon quoted in White to Lawrence, August 20, 1936, Frame 059111, Reel 27, Lawrence Papers.

50. White to Alexander J. Allen, October 22, 1936, Folder 13, Box 12, Princeton Physics Records; Lawrence to Henderson and White, October 24, 1936, Frame 027514, Henderson to Lawrence, October 25, 1936, Frame 027515, both in Reel 13, Lawrence Papers; Faculty Meeting Minutes, June 4, 1936, 112, November 30, 1936,

119, in "Department of Physics, Princeton University, Minute Book, 1928," Folder 9, Box 32, Princeton Physics Records; White to Gerald Kruger, January 13, 1937, White to Lawrence, March 10, 1937, both in Folder 13, Box 12, Princeton Physics Records.

51. Condon to Donald S. Villars, August 12, 1937, Folder: "Villars, Don S., #2," Box 125, Condon Papers; Condon to Lewis W. Chubb, July 26, 1937, 1, Folder: "WEC, #2," Box 126, Condon Papers; Condon to Smyth, July 26, 1937, 2, Folder 3, Box 3, Princeton Physics Records; D. C. McRoberts, "Contributors to Rubber Compounding Progress: The New Jersey Zinc Co.'s Personnel and Laboratory Facilities," *India Rubber World* 92 (July 1935): 31–32, 33.

52. E. U. Condon and Ralph Havens, "Theoretical Discussion of Heat Flow in Briquets Undergoing Reduction," September 9, 1937, 1, Folder: "The New Jersey Zinc Company (of Pennsylvania), Heat Flow in Briquets," Box 94, Condon Papers.

CHAPTER 2. SCIENCE IN THE STEEL CITY

1. Faculty Meeting Minutes, December 3, 1935, 98, in "Department of Physics, Princeton University, Minute Book, 1928," Folder 9, Box 32, Princeton Physics Records; Henderson to Lawrence, December 17, 1935, Frame 027506, Reel 13, Lawrence Papers; Henderson to Johnston, January 8, 1936, A. H. Warren to Henderson, January 27, 1936, 2, Warren to White, February 20, 1936, all in Folder 3, Box 5, Princeton Physics Records.

2. "Because the University of Pittsburgh has no equipment and no background of research in the field of nuclear physics," physics department chairman Elmer Hutchisson wrote in late 1937, "it seems wise not to develop the nuclear physics field, especially in the face of the feverish activity of so many other leading academic institutions." A close reading of the annual reports of the physics department at nearby Carnegie Institute of Technology (CIT) confirms that a similar scenario played out there. Only the institute's new but separately administered research laboratory of molecular physics, founded in 1934, recorded some progress, but nothing on the scale achieved at Princeton or contemplated at Westinghouse. The laboratory, which employed scientists who had fled Nazi Germany, obtained a radium-beryllium source for low-voltage studies of neutron scattering during the 1935–36 academic year. Elmer Hutchisson to John C. Hubbard, December 23, 1937, 1, Folder: "Correspondence and Records, 1936–1944, 1 of 2," Box 1, Elmer Hutchisson Papers, Niels Bohr Library and Archives, American Institute of Physics, College Park, MD (hereafter cited as Hutchisson Papers); "3 Germans to Join Carnegie Faculty," *New York Times*, September 13, 1933, 8; "German Scholars Get Places Here," *New York Times*, January 28, 1934, 22; "German Atom Expert Heads New Laboratory," *Science News Letter* 25 (March 24, 1934): 184; Otto Stern, "Research Laboratory of Molecular Physics," June 30, 1936, 1, Folder 17, Box 13, Records of the College of Engineering and Science, University Archives, Car-

negie Mellon University, Pittsburgh, PA (hereafter cited as Carnegie Engineering and Science Records). The CIT physics department annual reports (through World War II) are located in Boxes 1 and 2.

3. James Parton, "Pittsburg," *Atlantic Monthly* 21 (January 1868): 19, 21; W. A. Hamor and A. S. Keller, "Natural Resources and Manufactures of Western Pennsylvania [part 1]," *Chemical and Metallurgical Engineering* 34 (July 1927): 426; Hamor and Keller, "Natural Resources and Manufactures of Western Pennsylvania [part 2]," *Chemical and Metallurgical Engineering* 34 (August 1927): 497; "The Middle Atlantic Region," *Index* 11 (May 1931): 95; Langdon White, "The Iron and Steel Industry of the Pittsburgh District," *Economic Geography* 4 (April 1928): 119–21, 128–33.

4. "Jones-Laughlin Steel to Be Reorganized," *New York Times*, December 6, 1922, 2; E. C. Barringer, "Five Interests Make Two-Thirds of Country's Iron and Steel," *Iron Trade Review* 85 (August 15, 1929): 386; F. J. Crolius, "Pittsburgh as a Tonnage Producing Center," *Blast Furnace and Steel Plant* 11 (October 1923): 516; Roy M. Fenton, "Heavy Machinery Makers Compared," *Magazine of Wall Street* 60 (July 31, 1937): 480; "United Celebrates Golden Anniversary," *Iron and Steel Engineer* 28 (December 1951): 124; "Pittsburgh as an Industrial Center," *Industrial and Engineering Chemistry* 14 (June 20, 1936): 247; "Life Goes On," *Fortune* 9 (January 1934): 43; "Opportunities in Expanding Industries," *Magazine of Wall Street* 57 (March 14, 1936): 653; "The Mellon-Owned Oil Company," *Barron's* 15 (May 6, 1935): 14; Phillip Dobbs, "An Industry in Itself, Pioneer Feels Full Weight of Business Letdown," *Magazine of Wall Street* 50 (June 11, 1932): 230–32.

5. "Second in Electrical Equipment," *Barron's* 16 (December 28, 1936): 14; Roy M. Fenton, "Two Electrical Equipment Giants Compared," *Magazine of Wall Street* 60 (June 19, 1937): 314; "The Corporation," *Fortune* 13 (March 1936): 61, 66; "Most U.S. Steel Corp. Subsidiary Plants Are Clustered about Pittsburgh," *Steel* 96 (March 25, 1935): 21. US Steel moved its corporate headquarters from New York City to Pittsburgh in January 1938. "Pittsburgh to Be Headquarters for U.S. Steel Corporation," *Blast Furnace and Steel Plant* 25 (December 1937): 1317–18.

6. W. A. Hamor, "Pittsburgh as a Chemical Research Center," *Journal of Industrial and Engineering Chemistry* 14 (September 1922): 765–68; *Plant and Product of the Mesta Machine Company* (Pittsburgh: Mesta Machine Company, 1919), 8–9; J. B. Morrow, "Research Displaces Rule-of-Thumb," *Coal Age* 32 (November 1927): 244–45; Charles Longenecker, "Jones & Laughlin Steel Corporation in the Service of the Country for Almost a Century," *Blast Furnace and Steel Plant* 29 (August 1941): 890–92.

7. "To Direct Research Work," *Iron Age* 119 (June 9, 1927): 1682; "James Bliss Austin Heads United States Steel Research Laboratory," *Bulletin of the American Ceramic Society* 26 (January 15, 1947): 2; "New Research Program at Pittsburgh Plate Glass," *Glass*

Industry 18 (December 1937): 411–14; Charles E. Skinner and R. W. E. Moore, "The New Westinghouse Research Building," *Electrical World* 71 (June 1, 1918): 1132–33; Margaret B. W. Graham and Bettye H. Pruitt, *R&D for Industry: A Century of Technical Innovation at Alcoa* (Cambridge: Cambridge University Press, 1990), chaps. 1–3.

8. See, for example, Alan G. Wikoff, "Some Research of General Interest in Progress at the Mellon Institute," *Chemical and Metallurgical Engineering* 28 (April 9, 1923): 625–32; E. W. Tillotson, "Researches in Glass at Mellon Institute," *Chemical and Metallurgical Engineering* 34 (April 1927): 232; and Lawrence W. Bass, "Solving Food Manufacturing Problems by Research Institute Method," *Food Industries* 2 (September 1930): 406–9. On the founding of the Mellon Institute, see Robert Kennedy Duncan, "Industrial Fellowships of the Mellon Institute," *Science* 39 (May 8, 1914): 672–78.

9. "Mellon-Owned Oil Company," 14; M. G. Van Voorhis, "Industrial Fellowship System Expands in Temple of Research," *National Petroleum News* 29 (June 23, 1937): 50–52; "Research in Pittsburgh," *Industrial and Engineering Chemistry* 14 (July 20, 1936): 282; "Laboratories of the Gulf Research & Development Corporation," *Review of Scientific Instruments* 6 (November 1935): 338–41.

10. Samuel M. Kintner, "The Engineer and Research," *Journal of Engineering Education* 19 (April 1929): 810; W. Uyterhoeven to Loeb, November 19, 1929, 1, Folder: "U Miscellaneous," Box 14, Loeb Papers; Samuel M. Kintner, "Making Research Profitable," *Manufacturing Industries* 14 (December 1927): 415.

11. For additional biographical details on Kintner, see "Samuel Montgomery Kintner," *Journal of Applied Physics* 8 (February 1937): 117–21.

12. "Westinghouse Air Brake," *Fortune* 15 (March 1937): 115–16; Grondahl quoted in "Advisory Council on Applied Physics Meeting," November 16, 1935, 47, Folder: "Conference on Industrial Physics, Pittsburgh, Nov. 16, 1935," Box 4, Henry A. Barton Papers, Niels Bohr Library and Archives, American Institute of Physics, College Park, MD (hereafter cited as Barton Papers); Scientific News and Notes, *Science* 52 (October 29, 1920): 404; Cattell and Cattell, *American Men of Science*, 6th ed., *s.v.* "Grondahl, Lars Olai"; L. O. Grondahl, "The Role of Physics in Modern Industry," *Science* 70 (August 23, 1929): 179.

13. Loeb to L. J. Neuman, February 12, 1935, 1, Folder: "Letters Written by Loeb, January–March 1935," Box 17, Loeb Papers; Scientific News and Notes, *Science* 72 (September 26, 1930): 316; Austin M. Cravath to Loeb, May 18, 1930, 2, June 12, 1930, 4, Folder: "Cravath, Austin Melville, 1900–," Box 4, Loeb Papers; Loeb to Lawrence, October 5, 1955, with "Ph.D. Students, Leonard B. Loeb, The University of California, Berkeley, California" (undated attachment), Frames 038155, 038157, Reel 17, Lawrence Papers; Cravath to Loeb, January 17, 1932, 5, Folder: "Cravath, Austin Melville, 1900–," Box 4, Loeb Papers; "Union Switch & Signal Company," February 12, 1934,

7, Accession No. 1810, Folder 7, Box 1130, Manuscripts and Archives, Hagley Museum and Library, Wilmington, DE.

14. "Material for the AIP Project on the History of Recent Physics, Prepared by Arthur E. Ruark for Dr. James W. King," August 1963, A-2, B-11, Arthur E. Ruark File, Biographical Material Files, Niels Bohr Library and Archives, American Institute of Physics, College Park, MD; Morris Muskat to Paul S. Epstein, November 7, 1930, 2, 4, Folder 75, Box 5, Paul S. Epstein Papers, Archives and Special Collections, California Institute of Technology, Pasadena, CA.

15. See Leonard S. Reich, "Irving Langmuir and the Pursuit of Science and Technology in the Corporate Environment," *Technology and Culture* 24 (April 1983): 199–221; and Russo, "Fundamental Research at Bell Laboratories."

16. See Lillian Hoddeson, "The Entry of the Quantum Theory of Solids into the Bell Telephone Laboratories, 1925–40: A Case Study of the Industrial Application of Fundamental Science," *Minerva* 18 (Autumn 1980): 422–47; and Hoddeson, "The Discovery of the Point Contact Transistor," *Historical Studies in the Physical Sciences* 12, no. 1 (1981): 41–76.

17. Overton Luhr to Loeb, April 14, 1935, 1, Folder: "Luhr, Overton," Box 9, Loeb Papers; Wise, *Willis R. Whitney, General Electric, and the Origins of U.S. Industrial Research*, 159–61, 229–31, 252–53.

18. Henry A. Barton to Hutchisson, October 15, 1935, October 21, 1935, Hutchisson to Barton, October 22, 1935, all in Folder: "Journal of Applied Physics—Dr. Elmer Hutchisson, 1935 to 1940," Box 7, Barton Papers; Barton to Compton, October 17, 1935, Compton to Barton, October 18, 1935, both in Folder: "Compton, Karl T., 1935, July–December," Box 3, Barton Papers; E. Hutchisson, "Conference on Industrial Physics," *Review of Scientific Instruments* 6 (December 1935): 381–82; Henry A. Barton, "Report of Conference on Applied Physics," *Review of Scientific Instruments* 7 (March 1936): 113.

19. "The American Institute of Physics," *Science* 74 (November 20, 1931): 508–9; Henry A. Barton, "Advisory Council on Applied Physics," *Review of Scientific Instruments* 6 (December 1935): 383–84; "Advisory Council on Applied Physics of the American Institute of Physics, Report of Meeting, New York, October 28, 1936," *Journal of Applied Physics* 8 (February 1937): 99–100; Henry A. Barton, "New Journal of Applied Physics," *Review of Scientific Instruments* 7 (November 1936): 399; "Report of the Director for the Year 1936," February 24, 1937, 3–7, Folder: "AIP, 1937," Box 1, Barton Papers; Spencer R. Weart, "The Physics Business in America, 1919–1940: A Statistical Reconnaissance," in *The Sciences in the American Context: New Perspectives*, ed. Nathan Reingold (Washington, DC: Smithsonian Institution Press, 1979), 301–5, 318–26.

20. Foote, Dushman, and Hull quoted in Barton, "Report of Conference on Applied Physics," 119–20, 122.

21. Henry A. Barton, "A Conference on Applied Physics," *Review of Scientific Instruments* 6 (February 1935): 32. The recommendation to establish the AIP advisory council on applied physics originated in this earlier joint meeting. See Barton, "Advisory Council on Applied Physics," 383.

22. See Kline and Lassman, "Competing Research Traditions in American Industry."

23. Kintner, "Making Research Profitable," 418; "Personalities in Research," *Scientific American* 156 (January 1937): 3.

24. J. Slepian, X-Ray Tube, US Patent 1,645,304, filed April 1, 1922, and issued October 11, 1927; Donald W. Kerst, "Development of the Betatron and Applications of High Energy Betatron Radiations," *American Scientist* 35 (January 1948): 57–59. See also on Slepian, "Personalities in Science," *Scientific American* 165 (July 1941): 3.

25. Samuel M. Kintner to Chubb, May 10, 1935, Chubb to Kintner, May 20, 1935, both in Folder 1, Box 178, Records of the Westinghouse Electric Corporation, Thomas and Katherine Detre Library and Archives, Senator John Heinz History Center, Pittsburgh, PA (hereafter cited as Westinghouse Records).

26. Paul D. Foote to Barton, October 25, 1932, September 18, 1934, Folder: "Foote, Paul D., 1931–39," Box 5, Barton Papers; Cravath to Loeb, January 17, 1932, 2–5, May 17, 1932, 1, Folder: "Cravath, Austin Melville, 1900–," Box 4, Loeb Papers; Westinghouse Electric and Manufacturing Company, *55th Annual Report, 1940*, 16, copy preserved at the Library of Congress; Schatz, *Electrical Workers*, 60–61; "Personnel Report for June," July 5, 1933, 2, "Personnel Report for June," June 29, 1934, 2, both in "Personnel Rosters, January 1927 to December 1940 (Inclusive)," Folder 9, Box 165, Westinghouse Records; Wise, *Willis R. Whitney, General Electric, and the Origins of U.S. Industrial Research*, 301; Chubb to Kintner, May 20, 1935, Folder 1, Box 178, Westinghouse Records.

27. R. P. Jackson to O. H. Eschholz, July 18, 1935, Eschholz to Distribution, July 22, 1935, Harvey C. Rentschler to Eschholz, July 31, 1935, all in Folder 1, Box 178, Westinghouse Records; Emilio Segrè, *Enrico Fermi: Physicist* (Chicago: University of Chicago Press, 1970), chap. 3; Simone Turchetti, "The Invisible Businessman: Nuclear Physics, Patenting Practices, and Trading Activities in the 1930s," *Historical Studies in the Physical and Biological Sciences* 37, no. 1 (September 2006): 160–64. See also Gerald Holton, "Striking Gold in Science: Fermi's Group and the Recapture of Italy's Place in Physics," *Minerva* 12 (April 1974): 159–98.

28. Rentschler to Eschholz, July 31, 1935, 2, Folder 1, Box 178, Westinghouse Records; Jackson to Lawrence, October 15, 1935, Frame 058741, Reel 27, Lawrence Papers; Heilbron and Seidel, *Lawrence and His Laboratory*, 184–87.

29. Dayton L. Ulrey to Chubb, n.d., Victor S. Beam to Eschholz, August 9, 1935, both in Folder 1, Box 178, Westinghouse Records. The Westinghouse X-Ray Company

had been founded in 1930 through the acquisition and merger of two existing firms: the Wappler Electric Company of Long Island City, New York, and the American X-Ray Corporation, located in Chicago. "Westinghouse Forms X-Ray Company," *Radiology* 15 (September 1930): 408–9.

30. Chubb and Rentschler to Kintner, September 11, 1935, Chubb to Kintner, September 23, 1935, Kintner to David S. Youngholm, October 11, 1935, all in Folder 1, Box 178, Westinghouse Records; Turchetti, "Invisible Businessman," 164.

31. Charles M. Slack to Ladenburg, October 28, 1935, Ladenburg to Slack, November 2, 1935, Slack to Ladenburg, November 11, 1935, all in Folder 8, Box 9, Princeton Physics Records; Chubb to Kintner, November 29, 1935, Folder 1, Box 178, Westinghouse Records; Beam to Eschholz, January 17, 1936, Kintner to Chubb, n.d., Eschholz to Beam, January 22, 1936, all in Folder 2, Box 178, Westinghouse Records.

32. Chubb to Kintner, October 29, 1935, Jackson to Chubb, October 15, 1935, Jackson, "Dr. E. O. Lawrence, Cyclotron," July 18, 1935, 3, all in Folder 1, Box 178, Westinghouse Records.

33. Ulrey to Lawrence, October 23, 1935, Frame 058744, Lawrence to Ulrey, November 6, 1935, Frame 058746; Ulrey to Lawrence, November 27, 1935, Frame 058747, all in Reel 27, Lawrence Papers.

34. Ulrey to Smyth, October 23, 1935, Smyth to Ulrey, October 25, 1935, 1, Ulrey to Smyth, October 29, 1935, all in Folder 6, Box 11, Princeton Physics Records; Ulrey to White, October 30, 1935, Ulrey to Chubb, December 23, 1935, 2, both in Folder 1, Box 178, Westinghouse Records; White to Lawrence, November 1, 1935, Frame 059102, November 21, 1935, Frame 059106, Reel 27, Lawrence Papers.

35. Ulrey to Louis N. Ridenour, February 19, 1936, Folder 2, Box 178, Westinghouse Records; Ulrey to Chubb, December 23, 1935, 2, Folder 1, Box 178, Westinghouse Records; Chubb to Bitter, February 7, 1936, Folder: "Westinghouse Correspondence, 1936," Box 1, Francis Bitter Papers, Institute Archives and Special Collections, Massachusetts Institute of Technology, Cambridge, MA.

36. John A. Fleming to Merle A. Tuve, June 20, 1933, 2, Folder: "Letters, 1933," Box 8, Merle A. Tuve Papers, Manuscript Division, Library of Congress, Washington, DC (hereafter cited as Tuve Papers); Fleming to William H. Wells, June 17, 1933, Untitled Folder, Box 15, Tuve Papers; Tuve, "Report for May 1934," June 9, 1934, Folder: "Monthly Bureau Reports-M.A.T.," Box 15, Tuve Papers. On Tuve and the origins of high-voltage research at the Carnegie Institution of Washington, see Thomas D. Cornell, "Merle A. Tuve and His Program of Nuclear Studies at the Department of Terrestrial Magnetism: The Early Career of a Modern American Physicist" (PhD diss., Johns Hopkins University, 1986).

37. Fleming to John T. Tate, May 14, 1934, Tate to Tuve, May 22, 1934, Fleming to Tate, August 13, 1934, all in Folder: "Lab Letters, 1934," Box 13, Tuve Papers; Tuve,

"Department of Terrestrial Magnetism, Carnegie Institution of Washington," August 26, 1935, 1, Untitled Folder, Box 15, Tuve Papers.

38. Smyth promised to support White if the National Research Council did not renew his postdoctoral fellowship for a second year. Ridenour joined the Princeton physics faculty as a full-time instructor in January 1936. White to Lawrence, Frame 059103, Reel 27, Lawrence Papers; Faculty Meeting Minutes, January 7, 1936, 100, January 8, 1936, 101, November 30, 1936, 119, in "Department of Physics, Princeton University, Minute Book, 1928," Folder 9, Box 32, Princeton Physics Records; Smyth to Dodds, June 3, 1936, Dodds to Smyth, June 11, 1936, both in Folder 4, Box 134, Dodds Papers; "Scholarship, Prize, Fellowship Appointment," September 29, 1934, October 19, 1935, Folder: "Wells, William H.," Box 133, Personnel Records, Archives and Special Collections, University of Minnesota, Minneapolis, MN; "University of Minnesota Board of Regents Minutes," January 20, 1931, 828, copy retrieved on February 21, 2017, from the Libraries Digital Conservancy, Archives and Special Collections, University of Minnesota, http://hdl.handle.net/11299/45429 (paper copy also in the author's possession); Fleming to Ulrey, October 28, 1935, 1, Folder: "Lab Letters, 1935," Box 14, Tuve Papers.

39. "Physics Symposium at Cornell," *Review of Scientific Instruments* 7 (September 1936): 361; Tuve to Lawrence, January 7, 1936, Folder: "Lab Letters, 1936," Box 16, Tuve Papers; Fleming to John C. Merriam, May 28, 1936, Merriam, Handwritten Note, May 29, 1936, both in Untitled Folder, Box 15, Tuve Papers. On Tuve's proposal for the 20 MeV pressure installation, see Tuve, "A Full-Scale Equipment for Researches in High-Energy Physics," August 20, 1935 (revised September 30, 1935), Untitled Folder, Box 15, Tuve Papers.

40. Wells to Chubb, August 7, 1936, Wells, "Report and Recommendations on Nuclear Physics," n.d., 5–11, both in Folder 2, Box 178, Westinghouse Records.

41. Wells, "Report and Recommendations on Nuclear Physics," 6, 15–16, Folder 2, Box 178, Westinghouse Records; Heilbron and Seidel, *Lawrence and His Laboratory*, 27, 184–99, 314.

42. Chubb to Rentschler, August 5, 1936, Rentschler to Chubb, August 6, 1936, Chubb, "Memorandum on Nuclear Physics," August 13, 1936, 1–2, all in Folder 2, Box 178, Westinghouse Records. Fermi did not obtain a US patent until July 1940. See Turchetti, "Invisible Businessman," 170–71.

43. Chubb, "Memorandum on Nuclear Physics," August 13, 1936, 2, Folder 2, Box 178, Westinghouse Records.

44. See Larry Owens, "MIT and the Federal 'Angel': Academic R&D and Federal-Private Cooperation before World War II," *Isis* 81 (June 1990): 189–213.

45. See, for example, "Disks Spinning in Vacuum May Replace Present Generators," *Science News Letter* 24 (September 23, 1933): 202; "Huge Power Plan Gets Fed-

eral Aid," *New York Times,* December 5, 1933, 41; "Revolutionary Method of Power Transmission Urged," *Science News Letter* 26 (December 29, 1934): 408; and Watson Davis, "A New Plan for the Production and Transmission of Electrical Power," *Science,* Supplement 81 (January 11, 1935): 8.

46. Chubb, "Memorandum on Nuclear Physics," August 13, 1936, 2–3, Folder 2, Box 178, Westinghouse Records; Wise, *Willis R. Whitney, General Electric, and the Origins of U.S. Industrial Research,* 214. On Steinmetz, see Ronald R. Kline, *Steinmetz: Engineer and Socialist* (Baltimore: Johns Hopkins University Press, 1992).

47. "Samuel M. Kintner, Engineer, 64, Dead," *New York Times,* September 29, 1936, 27; M. W. Smith, "The Importance of Research and Development in Maintaining Technical Progress," *Electrical Engineering* 57 (December 1938): 484.

48. Marvin W. Smith to Ralph Kelly, September 16, 1936 (attached memorandum: Smith, "Nuclear Physics," n.d., 1), Chubb, "Memorandum on Nuclear Physics," August 13, 1936, 3, all in Folder 2, Box 178, Westinghouse Records.

49. Wells to Joseph Slepian, August 17, 1936, Chubb, "Memorandum on Nuclear Physics," September 8, 1936, Smith to Kelly, September 16, 1936, G. W. Penney, "Memorandum on Nuclear Physics," n.d. (attached to letter from Penney to Smith, September 11, 1936), Smith to Roscoe Seybold, September 22, 1936, Seybold to Smith, October 16, 1936, all in Folder 2, Box 178, Westinghouse Records.

50. "Scientists Unleash Largest Atom Attacking Machine," *Science News Letter* 24 (December 2, 1933): 364.

51. Watson Davis, "Experimenting with Millions of Volts," *Science News Letter* 19 (May 23, 1931): 326–27; Robert J. Van de Graaff et al., "Report on the Nuclear Research Project at Round Hill for the Year Ending Jan. 1, 1935," 1, 3, Folder: "Van de Graaff, R. J., Jan.–April 1935," Box 229, Compton Papers; "The Round Hill High Voltage-Nuclear Disintegration Project," June 21, 1935, 7, Folder: "Round Hill, June–Nov. 1935," Box 187, Compton Papers; R. J. Van de Graaff et al., "The Electrostatic Production of High Voltage for Nuclear Investigations," *Physical Review* 43 (February 1, 1933): 156–57.

52. "Report on the M.I.T. Nuclear Research Project at Round Hill for the Year Ending Dec. 31, 1935," 3, Folder: "Van de Graaff, R. J., May–Dec. 1935," Box 229, Compton Papers; "Report to the Research Corporation on the M.I.T. High Voltage Nuclear Research Project for the Year 1936," January 9, 1937, 1–2, "Report to the Research Corporation on the M.I.T. High Voltage Nuclear Research Project for the Year 1937," 1–2, both in Folder: "Van de Graaff, R. J., 1937," Box 229, Compton Papers; Wells, "Report and Recommendations on Nuclear Physics," 9, Folder 2, Box 178, Westinghouse Records; Compton to Alfred L. Loomis, May 27, 1937, 2–4, Folder: "Round Hill, 1937," Box 187, Compton Papers.

53. Condon to Chubb, January 29, 1941, 7, Folder: "WEC, #2," Box 126, Condon Papers; W. H. Wells et al., "Design and Preliminary Performance Tests of the West-

inghouse Electrostatic Generator," *Physical Review* 58 (July 15, 1940): 162; W. H. Wells, "Production of High Energy Particles," *Journal of Applied Physics* 9 (November 1938): 683–86; Livingston and Blewett, *Particle Accelerators*, 35–36; Wells, "Report and Recommendations on Nuclear Physics," 9, Folder 2, Box 178, Westinghouse Records; Fleming to Merriam, October 1, 1935, Tuve, "A Full-Scale Equipment for Researches in High-Energy Physics," August 20, 1935 (revised September 20, 1935), 4, both in Untitled Folder, Box 15, Tuve Papers; Tuve, "Report for May 1937," June 7, 1937, Folder: "Monthly Report—M.A.T.," Box 15, Tuve Papers; "Why Science Seeks to Smash Atom," *Washington Post*, August 15, 1937, A6.

54. Wells, "Memorandum on Public Relations Aspect of the High Voltage and Nuclear Physics Research Program," September 24, 1936, 1–2, Folder 2, Box 178, Westinghouse Records. The company limited the voltage to 5 MeV in public announcements. Wells to W. W. Rodgers, November 4, 1936, 1, Folder 2, Box 178, Westinghouse Records. See also "Cracking the Atom," *Daily Journal of Commerce* [Portland, OR], November 18, 1936, 2; "Big Steel Bulb to Aid Science," *Boston Globe*, April 13, 1937, 13; "Describes Feats of Atom Smasher," *New York Times*, June 3, 1937, 27; "Atom Smasher," *Chicago Daily Tribune*, July 29, 1937, 9; L. W. Chubb, "Exploring the Atom," *Scientific Monthly* 45 (September 1937): 285–87; and "Atom Smashing Equipment," *Journal of Applied Physics* 8 (November 1937): 758.

55. Chubb to Smith, November 4, 1936, Folder 2, Box 178, Westinghouse Records; Wise, *Willis R. Whitney, General Electric, and the Origins of U.S. Industrial Research*, 302, 304–5.

56. Saul Dushman to Lawrence, January 28, 1935, Frame 023458, Reel 11, Lawrence Papers; Frederick Seitz, *On the Frontier: My Life in Science* (New York: AIP Press, 1994), 95–96, 103; Frederick Seitz to Eugene Wigner, November 23, 1937, 1, Folder 2, Box 83, Eugene P. Wigner Papers, Rare Books and Special Collections, Princeton University, Princeton, NJ (hereafter cited as Wigner Papers).

57. Hoddeson, "Entry of the Quantum Theory of Solids into the Bell Telephone Laboratories," 441–45; William B. Shockley to Tuve, June 5, 1940, 1, "Summary and Conclusion," n.d., both in Folder: "Tuve Letters, 1940," Box 19, Tuve Papers.

58. George S. Sangdahl to C. G. Bunnell and L. C. Tomer, November 28, 1936, Folder 2, Box 178, Westinghouse Records; Westinghouse Electric and Manufacturing Company, *55th Annual Report, 1940*, 16; Alfred D. Chandler Jr., *Strategy and Structure: Chapters in the History of the American Industrial Enterprise* (Cambridge, MA: MIT Press, 1962), 363–68, 399n1; Ernest Dale, *The Great Organizers* (New York: McGraw-Hill, 1960), 144–45, 150–52, 161–62.

59. Smith to Frank W. Merrick, March 17, 1937, 1, Folder 3, Box 178, Westinghouse Records.

60. L. W. Chubb, "New Products from Research in the Electrical Industry," *Edison*

Electric Institute Bulletin 5 (July 1937): 294; "Organizes New Products Division," *Steel* 101 (August 16. 1937): 35.

61. Smith to Merrick, March 17, 1937, 2, Folder 3, Box 178, Westinghouse Records; Hounshell and Smith, *Science and Corporate Strategy*, chaps. 12–13. Smith quoted in his proposal to Merrick excerpts from an article Stine wrote that chronicled the origins and growth of DuPont's fundamental research program. See C. M. A. Stine, "The Place of Fundamental Research in an Industrial Research Organization," *Transactions of the American Institute of Chemical Engineers* 32 (June 29, 1936): 127–37, esp. 127, 131.

62. Hounshell and Smith, *Science and Corporate Strategy*, 225; Smith to Merrick, March 17, 1937, 1–2, Folder 3, Box 178, Westinghouse Records.

63. Emerson B. Roberts to Tuve, February 9, 1937, Tuve to Roberts, February 23, 1937, 2, both in Folder: "1933–1939," Box 3, Tuve Papers.

64. Kevles, *Physicists*, 303; Lee A. DuBridge to Alan C. Valentine, n.d., Folder: "36/37," Box 233, DuBridge Papers; Smith to Merrick, March 17, 1937, 2–3, Folder 3, Box 178, Westinghouse Records.

65. "Notes Taken in an Interview with Professor DuBridge on Thursday Evening, April 8," n.d., Valentine to DuBridge, April 13, 1937, April 20, 1937, DuBridge to Valentine, May 11, 1937, all in Folder: "36/37," Box 233, DuBridge Papers; DuBridge to Condon, July 21, 1937, 1, Folder: "WEC, #2," Box 126, Condon Papers.

66. Slepian to Tuve, May 3, 1937, Tuve to Slepian, May 6, 1937, Tuve to Fleming, May 10, 1937, all in Folder: "1933–1939," Box 3, Tuve Papers; Tuve to Robert Brode, February 1, 1939, 1, Folder: "APO-1939," Box 18, Tuve Papers; "Memorandum of Conversation with Dr. Fleming, and Later with Fleming, Tuve, and Breit," May 4, 1937, Folder 24, Box 4, Records of the Department of Terrestrial Magnetism, Carnegie Institution for Science, Washington, DC (hereafter cited as DTM Records); Tuve to Slepian, May 7, 1937, 1, Folder 10, Box 8, DTM Records.

67. Tuve to Slepian, May 7, 1937, 2, Folder 10, Box 8, DTM Records; Roberts to Tuve, May 13, 1937, Folder: "1933–1939," Box 3, Tuve Papers; Condon, "Suggested List of Ten Witnesses to Ask to Appear on Condon's Behalf," April 5, 1948, 2, Folder: "Security Investigations, 1948, #3," Box 108, Condon Papers; DuBridge to Condon, July 21, 1937, 1, Folder: "WEC, #2," Box 126, Condon Papers; Westinghouse Research Laboratories, "Work and Progress Report," July–September 1937, 30, Westinghouse Quarterly Research Reports, Accession No. 2017.0312, Thomas and Katherine Detre Library and Archives, Senator John Heinz History Center, Pittsburgh, PA (hereafter cited as Westinghouse Research Reports).

CHAPTER 3. ATOM SMASHING AT EAST PITTSBURGH

1. Loeb to Condon, December 31, 1937, 1, Folder: "Loeb, Leonard B., #1," Box 82, Condon Papers; White to Lawrence, October 18, 1937, Frame 059125, Reel 27, Lawrence Papers.

2. Slepian to Condon, July 14, 1937, Chubb to Condon, July 21, 1937, 1, Condon to Chubb, July 26, 1937, 1, all in Folder: "WEC, #2," Box 126, Condon Papers; Condon to Smyth, July 26, 1937, 1, Folder 3, Box 3, Princeton Physics Records.

3. DuBridge to Condon, July 21, 1937, 2–3, Chubb to Condon, July 21, 1937, 2, Condon, "Division of Fundamental Research," n.d., all in Folder: "WEC, #2," Box 126, Condon Papers.

4. Condon to Chubb, August 6, 1937, 1–2, Folder: "WEC, #2," Box 126, Condon Papers; John A. Hipple, "Development of a Portable Mass Spectrometer for Gas Analysis and Research," April 1, 1941, 3, Folder: "WEC, Mass Spectrometer for Gas Analysis and Research, #1," Box 127, Condon Papers; Walker Bleakney to G. G. Oberfell, June 14, 1937, Folder 5, Box 2, Princeton Physics Records; Bleakney, "Report to the Chairman," July 2, 1937, 3, Folder 15, Box 12, Princeton Physics Records.

5. Condon to Chubb, August 6, 1937, 2, Folder: "WEC, #2," Box 126, Condon Papers; Ladenburg to Ralph B. Barnes, March 17, 1933, Folder 4, Box 19, Princeton Physics Records; Barnes to Smyth, June 27, 1935, 2, Folder 15, Box 12, Princeton Physics Records; Condon to Shortley, June 9, 1936, 1, Folder: "Cambridge University Press, #4," Box 14, Condon Papers; Adams to Dodds, February 15, 1935, 5, Barnes to Dodds, May 25, 1936, both in Folder 4, Box 134, Dodds Papers; Yakov M. Rabkin, "Technological Innovation in Science: The Adoption of Infrared Spectroscopy by Chemists," *Isis* 78 (March 1987): 38; Anthony S. Travis, *Dyes Made in America, 1915–1980: The Calco Chemical Company, American Cyanamid, and the Raritan River* (Jerusalem: Hexagon Press and the Sidney M. Edelstein Center for the History and Philosophy of Science, Technology, and Medicine, Hebrew University, 2004), 101–3.

6. Condon to Chubb, August 6, 1937, 3, Folder: "WEC, #2," Box 126, Condon Papers; Westinghouse Research Laboratories, "Work and Progress Report," July–September 1937, 30, Westinghouse Research Reports.

7. Condon to Smyth, September 26, 1937, 1, Folder 3, Box 3, Princeton Physics Records; Condon to Dodds, September 27, 1937, 4–5, Folder 5, Box 134, Dodds Papers. At Princeton in the fall of 1936, physics instructors Malcolm Henderson, Louis Ridenour, and Milton White each received a $2,250 salary. Faculty Meeting Minutes, November 30, 1936, 119, in "Department of Physics, Princeton University, Minute Book, 1928," Folder 9, Box 32, Princeton Physics Records.

8. Westinghouse Research Laboratories, "Work and Progress Report," October–December 1937, 30, Westinghouse Research Reports; Hounshell and Smith, *Science and Corporate Strategy*, 292–93.

9. Condon to Chubb, January 29, 1941, 2–3, Folder: "WEC, #2," Box 126, Condon Papers.

10. Condon to Dodds, November 11, 1937, Folder 5, Box 134, Dodds Papers; Condon to Chubb, January 29, 1941, 1–3, Folder: "WEC, #2," Box 126, Condon Papers; Condon to Loeb, December 24, 1933, 4, Folder: "Condon, Edward Uhler, 1902–1974, 20 Letters, 1926–1933," Box 4, Loeb Papers.

11. Smyth to the President, November 26, 1937, Folder: "Condon, Edward Uhler," Princeton Faculty Files; Condon to Dodds, September 27, 1937, 1–2, Dodds to Condon, November 22, 1937, Condon to Dodds, December 1, 1937, December 3, 1937, all in Folder 5, Box 134, Dodds Papers; Dodds to Condon, December 2, 1937, Folder: "Princeton University, #4," Box 104, Condon Papers.

12. See, for example, "The Westinghouse Research Fellowships," *Science* 86 (December 31, 1937): 605–6; "Westinghouse Research Fellowships," *Journal of Applied Physics* 9 (January 1938): 34–35; "Ten Will 'Explore' into Pure Science," *New York Times*, December 19, 1937, 49; and "Westinghouse Will Endow Scientists," *Los Angeles Times*, December 20, 1937, 5. A copy of the original fellowship announcement is in Folder: "WB-WG, Miscellaneous," Box 14, Loeb Papers.

13. Westinghouse Research Laboratories, "Work and Progress Report," October–December 1937, 30, January–March 1938, 29, Westinghouse Research Reports; Financial News of the Week, *Annalist* 52 (August 3, 1938): 181. Business conditions began to improve during the last quarter (October through December) of 1938. See Stanley Devlin, "Undiscounted Earnings and Prospects," *Magazine of Wall Street* 63 (December 3, 1938): 212.

14. H. F. MacLane to S. R. Shave, October 28, 1937, MacLane to Smith, November 2, 1937, Smith to MacLane, November 4, 1937, all in Folder 3, Box 178, Westinghouse Records; Condon to DuBridge, May 19, 1938, 1, Folder: "DuBridge, Lee Alvin, #1," Box 61, Condon Papers.

15. Condon to DuBridge, May 19, 1938, 1, Folder: "DuBridge, Lee Alvin, #1," Box 61, Condon Papers; Condon to Smyth, May 5, 1938, 1, Folder: "Princeton University, #5," Box 104, Condon Papers; Condon to Birge, November 30, 1927, 2, Folder: "Condon, Edward Uhler, 1902–1974," Box 6, Birge Papers; DuBridge to Condon, May 23, 1938, Folder: "DuBridge, Lee Alvin, #1," Box 61, Condon Papers; Condon to Dodds, May 5, 1938, Dodds to Condon, May 9, 1938, both in Folder 5, Box 134, Dodds Papers; Smyth to Condon, May 18, 1938, Folder: "Princeton University, #5," Box 104, Condon Papers.

16. Loeb to Condon, July 10, 1928, Folder: "Letters Written by Loeb, 1928," Box 16, Loeb Papers; Condon to Loeb, July 31, 1928, 1, December 24, 1933, 3, Folder: "Condon, Edward Uhler, 1902–1974, 20 Letters, 1926–1933," Box 4, Loeb Papers.

17. Smyth to Louis A. Turner, December 22, 1937, 1, Folder 3, Box 3, Princeton Physics Records; White interview, 113–14.

18. Condon to DuBridge, May 19, 1938, 1, Folder: "DuBridge, Lee Alvin, #1," Box 61, Condon Papers; Smyth to Turner, December 22, 1937, 1, February 15, 1938, 1, Folder

3, Box 3, Princeton Physics Records; Smyth to the President, November 26, 1937, December 11, 1939, Folder: "Bleakney, Walker," Princeton Faculty Files; Smyth to Robert K. Root, December 12, 1939, 1, Folder 1, Box 4, Princeton Physics Records; Condon to Loeb, August 15, 1938, 1, Folder: "Condon, Edward Uhler, 26 Letters, 1934–52," Box 4, Loeb Papers; Westinghouse Research Laboratories, "Work and Progress Report," January–March 1938, 29, Westinghouse Research Reports; "Research Fellowship Appointments Announced," *Review of Scientific Instruments* 9 (June 1938): 206.

19. Condon to Chubb, August 15, 1938, 1–2, Folder 4, Box 178, Westinghouse Records; Smyth to Turner, February 15, 1938, 1, Folder 3, Box 3, Princeton Physics Records.

20. Condon to Loeb, December 24, 1933, 4–5, Folder: "Condon, Edward Uhler, 1902–1974, 20 Letters, 1926–1933," Box 4, Loeb Papers; Lawrence to DuBridge, June 1, 1939, Frame 049917, Reel 23, Lawrence Papers; F. Bloch, "Leonard Isaac Schiff, 1915–1971," in *Biographical Memoirs* (Washington, DC: National Academy Press, 1983), 54:302–4.

21. Condon to Chubb, October 10, 1938, J. K. Hodnette to Condon, October 27, 1938, both in Folder: "WEC, Research Memorandums," Box 127, Condon Papers.

22. Condon quoted in "University of Pittsburgh, Proposal for an Expanded Program of Research on Metals," n.d., 2–3, Folder 253, Box 36, Rufus H. Fitzgerald Papers, Administrative Files, 1945–55, Archives and Special Collections, University of Pittsburgh, Pittsburgh, PA (hereafter cited as Fitzgerald Papers).

23. Condon to John G. Bowman, November 19, 1940, 1, 6–10, Folder 26, Box 4, John G. Bowman Papers, Administrative Files, 1921–45, Archives and Special Collections, University of Pittsburgh, Pittsburgh, PA (hereafter cited as Bowman Papers).

24. "Conference: Dean Crawford and Mr. Fitzgerald," March 6, 1939, 1, "Conference: Dr. Condon, Dr. Hutchisson, Dr. King, Dr. Crawford, and Mr. Fitzgerald," March 8, 1939, "Conference: Mr. Weber and Mr. Fitzgerald," March 9, 1939, all in Folder 224, Box 27, Fitzgerald Papers; Condon to Bowman, November 19, 1940, 1, Folder 26, Box 4, Bowman Papers.

25. Condon to Loeb, August 15, 1938, 1, Folder: "Condon, Edward Uhler, 26 Letters, 1934–52," Box 4, Loeb Papers; Condon to Loeb, March 6, 1941, Folder: "Loeb, Leonard B., #4," Box 82, Condon Papers.

26. *Research Progress*, no. 1, n.d., 3, Folder 5, Box 172, Westinghouse Records; Condon to Chubb, January 29, 1941, 14–16, Folder: "WEC, #2," Box 126, Condon Papers.

27. Paul Hoch, "The Development of the Band Theory of Solids, 1933–1960," in *Out of the Crystal Maze: Chapters from the History of Solid-State Physics*, ed. Lillian Hoddeson et al. (New York: Oxford University Press, 1992), 183–87; Condon, "Frederick Seitz," February 24, 1934, Folder 10, Box 10, Princeton Physics Records; DuBridge quoted in "The Structure of Solids," *Science*, Supplement 87 (March 4, 1938): 10;

Seitz, *On the Frontier*, 96; DuBridge to William E. Weld, January 14, 1935, Folder: "34/35," Box 233, DuBridge Papers; DuBridge to Leonard Carmichael, March 16, 1937, 2, Folder: "36/37," Box 233, DuBridge Papers; Harnwell to Paul H. Musser, February 24, 1938, March 4, 1938, Folder 18, Box 3, Gaylord P. Harnwell Papers, University Archives, University of Pennsylvania, Philadelphia, PA (hereafter cited as Harnwell Papers); Harnwell to Lyman J. Briggs, December 3, 1938, 1, Folder 2, Box 6, Harnwell Papers; Seitz to Condon, n.d. (written sometime during the second half of December 1938), Folder: "Frederick Seitz, #2," Box 110, Condon Papers.

28. Condon to Seitz, December 14, 1938, 1, Folder: "Frederick Seitz, #1," Box 110, Condon Papers; Scientific News and Notes, *Science* 86 (August 27, 1937): 193; "Summer Session, Physics of Metals and Courses in Physics, University of Pittsburgh," 1938, 3–4, Folder: "Correspondence and Records, 1936–1944, 1 of 2," Box 1, Hutchisson Papers; Condon to Chubb, January 29, 1941, 16–17, Folder: "WEC, #2," Box 126, Condon Papers; "University of Pittsburgh Department of Physics, a 10-Year Plan of Development," n.d., 1–2, 4 (attached to letter from Hutchisson to Stanton C. Crawford, September 30, 1937), Folder 144, Box 19, Bowman Papers; "Annual Report to the Dean of the College from the Head of the Physics Department for the School Year 1937–38 at the University of Pittsburgh," n.d., 3, Folder: "Correspondence and Records, 1936–1944, 1 of 2," Box 1, Hutchisson Papers.

29. Smith to Thomas S. Gates, October 24, 1938, Harnwell to Chester E. Tucker, October 25, 1938, both in Folder 18, Box 3, Harnwell Papers; Condon to Seitz, December 14, 1938, Folder: "Seitz, Frederick, #1," Box 110, Condon Papers; Condon to Seitz, January 10, 1939, 1, Folder: "Seitz, Frederick, #2," Box 110, Condon Papers.

30. Frederick Seitz, *The Modern Theory of Solids* (New York: McGraw-Hill, 1940), vi; Spencer R. Weart, "The Solid Community," in *Out of the Crystal Maze*, ed. Hoddeson et al., 628; Westinghouse Research Laboratories, "Work and Progress Report," April–June 1939, 28, Westinghouse Research Reports; Thomas A. Read, "Research Program for 1941," n.d., 1, Folder: "Bell Type Generator, #1," Box 11, Condon Papers; H. Walther, "Internal Friction in Solids," *Scientific Monthly* 41 (September 1935): 275.

31. Condon to Chubb, January 29, 1941, 16, Folder: "WEC, #2," Box 126, Condon Papers; Westinghouse Research Laboratories, "Work and Progress Report," July–September 1939, 32, July–September 1940, 32, October–December 1940, 29, Westinghouse Research Reports; Ernest Braun, "Mechanical Properties of Solids," in *Out of the Crystal Maze*, ed. Hoddeson et al., 317–37; Seitz, *On the Frontier*, 135.

32. Frederick Seitz and T. A. Read, "Theory of the Plastic Properties of Solids. I," *Journal of Applied Physics* 12 (February 1941): 100; "American Institute of Physics," *Review of Scientific Instruments* 7 (August 1936): 321; Barton, "New Journal of Applied Physics," 399; "Editorial Office of the J.A.P.," *Journal of Applied Physics* 8 (September 1937): 614.

33. Seitz, *On the Frontier*, 135; L. O. Grondahl, "An Adventure in Research: Copper-Oxide Rectifiers and Their Applications," *American Physics Teacher* 4 (September 1936): 105–7. Grondahl described the new device for the first time at a meeting of the American Physical Society in Washington, DC, in April 1926. See L. O. Grondahl, "A New Type of Contact Rectifier," abstract, *Physical Review* 27 (June 1926): 813.

34. Ernest Braun, "Selected Topics from the History of Semiconductor Physics and Its Applications," in *Out of the Crystal Maze*, ed. Hoddeson et al., 443–46; Westinghouse Research Laboratories, "Work and Progress Report," October–December 1936, 1, Westinghouse Research Reports; Condon to Philip M. Morse, January 18, 1939, 2, Folder: "Condon, Edward U.," Box 6, Philip M. Morse Papers, Institute Archives and Special Collections, Massachusetts Institute of Technology, Cambridge, MA (hereafter cited as Morse Papers).

35. Seitz to Condon, January 12, 1940, 1, Condon to Seitz, January 25, 1940, 1, Seitz to Condon, January 28, 1940, 1, Chubb to Seitz, February 5, 1940, all in Folder: "Seitz, Frederick, #4," Box 110, Condon Papers.

36. Seitz to Condon, March 9, 1940, 1, Condon to Seitz, March 11, 1940, 1, both in Folder: "Seitz, Frederick, #4," Box 110, Condon Papers; Harnwell to Musser, March 6, 1940, 1, Folder: "College (Physics Department), III, 1935–1940," Box 11, Records of the Office of the President, 1930–44 (Thomas S. Gates), University Archives, University of Pennsylvania, Philadelphia, PA (hereafter cited as Gates Papers); Condon to Chubb, January 29, 1941, 17, Folder: "WEC, #2," Box 126, Condon Papers; Seitz to Condon, January 26, 1942, Condon to Seitz, February 10, 1942, both in Folder: "Seitz, Frederick, #6," Box 110, Condon Papers.

37. Condon to Chubb, January 29, 1941, 14, Folder: "WEC, #2," Box 126, Condon Papers.

38. "Prof. Harry S. Hower," *New York Times*, October 11, 1941, 17; Seitz to Condon, March 13, 1942, Condon to Seitz, March 17, 1942, Seitz to Condon, April 6, 1942, 1, Condon, "Memo of Conversation with Seitz," August 8, 1942, all in Folder: "Seitz, Frederick, #6," Box 110, Condon Papers; Seitz, *On the Frontier*, 123–24.

39. Condon to Seitz, January 9, 1941, Seitz to Condon, January 14, 1941, February 10, 1941, February 26, 1941, all in Folder: "Seitz, Frederick, #5," Box 110, Condon Papers; Seitz to Condon, July 7, 1942, Folder: "Seitz, Frederick, #7," Box 110, Condon Papers; Frederick Seitz, "Annual Report of the Department of Physics," June 30, 1943, 1–3, 8, Folder 23, Box 1, Carnegie Engineering and Science Records.

40. Smith to Chubb, February 23, 1938, Chubb to Smith, February 24, 1938, both in Folder 4, Box 178, Westinghouse Records.

41. Smith to Chubb, June 13, 1938, Chubb to Smith, June 14, 1938, both in Folder 4, Box 178, Westinghouse Records; Tuve to Fleming, February 18, 1938, Folder: "Tuve," Box 17, Tuve Papers.

42. Condon to Loeb, June 8, 1938, 2, Folder: "Condon, Edward Uhler, 26 Letters, 1934–52," Box 4, Loeb Papers; Condon to Gregory Breit, September 6, 1938, Folder 163 (Reel 3), Box 5, Gregory Breit Papers (Microfilm), Niels Bohr Library, American Institute of Physics, College Park, MD (hereafter cited as Breit Papers); Chubb to A. M. Fleming, July 2, 1937, 2–3, Folder 3, Box 178, Westinghouse Records; Here and There, *Journal of Applied Physics* 8 (January 1937): 73; Loeb to Condon, December 31, 1937, 1, Folder: "Loeb, Leonard B., #1," Box 82, Condon Papers; Coolidge quoted in Seitz to Condon, n.d., 2, Folder: "Seitz, Frederick, #1," Box 110, Condon Papers.

43. Condon to Chubb, January 29, 1941, 7–8, Folder: "WEC, #2," Box 126, Condon Papers; Westinghouse Research Laboratories, "Work and Progress Report," October–December 1938, 26, Westinghouse Research Reports; Chubb to Smith, July 11, 1939, Folder 4, Box 178, Westinghouse Records.

44. Chubb, "Memorandum on Nuclear Physics," September 8, 1936, 1, Folder 2, Box 178, Westinghouse Records; Smith to George H. Bucher, July 12, 1939, Folder 4, Box 178, Westinghouse Records. Upon Bucher's promotion, Merrick became vice-chairman of the company's board of directors. "Bucher Promoted by Westinghouse," *New York Times*, February 24, 1938, 31.

45. Condon et al., "First Report on Nuclear Physics Investigations," Research Report R-94236-A (Summary), September 11, 1939, 1, Folder: "WEC. First Report on Nuclear Physics Investigations, #1," Box 127, Condon Papers; Condon et al., "First Report on Nuclear Physics Investigations," Research Report R-94236-A, September 11, 1939, 69–70, Folder: "WEC. First Report on Nuclear Physics Investigations, #6," Box 127, Condon Papers; Chubb to Smith, December 28, 1939, 1, Wells, "Trip to Department of Terrestrial Magnetism, Washington, D.C.," November 7, 1939, 2, both in Folder 4, Box 178, Westinghouse Records.

46. Condon et al., "First Report on Nuclear Physics Investigations," 73–74, Folder: "WEC. First Report on Nuclear Physics Investigations, #7," Box 127, Condon Papers; Robert D. Potter, "Some Papers Read at the New York Meeting of the American Physical Society," *Science*, Supplement 91 (March 1, 1940): 10; Chubb to Smith, December 28, 1939, Folder 4, Box 178, Westinghouse Records; Westinghouse Research Laboratories, "Work and Progress Report," April–June 1940, 33, Westinghouse Research Reports; Condon to Chubb, January 29, 1941, 10, Folder: "WEC, #2," Box 126, Condon Papers; T. H. Osgood, "Physics in 1940," *Journal of Applied Physics* 12 (February 1941): 84–85; R. O. Haxby et al., "Photo-Fission of Uranium and Thorium," *Physical Review* 58 (July 1, 1940): 92. For a complete discussion of the results, see R. O. Haxby et al., "Photo-Fission of Uranium and Thorium," *Physical Review* 59 (January 1, 1941): 57–62.

47. "Mightiest Atom Smasher at East Pittsburgh, PA," *Life*, August 30, 1937, 36–37; George E. Pendray to Chubb, August 27, 1937, Folder 3, Box 178, Westinghouse Records; "G. E. Pendray, 86, Rocket Proponent," *New York Times*, September 20, 1987,

60; "Westinghouse Electric," *Fortune* 17 (February 1938): 55; Edward U. Condon, "Sharpshooting at the Atom," *Popular Mechanics* 74 (July 1940): 1–5, 143A–44A; Condon to Chubb, January 29, 1941, 13–14, Folder: "WEC, #2," Box 126, Condon Papers.

48. See, for example, "At Westinghouse," *Time*, February 12, 1940, 44–45; "New Method of Splitting Uranium Atoms Discovered," *Los Angeles Times*, July 2, 1940, 9; "New Way to Split Uranium and Release Energy Found," *Science News Letter* 38 (July 6, 1940): 3–4; William S. Barton, "Our: Progress in the World of Science," *Los Angeles Times*, March 17, 1940, I16; and Waldemar Kaempffert, "Science in the News," *New York Times*, February 9, 1941, D5.

49. Condon to Morse, January 18, 1939, 1, Folder: "Condon, Edward U.," Box 6, Morse Papers; Westinghouse Research Laboratories, "Work and Progress Report," April–June 1939, 28, Westinghouse Research Reports.

CHAPTER 4. NEW PRODUCTS FOR NEW MARKETS

1. Westinghouse Research Laboratories, "Work and Progress Report," April–June 1939, 28, Westinghouse Research Reports; Condon to Seitz, March 21, 1939, 1, Folder: "Seitz, Frederick, #2," Box 110, Condon Papers.

2. "Westinghouse Research Grant," *Journal of Applied Physics* 9 (January 1938): 35; "Conditioned Heated Air Induces Fever That Cures," *Science News Letter* 35 (February 11, 1939): 90; T. Lyle Hazlett and William W. Hummel, *Industrial Medicine in Western Pennsylvania, 1850–1950* (Pittsburgh: University of Pittsburgh Press, 1958), ix, 171–73; Condon to Chubb, February 21, 1939, 1, Folder 4, Box 178, Westinghouse Records.

3. Condon to Chubb, February 21, 1939, 1–2, Folder 4, Box 178, Westinghouse Records; Heilbron and Seidel, *Lawrence and His Laboratory*, 240–41, 281–98; "Cyclotron to Be Built by the Massachusetts Institute of Technology," *Science*, Supplement 88 (July 15, 1938): 6; Tuve, "Report of April 1939," May 16, 1939, 2, "Report for July 1940," August 8, 1940, both in Folder: "Extra Copies, Mss. & Reports," Box 21, Tuve Papers; Fleming to Lawrence, May 27, 1939, Frames 009282–83, Reel 5, Lawrence Papers; Tuve to Breit, June 4, 1939, Folder 978 (Reel 14), Box 24, Breit Papers; DuBridge to Lawrence, June 21, 1939, Frame 049925, Reel 23, Lawrence Papers; DuBridge to Lawrence, November 20, 1939, Frame 019050, March 9, 1940, Frame 019052, Reel 9, Lawrence Papers; Donald Cooksey to Charles Seymour, March 26, 1940, Frame 059671, Reel 27, Lawrence Papers.

4. Condon to Chubb, February 21, 1939, 2–3, Folder 4, Box 178, Westinghouse Records; "$15,000,000 Estate Will Go to Charity," *New York Times*, June 21, 1927, 28; Alexander Silverman to Bowman, October 3, 1935, Folder 80, Box 9, Bowman Papers; "A Brief Resume of the Buhl Foundation Research Project in Chemistry, Physics, and Psychology at the University of Pittsburgh, 1936 to February 1943," 1, Folder 1, Box 1, Records of the Division of Research in the Natural Sciences, Archives and Special

Collections, University of Pittsburgh, Pittsburgh, PA (hereafter cited as DRNS Records); "Report to the Buhl Foundation, Research Project in Biochemistry," May 20, 1939, 1, 13–15, Folder 253, Box 36, Fitzgerald Papers.

5. Condon to Chubb, February 21, 1939, 2–3, Folder 4, Box 178, Westinghouse Records; "New Appointments," *Journal of Applied Physics* 10 (April 1939): 267; Smith to Chubb, February 23, 1939, Folder 4, Box 178, Westinghouse Records.

6. "Annual Report to the Dean of the Graduate School from the Head of the Department of Physics, 1938–1939," n.d., 6, Folder 168, Box 18, DRNS Records; Hazlett and Hummel, *Industrial Medicine in Western Pennsylvania*, 174; Hutchisson, "Research in Nuclear Physics," n.d., 2 (attached to note from Hutchisson to Charles G. King, February 27, 1941), Folder 46, Box 6, DRNS Records; Allen to Lawrence, January 28, 1937, Frames 002403–4, Allen to Cooksey, September 22, 1937, Frame 002408, Allen to Lawrence, April 8, 1938, Frame 002423, Allen to Lawrence and Cooksey, June 30, 1939, Frame 002481, Allen to Cooksey, July 29, 1939, Frame 002486, all in Reel 2, Lawrence Papers; "The Biochemical Research Foundation of the Franklin Institute," *Science* 83 (January 10, 1936): 26; "New 'Atom Smasher' Shown," *New York Times*, February 4, 1938, 6.

7. "Scientists Pool Resources in $60,000 Research Plan," *Science News Letter* 36 (August 19, 1939): 120; Hutchisson, "Annual Report to the Dean of the Graduate School from the Head of the Department of Physics, 1938–1939," n.d., 8, Folder 168, Box 18, DRNS Records; King, "Outline of Plan for the Administration of the Buhl Foundation Grant in Support of Fundamental Research at the University of Pittsburgh," n.d., 1 (attached to letter from Bowman to Rufus H. Fitzgerald, July 13, 1939), "Progress Report to the Buhl Foundation, Research Project in Chemistry, Physics, and Biology, University of Pittsburgh," November 1, 1939, 1, 6–7, both in Folder 253, Box 36, Fitzgerald Papers; Hutchisson to Lawrence, May 17, 1939, Frame 052874, Lawrence to Hutchisson, May 20, 1939, Frames 052875–76, Samuel J. Simmons to Lawrence, September 26, 1939, Frames, 052879–80, all in Reel 24, Lawrence Papers; "Proposed University of Pittsburgh Cyclotron, Summary of Report," n.d., "Proposed University of Pittsburgh Cyclotron," n.d., 6–9 (attached to letter from Hutchisson to Fitzgerald, October 26, 1939), both in Folder 253, Box 36, Fitzgerald Papers.

8. Stanley Van Voorhis to Cooksey, June 11, 1939, Frame 057436, Reel 26, Lawrence Papers; Simmons to "Bob" (unknown recipient), January 11, 1940, Frames 052888–89, Simmons to Lawrence, April 18, 1940, Frame 052892, both in Reel 24, Lawrence Papers; Allen to Lawrence, October 23, 1940, Frame 002502, Reel 2, Lawrence Papers; "Proposed University of Pittsburgh Cyclotron," 13, Folder 253, Box 36, Fitzgerald Papers.

9. "Research Group Dinner Meeting at the Fitzgerald Home," October 16, 1939, Folder 224, Box 27, Fitzgerald Papers; "Progress Report to the Buhl Foundation, Re-

search Project in Chemistry, Physics, and Biology, University of Pittsburgh," January 31, 1940, 7, "Progress Report to the Buhl Foundation, Research Project in Chemistry, Physics, and Biology, University of Pittsburgh, July 1 to October 1, 1940," 11, both in Folder 253, Box 36, Fitzgerald Papers; Allen to Lawrence, October 29, 1939, Frame 002491, Reel 2, Lawrence Papers.

10. Simmons to Helen Griggs, September 24, 1939, Frame 052878, Simmons to Lawrence, April 18, 1940, Frame 052892, both in Reel 24, Lawrence Papers; Allen to Lawrence, October 23, 1940, Frame 002502, Reel 2, Lawrence Papers; Sarah Mellon Scaife to William S. McEllroy, December 26, 1940, Folder 49, Box 6, Bowman Papers; G. S. Rupp to Bowman, June 27, 1942, McEllroy to Bowman, July 3, 1942, both in Folder 104, Box 13, Bowman Papers.

11. Condon, "Artificial Radio-Active Materials as a New Product," December 21, 1939, 1, 8–9, Folder: "WEC, Radio-Active Materials," Box 127, Condon Papers.

12. Heilbron and Seidel, *Lawrence and His Laboratory*, 192–93; Condon, "Artificial Radio-Active Materials as a New Product," December 21, 1939, 2–3, Folder: "WEC, Radio-Active Materials," Box 127, Condon Papers; Tuve to Shockley, May 13, 1940, 1–2, Folder: "Tuve Letters, 1940," Box 19, Tuve Papers.

13. Condon, "Artificial Radio-Active Materials as a New Product," December 21, 1939, 4–7, 10, Folder: "WEC, Radio-Active Materials," Box 127, Condon Papers.

14. Condon to Loeb, January 4, 1940, January 22, 1940, Folder: "Condon, Edward Uhler, 26 Letters, 1934–1952," Box 4, Loeb Papers; Lawrence to Condon, April 13, 1940, Frame 011621, Condon to Lawrence, March 29, 1940, Frames 011619–20, both in Reel 6, Lawrence Papers; Cooksey to Loomis, May 10, 1940, Frame 157288, May 14, 1940, Frames 157290, 157292, Lawrence to Loomis, May 20, 1940, Frames 157300–301, all in Reel 70, Lawrence Papers.

15. "University of Pittsburgh Summer Session, Physics of Metals, 1940," 2, Folder: "Correspondence and Records, 1936–1944, 1 of 2," Box 1, Hutchisson Papers; DuBridge to Lawrence, August 28, 1940, Frame 019065, Reel 9, Lawrence Papers; Condon, "Artificial Radio-Active Materials as a New Product," December 21, 1939, 4–6, Folder: "WEC, Radio-Active Materials," Box 127, Condon Papers; Condon to Chubb, January 29, 1941, 9, Folder: "WEC, #2," Box 126, Condon Papers.

16. Cattell and Cattell, *American Men of Science*, 6th ed., s.v. "Spooner, Thomas"; Chubb, Thomas Spooner, and Slepian to Smith, February 15, 1939, 1-2, Folder 19, Box 163, Westinghouse Records.

17. Chubb, Spooner, and Slepian to Smith, February 15, 1939, 2, 9, Folder 19, Box 163, Westinghouse Records

18. DuBridge to Carmichael, March 17, 1938, 1, Folder: "37/38," Box 233, DuBridge Papers; DuBridge to Lawrence, March 19, 1938, Frames 019032–33, Reel 9, Lawrence Papers; DuBridge to Lawrence, April 4, 1938, Frame 049845, DuBridge to Van

Voorhis, May 18, 1938, Frames 049850–51, DuBridge to Lawrence, June 1, 1939, Frame 049918, all in Reel 23, Lawrence Papers; DuBridge to Valentine, May 5, 1939, Folder: "38/39," Box 233, DuBridge Papers.

19. Lawrence to DuBridge, June 14, 1939, Frame 049922, DuBridge to Lawrence, June 21, 1939, Frame 049924, both in Reel 23, Lawrence Papers.

20. DuBridge to Lawrence, August 28, 1940, Frame 019065, Reel 9, Lawrence Papers.

21. DuBridge to Lawrence, August 28, 1940, Frames 019065–66, Lawrence to DuBridge, September 2, 1940, Frame 019067, both in Reel 9, Lawrence Papers.

22. DuBridge to Lawrence, September 10, 1940, Frame 019070, Reel 9, Lawrence Papers; Condon to Chubb, January 29, 1941, 9–10, 12, Folder: "WEC, #2," Box 126, Condon Papers.

23. Smith to Lawrence, July 31, 1940, Frame 058758, Lawrence to Smith, July 29, 1940, Frame 058757, both in Reel 27, Lawrence Papers. GE got the order from Berkeley at the end of 1940. "General Electric Co. to Build Machinery of Big Atom Smasher," *Chicago Daily Tribune*, December 8, 1940, B5; Smith to Lawrence, August 13, 1940, Frame 058761, Reel 27, Lawrence Papers.

24. "Devotes $2,000,000 to Cancer Research," *New York Times*, June 9, 1932, 13; "Donner Buys Home in Switzerland," *New York Times*, October 2, 1938, 56; Eugene Pendergrass to Gates, January 15, 1940, 1, Folder: "Bicentennial Celebration (Donner Cyclotron), I, 1935–1940," Box 11, Gates Papers; R. L. S. Doggett to DuBarry, November 21, 1945, Folder: "Betatron, II, 1945–1950," Box 28, Records of the Office of the President, 1945–55 (George W. McClelland), University Archives, University of Pennsylvania, Philadelphia, PA (hereafter cited as McClelland Papers); Harnwell to Coolidge, January 30, 1940, Folder 19, Box 3, Harnwell Papers; "Report Submitted by A. L. Hughes," n.d. (attached to letter from Arthur L. Hughes to Sherwood Moore, October 11, 1938), 12–13, Folder 18, Box 3, Harnwell Papers.

25. Heilbron and Seidel, *Lawrence and His Laboratory*, 359–60, 389–404; Ridenour to John H. Lawrence, February 3, 1940, Frames 046533–34, John H. Lawrence to Ridenour, February 6, 1940, Frames 046535–36, Raymond E. Zirkle to Lawrence, February 14, 1940, Frames 046540–41, Lawrence to Zirkle, February 20, 1941, Frame 046543, Harnwell to Lawrence, February 21, 1940, Frames 046544–45, all in Reel 21, Lawrence Papers.

26. Harnwell to Lawrence, June 10, 1940, Frame 046554, Reel 21, Lawrence Papers; Van Voorhis to Lawrence, June 21, 1940, Frame 057447, Reel 26, Lawrence Papers; Condon to Lawrence, September 18, 1940, Frame 011626, Lawrence to Condon, September 20, 1940, Frame 011627, both in Reel 6, Lawrence Papers; Harnwell to Edwin M. Chance, October 4, 1940, Folder 19, Box 3, Harnwell Papers; Alfred N. Richards to Lawrence, September 26, 1940, Frame 046571, Lawrence to Richards, October 3, 1940, Frame 046573, both in Reel 21, Lawrence Papers.

27. L. D. Canfield to Victor Hicks, March 26, 1937, Hicks to Loeb, March 31, 1937, 1, April 26, 1937, all in Folder: "Hicks, Victor, 1904–," Box 7, Loeb Papers; Canfield to C. S. Eason, November 28, 1940, 1, Folder: "Bicentennial Celebration (Donner Cyclotron), II, 1940–1945," Box 18, Gates Papers; Harnwell to Tuve, March 5, 1940, Tuve to Harnwell, March 8, 1940, both in Folder 19, Box 3, Harnwell Papers; Harnwell to DuBarry, October 31, 1940, December 3, 1940, Pendergrass to DuBarry, December 5, 1940, all in Folder: "Bicentennial Celebration (Donner Cyclotron), II, 1940–1945," Box 18, Gates Papers.

28. W. H. Branch to Harnwell, December 6, 1940, Branch to DuBarry, January 3, 1941, both in Folder: "Bicentennial Celebration (Donner Cyclotron), II, 1940–1945," Box 18, Gates Papers.

29. "Stettinius Quits $100,000-a-Year Steel Job to Give Full Time to National Defense," *New York Times*, June 5, 1940, 1; DuBarry to William H. Donner, January 14, 1941, Gates to Donner, January 15, 1941, DuBarry to Donner, January 22, 1941, DuBarry to J. J. O'Neill, November 3, 1944, all in Folder: "Bicentennial Celebration (Donner Cyclotron), II, 1940–1945," Box 18, Gates Papers; Doggett to DuBarry, November 21, 1945, Folder: "Betatron, II, 1945–1950," Box 28, McClelland Papers. In 1946, the university dropped plans to use the Donner bequest to acquire a cyclotron and chose instead to purchase a 20 MeV x-ray source called the betatron, which the Allis-Chalmers Manufacturing Company built and installed two years later. Pendergrass to Richards, January 3, 1946, Richards to DuBarry, February 12, 1946, John Gammell to DuBarry, May 3, 1946, all in Folder: "Betatron, I, 1945–1950," Box 28, McClelland Papers; DuBarry to Doggett, June 20, 1946, "Executive Committee," February 23, 1948, both in Folder: "Betatron, II, 1945–1950," Box 28, McClelland Papers; "New Weapons Foreseen," *New York Times*, December 7, 1948, 3.

30. DuBridge to Lawrence, September 10, 1940, Frame 019070, Reel 9, Lawrence Papers.

31. Travis, *Dyes Made in America*, 1, 3–4, 6–7, 74–79, 108–13; "American Cyanamid," *Fortune* 22 (September 1940): 68.

32. Heilbron and Seidel, *Lawrence and His Laboratory*, 205, 477; Loomis to Lawrence, May 28, 1940, Frame 157316, Reel 70, Lawrence Papers; Lawrence to William B. Bell, June 21, 1940, Frame 002932, Bell to Lawrence, June 28, 1940, Frame 002933, both in Reel 2, Lawrence Papers.

33. Bell to Lawrence, July 27, 1940, Frame 002937, Moses L. Crossley to Lawrence, August 2, 1940, Frame 002938, Bell to John H. Lawrence, August 2, 1940, Frame 002939, Stanley D. Beard to Lawrence, August 13, 1940, Frame 002941, all in Reel 2, Lawrence Papers.

34. Barnes to Lawrence, August 29, 1940, Frames 002943–44, Bell to John H. Lawrence, August 30, 1940, Frame 002946, both in Reel 2, Lawrence Papers; DuBridge to

Lawrence, September 10, 1940, Frame 019070, Reel 9, Lawrence Papers; Canfield to Eason, November 28, 1940, 1, Folder: "Bicentennial Celebration (Donner Cyclotron), II, 1940–1945," Box 18, Gates Papers; Condon to Chubb, January 29, 1941, 12, Folder: "WEC, #2," Box 126, Condon Papers.

35. Condon to Chubb, January 29, 1941, 12, Folder: "WEC, #2," Box 126, Condon Papers; Bell to Lawrence, October 4, 1940, Frame 002949, Reel 2, Lawrence Papers; Poillon to Lawrence, September 20, 1940, Frames 049359–61, Lawrence to Poillon, September 24, 1940, Frames 049362–63, both in Reel 23, Lawrence Papers.

36. Beard to Lawrence, October 10, 1940, Frame 002950, Barnes to Lawrence, February 20, 1941, Frame 002970, Bell to Lawrence, March 22, 1941, Frames 002973–74, Lawrence to Bell, April 7, 1941, Frame 002977, May 13, 1941, Frame 002979, Bell to Lawrence, May 15, 1941, Frame 002981, May 27, 1941, Frame 002984, June 5, 1941, Frame 002986, all in Reel 2, Lawrence Papers; Biographical Statement for Joseph W. Kennedy, n.d., Folder: "Kennedy, Joseph," Box K1, Biographical Faculty Files, Records of the Office of Public Affairs, University Archives, Washington University, St. Louis, MO.

37. Bell to Lawrence, July 18, 1941, Frame 002997, August 1, 1941, Frame 003007, August 10, 1941, Frame 003013, Ralph W. G. Wyckoff to Lawrence, August 11, 1941, Frame 003014, Bell to Lawrence, August 30, 1941, Frames 003020–22, Barnes to Lawrence, September 16, 1941, Frame 003031, "Personnel Record of Raymond L. Libby," n.d. (attached to letter from Bell to Lawrence, May 31, 1948), Frames 003264–65, all in Reel 2, Lawrence Papers.

38. Henry E. Guerlac, *Radar in World War II, Sections A–C*, vol. 8 of *The History of Physics, 1800–1950* (New York: Tomash Publishers and the American Institute of Physics, 1987), 260; Condon to Chubb, January 29, 1941, 4–6, 13, Folder: "WEC, #2," Box 126, Condon Papers; Westinghouse Research Laboratories, "Work and Progress Report," July–September 1941, 33, January–March 1942, 33, Westinghouse Research Reports.

39. Bleakney to Smyth, June 24, 1935, Folder 15, Box 12, Princeton Physics Records; Bleakney to Harold C. Urey, May 20, 1936, Folder 4, Box 2, Princeton Physics Records; Poillon to Bleakney, July 6, 1937, May 11, 1938, Bleakney, "Quantitative Analysis with the Mass Spectrometer" n.d., 4, all in Folder: "Mass Spectrometer Patent, Research Corp.," Box 20, Walker Bleakney Papers, Rare Books and Special Collections, Princeton University, Princeton, NJ (hereafter cited as Bleakney Papers); Bleakney to Poillon, July 13, 1937, Bleakney, "Resume of Patent Possibilities of the Mass Spectrograph," July 13, 1937, 3–4, Poillon to Bleakney, July 16, 1937, all in Folder 5, Box 2, Princeton Physics Records; Poillon to Smyth, August 12, 1938 (plus attached draft letter to oil companies, n.d.), Folder 3, Box 6, Princeton Physics Records. On the founding and early history of the Research Corporation, see Thomas D. Cornell, *Establishing Research*

Corporation: A Case Study of Patents, Philanthropy, and Organized Research in Early Twentieth-Century America (Tucson, AZ: Research Corporation, 2004).

40. Archibald F. Meston to Carroll L. Wilson, September 29, 1938, Wilson to Bleakney, June 30, 1939, Sylvain Pirson to Wilson, April 13, 1939, Wilson to Pirson, June 5, 1939, Pirson to Wilson, February 23, 1940, Wilson to Pirson, February 26, 1940, all in Folder: "Mass Spectrometer Patent, Research Corp.," Box 20, Bleakney Papers; Smyth to Poillon, October 8, 1938, Poillon to Smyth, October 10, 1938, both in Folder 3, Box 6, Princeton Physics Records; Poillon to Bleakney, October 21, 1939, Wilson to Bleakney, December 31, 1939, 1, William T. Thom to Hendrix Rowell, March 4, 1940, all in Folder "Mass Spectrometer Patent, Research Corp.," Box 20, Bleakney Papers.

41. Ira H. Cram to Wilson, February 27, 1940, Wilson to Cram, March 7, 1940, April 9, 1940, Cram to Wilson, May 17, 1940, Blaine B. Wescott to Wilson, May 9, 1940, Wilson to Bleakney, May 13, 1940, 1, June 30, 1940, all in Folder: "Mass Spectrometer Patent, Research Corp.," Box 20, Bleakney Papers; W. Bleakney, Focusing and Separation of Charged Particles, US Patent 2,221,467, filed December 27, 1938, and issued November 12, 1940.

42. Irvin Stewart, *Organizing Scientific Research for War: The Administrative History of the Office of Scientific Research and Development* (Boston: Little, Brown, 1948), 10; "The National Defense Research Committee," *Science* 92 (November 22, 1940): 484; Scientific Notes and News, *Science* 93 (June 13, 1941): 565; Smyth to Dodds, July 3, 1941 (plus attachment, "Princeton University, Department of Physics," n.d.), Folder 5, Box 134, Dodds Papers; Wilson to Bleakney, December 31, 1940, June 30, 1941, Folder: "Mass Spectrometer Patent, Research Corp.," Box 20, Bleakney Papers.

43. "Record-Breaking Magnetic Field Employed in New Mass Spectrometer," *Electronics* 13 (May 1940): 40, 44; Condon to Chubb, January 29, 1941, 18–19, Folder: "WEC, #2," Box 126, Condon Papers; Westinghouse Research Laboratories, "Work and Progress Report," January–March 1938, 29, July–September 1939, 31, October–December 1939, 30, Westinghouse Research Reports.

44. Westinghouse Research Laboratories, "Work and Progress Report," January–March 1940, 34, Westinghouse Research Reports; Condon to Chubb, January 29, 1941, 11, Folder: "WEC, #2," Box 126, Condon Papers; A. O. Nier et al., "Nuclear Fission of Separated Uranium Isotopes," *Physical Review* 57 (March 15, 1940): 546; A. O. Nier et al., "Further Experiments on Fission of Separated Uranium Isotopes," *Physical Review* 57 (April 15, 1940): 748; K. H. Kingdon et al., "Fission of the Separated Isotopes of Uranium," *Physical Review* 57 (April 15, 1940): 749.

45. Condon to Lawrence, March 29, 1940, Frame 011619, Reel 6, Lawrence Papers; Westinghouse Research Laboratories, "Work and Progress Report," April–June 1940, 33, Westinghouse Research Reports.

46. See Thom to Rowell, March 4, 1940, Folder: "Mass Spectrometer Patent,

Research Corp.," Box 20, Bleakney Papers; and H. Hoover Jr. and H. Washburn, "A Preliminary Report on the Application of Mass Spectrometry to Problems in the Petroleum Industry," *Petroleum Technology* 3 (May 1940): 1–7.

47. Howard W. Blakeslee, "Herbert Hoover Jr. Uses Atom-Throwers to Find Oil," *Atlanta Constitution*, February 29, 1940, 16; "Consolidated Engineering Corp.," *Commercial and Financial Chronicle* 161 (June 25, 1945): 2784; "Electronic Unit to Sell Shares," *Los Angeles Times*, May 25, 1945, A6; Condon to Chubb, January 29, 1941, 19, Folder: "WEC, #2," Box 126, Condon Papers; Pirson to Wilson, February 23, 1940, Wilson to Bleakney, April 9, 1940, May 13, 1940, 2, all in Folder: "Mass Spectrometer Patent, Research Corp.," Box 20, Bleakney Papers.

48. Harold W. Washburn to Lawrence, February 20, 1940, Frame 058396, Reel 27, Lawrence Papers; Herbert Hoover Jr. and Harold Washburn, "Analysis of Hydrocarbon Gas Mixtures by Mass Spectrometry," *California Oil World and Petroleum Industry* 34 (November 1941): 21; Cattell and Cattell, *American Men of Science*, 6th ed., *s.v.* "Washburn, Harold Williams."

49. Hipple, "Development of a Portable Mass Spectrometer for Gas Analysis and Research," April 1, 1941, 3–4, Folder: "WEC, Mass Spectrometer for Gas Analysis and Research, #1," Box 127, Condon Papers; David Stevenson, "Research Program," January 23, 1941, Folder: "Bell Type Generator, #1," Box 11, Condon Papers; Condon to Chubb, January 29, 1941, 19–20, Folder: "WEC, #2," Box 126, Condon Papers; Westinghouse Research Laboratories, "Work and Progress Report," April–June 1941, 33, Westinghouse Research Reports.

50. Prince M. Carlisle, "Chemical Gains to Feature Show," *New York Times*, November 30, 1941, F1. Descriptions of the portable mass spectrometer appeared in "Portable Atom-Sorter for Industrial Uses," *Science News Letter* 40 (July 5, 1941): 9; "Mass Spectrographic," *Industrial and Engineering Chemistry, Analytical Edition* 13 (October 15, 1941): 749–50; "Mass Spectrometer for Gas Analysis," *Review of Scientific Instruments* 13 (May 1942): 238–39; John A. Hipple, "Portable Mass Spectrometer," *Nature* 150 (July 25, 1942): 111–12; and Hipple, "Gas Analysis with the Mass Spectrometer," *Journal of Applied Physics* 13 (September 1942): 551–59.

51. Hipple, "Development of a Portable Mass Spectrometer for Gas Analysis and Research," April 1, 1941, 4, Folder: "WEC, Mass Spectrometer for Gas Analysis and Research #1," Box 127, Condon Papers; Condon to Chubb, January 29, 1941, 20, Folder: "WEC, #2," Box 126, Condon Papers.

52. Condon to Compton, August 10, 1942, 1, Folder: "Condon, Edward U., 1940–1952," Box 62, Compton Papers; "Robot Chemist to Aid Testing of Butadiene," *Chicago Daily Tribune*, August 8, 1943, A7; "The Production of Synthetic Rubber," *Science*, Supplement 98 (August 20, 1943): 8–9; Seymour Meyerson, "Reminiscences of the Early Days of Mass Spectrometry in the Petroleum Industry," *Organic Mass Spectrometry*

21 (April 1986): 199–201; Jaques Cattell, ed., *American Men of Science: A Biographical Directory*, 7th ed. (Lancaster, PA: Science Press, 1944), *s.v.* "Sweeney, William Joseph"; "Top Changes Made in Research Group," *New York Times*, October 2, 1944, 27.

53. Ernest Solomon and Louis C. Rubin, "Mass Spectrometric Gas Analysis," *American Gas Association Monthly* 27 (October 1945): 461; "Mass Spectrometer Analysis of Gasoline," *Petroleum Engineering* 14 (April 1943): 44; O. L. Roberts, "Quantitative Analysis by Mass Spectrometry," *Petroleum Engineer* 14 (May 1943): 109–14, 116; "Consolidated Engineering Corp.," 2784; Washburn to Lawrence, December 26, 1946, Frame 058402, Reel 27, Lawrence Papers; F. W. Karasek, "Harold Washburn: MS Pioneer," *Research/Development* 23 (June 1972): 26–27; Washburn to Lawrence, July 25, 1939, Frames 058392–93, Lawrence to Washburn, August 3, 1939, Frame 058394, both in Reel 27, Lawrence Papers.

54. John L. Binder to William E. Stephens, October 22, 1946, Folder 34, Box 1, William E. Stephens Papers, University Archives, University of Pennsylvania, Philadelphia, PA; John A. Hutcheson to Smith, August 27, 1946, Hutcheson to Chubb, January 30, 1947, Hutcheson to G. C. Saltman, February 11, 1947, all in Folder 6, Box 53, Westinghouse Records; Raymond J. Woodrow to Elliott R. Weyer, June 11, 1946, Weyer to Bleakney, March 27, 1947 (plus attached annual report, December 31, 1946), all in Folder: "Mass Spectrometer Patent, Research Corp.," Box 20, Bleakney Papers; Condon to Hipple, April 1, 1946, Hipple to Condon, April 9, 1946, both in Folder: "D/APP," Box 1, Records of the Office of the Director, Director's Correspondence, 1923–63, Record Group 167, Records of the National Institute of Standards and Technology, National Archives and Records Administration, College Park, MD (hereafter cited as NBS Director's Correspondence); Hipple to Condon, May 6, 1946, Folder: "D/AP," Box 1, NBS Director's Correspondence; Hipple to Condon, December 23, 1946, Folder: "Hipple, John A.," Box 76, Condon Papers.

55. J. W. Hinkley to Bleakney, April 7, 1954 (plus attached annual report, n.d.), H. Gordon Howe to Bleakney, March 15, 1955 (plus attached annual report, n.d.), Howe to Bleakney, April 16, 1956 (plus attached annual report, n.d.), all in Folder: "Mass Spectrometer Patent, Research Corp.," Box 20, Bleakney Papers. By the early 1960s, according to one estimate, CEC controlled two-thirds of the US mass spectrometer market. "Trade-in Plan Set Up for Mass Spectrometers," *Chemical and Engineering News* 41 (August 19, 1963): 23.

CHAPTER 5. WESTINGHOUSE AT WAR

1. "40% of Westinghouse Production for Defense," *Steel* 108 (January 6, 1941): 3; "Record Expansion for Westinghouse," *New York Times*, March 11, 1941, 35. On Westinghouse's wartime operations, see Robert O. Woodbury, *Battlefronts of Industry: Westinghouse in World War II* (New York: John Wiley, 1948).

2. "The Fifth Washington Conference on Theoretical Physics," *Science* 89 (February 24, 1939): 180; Richard G. Hewlett and Oscar E. Anderson Jr., *The New World, 1939–1946*, vol. 1 of *A History of the United States Atomic Energy Commission* (University Park: Pennsylvania State University Press, 1962), 10–11, 19–20, 24–26; Stewart, *Organizing Scientific Research for War*, 7–9.

3. Stewart, *Organizing Scientific Research for War*, 11; Chubb to Briggs, July 2, 1940, Folder: "Chubb, L. W., ID-900-C, Nuclear Fission, Westinghouse E&M Co.," Box 3, S-1 Files, Lyman J. Briggs Alphabetical Files, Record Group 227, Records of the Office of Scientific Research and Development, National Archives and Records Administration, College Park, MD (hereafter cited as Briggs S-1 Files).

4. Hewlett and Anderson, *New World*, 13–14, 22, 27–31; Briggs to Chubb, July 15, 1940, 1, Folder: "Chubb, L. W., ID-900-C, Nuclear Fission, Westinghouse E&M Co.," Box 3, Briggs S-1 Files.

5. Condon to Chubb, January 29, 1941, 11, Folder: "WEC, #2," Box 126, Condon Papers; Condon to Smith, March 20, 1941, 1–2, Folder: "Shoupp, W. E.," Box 111, Condon Papers; J. Slepian, Ionic Centrifuge, US Patent 2,724,056, filed June 19, 1942, and issued November 15, 1955; Hewlett and Anderson, *New World*, 30, 59, 96–97; Condon to Chubb, April 18, 1944, 3, Folder: "Chubb, L. W., #1," Box 16, Condon Papers.

6. Condon to Smith, March 20, 1941, 2, Folder: "Shoupp, W. E.," Box 111, Condon Papers; Hewlett and Anderson, *New World*, 32; Briggs to Condon, July 3, 1941, Breit to Condon, October 14, 1941, both in Folder: "Atomic Energy, #2a," Box 6, Condon Papers.

7. See Karl Stephan, "Experts at Play: Magnetron Research at Westinghouse, 1930–1934," *Technology and Culture* 42 (October 2001): 737–49.

8. "Westinghouse Research Fellowships," *Journal of Applied Physics* 10 (May 1939): 320; Westinghouse Research Laboratories, "Work and Progress Report," April–June 1939, 28, Westinghouse Research Reports.

9. On the invention of the klystron at Stanford and the tube's early commercial development at the Sperry Gyroscope Company, see Edward L. Gintzon, "The $100 Idea," *IEEE Spectrum* 12 (February 1975): 30–39.

10. John H. Bryant, "Microwave Technology and Careers in Transition: The Interests and Activities of Visitors to the Sperry Gyroscope Company's Klystron Plant in 1939–1940," *IEEE Transactions on Microwave Theory and Techniques* 38 (November 1990): 1546, 1550; Westinghouse Research Laboratories, "Work and Progress Report," July–September 1939, 32, October–December 1939, 30, Westinghouse Research Reports; Condon to William W. Hansen, September 27, 1939, Condon to Morse, September 27, 1939, both in Folder 14, Box 2, William W. Hansen Papers, Special Collections and University Archives, Stanford University, Stanford, CA (hereafter cited as Hansen Papers).

11. "Westinghouse Research Fellowships Granted," *Journal of Applied Physics* 11 (May 1940): 340; Bryant, "Microwave Technology and Careers in Transition," 1546; Westinghouse Research Laboratories, "Work and Progress Report," October–December 1939, 31, January–March 1940, 34, October–December 1940, 29, Westinghouse Research Reports; Condon to Hansen, November 1, 1939, 1, Folder 14, Box 2, Hansen Papers; Condon to Hansen, March 13, 1940, Folder 16, Box 2, Hansen Papers; Condon to Chubb, January 29, 1941, 22–23, Folder: "WEC, #2," Box 126, Condon Papers.

12. Condon, "Reminiscences of a Life in and out of Quantum Mechanics," 17; Condon to Hansen, November 1, 1939, 1, December 7, 1939, Folder 14, Box 2, Hansen Papers; Condon to Hansen, January 4, 1940, Folder 15, Box 2, Hansen Papers; "Memorandum on Conversation with W. W. Hansen and Subsequent Conversation with President Wilbur," June 3, 1940, Folder 12, Box 1, David L. Webster Papers (SC0131), Series D, Special Collections and University Archives, Stanford University, Stanford, CA; Gintzon, "$100 Idea," 39; Stuart W. Leslie, *The Cold War and American Science: The Military-Industrial-Academic Complex at MIT and Stanford* (New York: Columbia University Press, 1993), 163–64.

13. Stewart, *Organizing Scientific Research for War*, 12; vol. 1 of "Westinghouse in World War II, Lamp Division," n.d., 92, Folder 2, Box 206, Westinghouse Records; Guerlac, *Radar in World War II*, 15–16, 39, 247–50; Stephan, "Experts at Play," 739.

14. "Conference on Applied Nuclear Physics," *Science* 92 (July 26, 1940): 73–74; Condon to Compton, September 3, 1940, Vannevar Bush to Condon, October 14, 1940, both in Folder: "Exhibits in Behalf of Condon, #9," Box 69, Condon Papers. The Office of Scientific Research and Development absorbed the National Defense Research Committee in June 1941. Bush, now director of OSRD, appointed Condon to the same functionally equivalent position in the new organization. Stewart, *Organizing Scientific Research for War*, 35–38; Bush to Condon, June 28, 1941, Folder: "Exhibits in Behalf of Condon, #9," Box 69, Condon Papers.

15. William E. Shoupp, "The Radar Committee," n.d., 1, in vol. 2 of "Westinghouse in World War II, Research Laboratories," Folder 2, Box 217, Westinghouse Records. Because of its high peak power, the resonant cavity magnetron found widespread use during the war as the microwave generator in radar transmitters. The klystron, by contrast, could not match the power output of the magnetron, but the ability to be tuned across a broad frequency band made it especially well suited for use in radar receivers. For a comparison of these tubes, see "Where Radar Frequencies Come From," *Westinghouse Engineer* 6 (January 1946): 4–5. See also John W. Coltman, "Resonant Cavity Magnetron," *Westinghouse Engineer* 6 (November 1946): 172–75; and Sidney Krasik, "The Klystron—Radar-Receiver-Oscillator," *Westinghouse Engineer* 6 (November 1946): 176–79.

16. Guerlac, *Radar in World War II*, 256–60; DuBridge to Lawrence, November 1, 1940, Frames 088123–24, Reel 40, Lawrence Papers; Condon to Hansen, November 8, 1940, Folder 18, Box 2, Hansen Papers; Condon to Chubb, January 29, 1941, 22, Folder: "WEC, #2," Box 126, Condon Papers.

17. Smith to Distribution, January 13, 1941, Condon to Smith, February 17, 1941, both in vol. 2 of "Westinghouse in World War II, Research Laboratories," Folder 2, Box 217, Westinghouse Records; Westinghouse Research Laboratories, "Work and Progress Report," January–March 1941, 33, April–June 1941, 34, Westinghouse Research Reports.

18. Heilbron and Seidel, *Lawrence and His Laboratory*, 24–25, 89–93, 116–21, 495; Lawrence to Poillon, October 6, 1938, Frame 049282, Reel 23, Lawrence Papers; Lawrence to Poillon, April 27, 1939, Frame 053165, Poillon to Lawrence, May 5, 1939, Frame 053167, both in Reel 24, Lawrence Papers; Cooksey to Loomis, May 14, 1940, Frame 157291, Reel 70, Lawrence Papers; H. Hugh Willis to Lawrence, June 5, 1940, Frame 053175, Lauriston C. Marshall to Willis, June 8, 1940, Frame 053176, both in Reel 24, Lawrence Papers; Loomis to Lawrence, July 9, 1940, Frame 157338, July 16, 1940, Frame 157339, July 30, 1940, Frame 157345, Reel 70, Lawrence Papers.

19. Guerlac, *Radar in World War II*, 250, 253–56; Loomis to Lawrence, September 17, 1940, Frame 157356, Lawrence to Loomis, January 10, 1941, Frame 157388, December 9, 1940, Frame 157386, all in Reel 70, Lawrence Papers; Marshall to David E. Sloan, April 5, 1941, Frame 053203, Reel 24, Lawrence Papers.

20. Heilbron and Seidel, *Lawrence and His Laboratory*, 500–501; Condon to Compton, April 16, 1941, in vol. 2 of "Westinghouse in World War II, Research Laboratories," Folder 2, Box 217, Westinghouse Records. Cassen received the patent, the rights of which he assigned to Westinghouse, three years later, in February 1944. See B. Cassen, Super Voltage X-Ray Tube, US Patent 2,342,789, filed April 19, 1941, and issued February 29, 1944; and "Supervoltage X-Ray Tube Produces 'Hard' Rays," *Science News Letter* 45 (March 11, 1944): 169.

21. Marshall to Leonard F. Fuller, April 3, 1941, Frame 053049, Marshall to Sloan, April 5, 1941, Frames 053203–6, Marshall to Sloan, May 21, 1941, Frames 053209–10, all in Reel 24, Lawrence Papers; Condon to Smith, April 14, 1942, Folder: "WEC, #3," Box 126, Condon Papers; Westinghouse Research Laboratories, "Work and Progress Report," April–June 1941, 34, July–September 1941, 33, January–March 1942, 33, July–December 1942, 46, Westinghouse Research Reports; F. W. Boggs, "Development of Various Resnatron Jamming Tubes for OSRD-Division 15," n.d., 2, in vol. 2 of "Westinghouse in World War II, Research Laboratories," Folder 2, Box 217, Westinghouse Records. Lawrence and Loomis initially tried to obtain, on behalf of Sloan and Marshall, a collaborative arrangement similar to the one that Sperry had negotiated with Stanford to develop the klystron. Sperry agreed to receive the Berkeley contingent, but the facilities and manpower available at the company's San Carlos factory did not meet

Sloan and Marshall's expectations. See Lawrence to Loomis, May 25, 1940, Frame 157312, Lawrence to Loomis, May 27, 1940, Frame 157313, Loomis to Lawrence, May 28, 1940, Frame 157316, all in Reel 70, Lawrence Papers; Poillon to Marshall, June 28, 1940, Frame 053184, Lawrence to Poillon, July 1, 1940, Frame 053186, both in Reel 24, Lawrence Papers.

22. Westinghouse Research Laboratories, "Work and Progress Report," January–March 1940, 34, January–March 1942, 33, Westinghouse Research Reports; "Westinghouse Research Fellows," *Science* 95 (June 5, 1942): 571; "Fellowships in Wartime," *Review of Scientific Instruments* 13 (December 1942): 544.

23. "National Defense," *Review of Scientific Instruments* 12 (May 1941): 283; Westinghouse Research Laboratories, "Work and Progress Report," April–June 1941, 33–34, Westinghouse Research Reports; John W. Coltman, interview by Thomas C. Lassman, June 29–30, 2004, transcript, 7–9, 16, 25, 29, 37–45, 49, Niels Bohr Library and Archives, American Institute of Physics, College Park, MD.

24. Condon to Smith, January 2, 1942, 2, Folder: "WEC, #3," Box 126, Condon Papers; Henry A. H. Boot and John T. Randall, "Historical Notes on the Cavity Magnetron," *IEEE Transactions on Electron Devices*, ED-23 (July 1976): 724–29, esp. 724.

25. Westinghouse did not develop and manufacture its own three-centimeter wavelength magnetron. Instead, the company based its in-house production models on designs the Bell Telephone Laboratories provided. J. W. Coltman, "Magnetron Group Activities," n.d., 1–3, in vol. 2 of "Westinghouse in World War II, Research Laboratories," Folder 2, Box 217, Westinghouse Records.

26. Seitz to Condon, January 26, 1942, Condon to Seitz, February 10, 1942, both in Folder: "Seitz, Frederick, #6," Box 110, Condon Papers; Condon to Hans and Rose Bethe, July 26, 1943, 2, Folder: "Atomic Energy, University of Chicago, Metallurgical Laboratory, #1," Box 10, Condon Papers; Stephen J. Angello to J. A. McLeod, November 15, 1943, in chap. 1, vol. 2 of "Westinghouse in World War II, Research Laboratories," Folder 2, Box 217, Westinghouse Records. See also Lillian Hoddeson, "Research on Crystal Rectifiers during World War II and the Invention of the Transistor," *History and Technology* 11, no. 2 (1994): 121–30, esp. 121–26.

27. Boggs, "Development of Various Resnatron Jamming Tubes for OSRD-Division 15," 1, 3, in vol. 2 of "Westinghouse in World War II, Research Laboratories," Folder 2, Box 217, Westinghouse Records; "Story of Research: The Resnatron," *Westinghouse Engineer* 6 (March 1946): 47; "Radar Countermeasures," *Electronics* 19 (January 1946): 92–97, esp. 97.

28. Condon to Distribution, March 24, 1943, Folder: "Atomic Energy, #3," Box 6, Condon Papers; Condon, Written Recollections on Wartime Work at Los Alamos, 1970, 1–2, Folder: "Condon, E. U., Recollections, Los Alamos, 1943," Box 24, Condon Papers.

29. Leslie R. Groves, *Now It Can Be Told: The Story of the Manhattan Project* (New York: Harper, 1962), 154–55. On the army's role in the development of the atomic bomb, see Vincent C. Jones, *Manhattan: The Army and the Atomic Bomb*, in *United States Army in World War II, Special Studies* (Washington, DC: US Army Center of Military History, 1985).

30. "Notes on Meeting of March 6, 1943," n.d., 2, Folder: "J. Robert Oppenheimer, Feb.–Dec. 1943," Box 291, J. Robert Oppenheimer Papers, Manuscript Division, Library of Congress, Washington, DC (hereafter cited as Oppenheimer Papers); Condon, Written Recollections on Wartime Work at Los Alamos, 1970, 6–8, Folder: "Condon, E. U., Recollections, Los Alamos, 1943," Box 24, Condon Papers; "Research News Letter," April 23, 1943, 3, Folder 8, Box 171, Westinghouse Records.

31. "First Meeting of Planning Board, March 30, 1943," n.d., "Second Meeting of Planning Board, April 2, 1943," n.d., 1, "Third Meeting of Planning Board, April 8, 1943," n.d., 3–4, all in Folder: "J. Robert Oppenheimer, Feb.–Dec. 1943," Box 291, Oppenheimer Papers; Condon, Written Recollections on Wartime Work at Los Alamos, 1970, 9, Folder: "Condon, E. U., Recollections, Los Alamos, 1943," Box 24, Condon Papers; E. U. Condon, "The Los Alamos Primer," n.d., 1, Folder: "The Los Alamos Primer, #6," Box 83, Condon Papers. See also Robert Serber, *The Los Alamos Primer: The First Lectures on How to Build an Atomic Bomb*, ed. Richard Rhodes (Berkeley: University of California Press, 1992); and Lillian Hoddeson et al., *Critical Assembly: A Technical History of Los Alamos during the Oppenheimer Years, 1943–1945* (Cambridge: Cambridge University Press, 1993).

32. Condon to William E. Shoupp, April 7, 1943, Condon to Chubb, April 7, 1943, 1–2, Condon to Smith, April 7, 1943, all in Folder: "Atomic Energy, #3," Box 6, Condon Papers; Condon to Emilie Condon, April 7, 1943, 1–2, Folder: "Condon, Emilie Honzik, #3," Box 45, Condon Papers.

33. Condon to Emilie Condon, April 12, 1943, 2, April 20, 1943, 1–2, Folder: "Condon, Emilie Honzik, #3," Box 45, Condon Papers.

34. J. Robert Oppenheimer to Condon, March 11, 1943, Folder: "Oppenheimer, J. R., #5," Box 99, Condon Papers; Condon to Oppenheimer, April 26, 1943, 2, Folder: "Condon, E. U., (1943–Feb. 1948)," Box 27, Oppenheimer Papers; Condon to Emilie Condon, April 20, 1943, 2, April 25, 1943, 1, Folder: "Condon, Emilie Honzik, #3," Box 45, Condon Papers; Kai Bird and Martin J. Sherwin, *American Prometheus: The Triumph and Tragedy of J. Robert Oppenheimer* (New York: Knopf, 2005), 223–26.

35. Condon to Oppenheimer, May 3, 1943, Folder: "Condon, E. U., (1943–Feb. 1948)," Box 27, Oppenheimer Papers; Groves, *Now It Can Be Told*, 156; Hewlett and Anderson, *New World*, 71–78, 116–20.

36. Condon, Written Recollections on Wartime Work at Berkeley, 1970, 2–5, Folder: "Condon, E. U., Recollections, Berkeley, 1943," Box 24, Condon Papers; Hewlett and

Anderson, *New World*, 56–60, 91–96; University of California Radiation Laboratory, "Volume V, Book 1, Research and Development," April 24, 1945, 32 (attached to letter from Cooksey to H. A. Fidler, April 25, 1945), Folder: "History (Early), General–through World War II [Manhattan Project Report]," Box 5, Series 11, Subseries C, McMillan Papers; Bird and Sherwin, *American Prometheus*, 193, 174–76, 236–37.

37. Condon, Written Recollections of Wartime Work at Berkeley, 1970, 5–6, Folder: "Condon, E. U., Recollections, Berkeley, 1943," Box 24, Condon Papers; "S-1 Section O.S.R.D., Scientific Research & Development Personnel, Westinghouse Research Laboratory Personnel Assigned to University of California Radiation Laboratory," n.d., Folder: "Westinghouse Research Laboratory (Personnel Assigned to U. of Cal. Rad. Lab.)," Box 8, S-1 Materials, Manhattan Engineer District Biography Files, Record Group 227, Records of the Office of Scientific Research and Development, National Archives and Records Administration, College Park, MD (hereafter cited as OSRD Records); Condon to Emilie Condon, January 29, 1944, 1, Folder: "Condon, Emilie Honzik, #4," Box 45, Condon Papers; Hewlett and Anderson, *New World*, 141–67, 294–95.

38. Condon to Emilie Condon, April 4, 1944, 1, Folder: "Condon, Emilie Honzik, #4," Box 45, Condon Papers; Condon to Chubb, April 18, 1944, 1–3, Folder: "Chubb, L. W., #1," Box 16, Condon Papers. Condon's deepening commitment at Berkeley had prompted Chubb to transition William Shoupp from acting to full-time head of the electronics division at East Pittsburgh in November 1943. "Research News Letter," November 19, 1943, Folder 8, Box 171, Westinghouse Records.

39. Condon to Chubb, April 18, 1944, 4–6, Folder: "Chubb, L. W., #1," Box 16, Condon Papers.

40. Smith to Condon, April 27, 1944, Folder: "Atomic Energy, #5," Box 6, Condon Papers; Chubb to Condon, April 28, 1944, Folder: "Chubb, L. W., #1," Box 16, Condon Papers.

41. Lawrence's postwar research plans are discussed in Robert W. Seidel, "Accelerating Science: The Postwar Transformation of the Lawrence Radiation Laboratory," *Historical Studies in the Physical Sciences* 13, no. 2 (1983): 375–400, esp. 376–84.

42. Condon to Smith, October 9, 1944, 1–2, Folder: "Smith, M. W.," Box 112, Condon Papers.

43. Condon to Smith, October 9, 1944, 2–3, Folder: "Smith, M. W.," Box 112, Condon Papers; Rabi to Condon, September 15, 1944, 1, Folder: "Condon, Edward U., 1927–1970," Box 2, Rabi Papers.

44. Condon to Smith, October 9, 1944, 1, 3, Folder: "Smith, M. W.," Box 112, Condon Papers; Condon to Smith, March 20, 1941, 1, 3, Folder: "Shoupp, W. E.," Box 111, Condon Papers.

45. Condon to Smith, October 26, 1944, 1, Folder: "Smith, M. W.," Box 112, Condon Papers; Seidel, "Accelerating Science," 379.

46. Condon to Smith, n.d., 1, 4–5, Folder: "WEC, #3," Box 126, Condon Papers; "Dr. R. E. Hellmund of Westinghouse," *New York Times*, May 18, 1942, 15.

47. F. L. Bishop, "Report of the Dean of the School of Engineering of the University of Pittsburgh," June 6, 1911, 1–2, Folder 21, Box 2, Samuel B. McCormick Papers, Administrative Files, 1904–20, Archives and Special Collections, University of Pittsburgh, Pittsburgh, PA; Bishop, "Cooperative Work, School of Engineering, University of Pittsburgh," August 19, 1924, Volume: "Reports to the Chancellor from Deans and Other Officers, 1923–1924," Box 1, Deans' Reports to the Chancellor, Archives and Special Collections, University of Pittsburgh, Pittsburgh, PA (hereafter cited as Pittsburgh Deans' Reports); L. Sieg, "The Graduate School, Annual Report of the Dean, 1926–1927," n.d., 11, Folder 1, Box 1, Records of the Graduate School, Publications, Archives and Special Collections, University of Pittsburgh, Pittsburgh, PA (hereafter cited as Pittsburgh Graduate School Records); H. E. Dyche and R. E. Hellmund, "The Pitt-Westinghouse Graduate Program," *Electrical Engineering* 53 (January 1934): 103, 105.

48. Sieg, "The Graduate School, Annual Report of the Dean, 1929–1930," June 19, 1930, 3–4, 11–12, Folder 1, Box 1, Pittsburgh Graduate School Records; E. A. Holbrook, "Report of the Schools of Engineering and Mines for the Year Ending June 30, 1935," June 20, 1935, 5, Volume: "Reports to the Chancellor from Deans and Other Officers, 1934–1935," Box 3, Pittsburgh Deans' Reports.

49. Condon to Smith, n.d., 1, 5, 8–15, Folder: "WEC, #3," Box 126, Condon Papers.

50. "Aeronautical College Assured for U.C.L.A.," *Los Angeles Times*, June 10, 1943, A1; "U.C.L.A. Fills Dean's Chair," *Los Angeles Times*, September 23, 1944, A1; Condon to Emilie Condon, April 4, 1944, 2–3, Folder: "Condon, Emilie Honzik, #4," Box 45, Condon Papers; Condon to Smith, October 26, 1944, 2, Folder: "Smith, M. W.," Box 112, Condon Papers; "Recent Appointments," *Review of Scientific Instruments* 14 (November 1943): 348.

51. Hutcheson to Condon, October 19, 1944, 1, Folder: "Hutcheson, J. A.," Box 77, Condon Papers; Condon to Smith, October 26, 1944, 1–2, Folder: "Smith, M. W.," Box 112, Condon Papers.

52. Smith to Condon, November 15, 1944, Condon to Smith, January 26, 1945, 3, both in Folder: "Smith, M. W.," Box 112, Condon Papers; "Director of Laboratories Named by Westinghouse," *New York Times*, March 12, 1948, 35.

53. Condon to Loeb, February 26, 1945, 2–3, Folder: "Condon, Edward Uhler, 26 Letters, 1934–52," Box 4, Loeb Papers; Minutes of the Meeting of the Patent Board of the University of California, March 20, 1945, Folder 17, Box 28, Records of the Office of the President, Special Problem Folders, 1899–58 (CU-5), Series 4, University Archives, University of California, Berkeley, CA; Condon to Robert G. Sproul, January 10, 1945, 3, Folder: "408: 1945," Box 1945 (403–420), Records of the Office of the Presi-

dent, 1914–58 (CU-5), Series 2, University Archives, University of California, Berkeley, CA (hereafter cited as Sproul Papers); Condon to Smith, January 26, 1945, 1–3, Folder: "Smith, M. W.," Box 112, Condon Papers; Condon to Loeb, February 26, 1945, 2–5, Folder: "Condon, Edward Uhler, 26 Letters, 1934–52," Box 4, Loeb Papers; Condon to Sproul, March 4, 1945, 5–6, Folder: "408: 1945," Box 1945 (403–420), Sproul Papers.

54. Condon to Loeb, February 26, 1945, 6, Folder: "Condon, Edward Uhler, 26 Letters, 1934–52," Box 4, Loeb Papers; Loeb to Condon, February 25, 1945, February 28, 1945, Folder: "Loeb, Leonard B., #4," Box 82, Condon Papers; Loeb to Sproul, February 28, 1945, 2, Sproul to Loeb, March 8, 1945, both in Folder: "1945: 400-Physics," Box 1945 (400M–402), Sproul Papers.

55. Loeb to Sproul, March 13, 1945, Folder: "1945: 400-Physics," Box 1945 (400M–402), Sproul Papers; Loeb to Condon, March 13, 1945, Folder: "Loeb, Leonard B., #4," Box 82, Condon Papers; Condon to Loeb, April 3, 1945, Folder: "Condon, Edward Uhler, 26 letters, 1934–52," Box 4, Loeb Papers; Condon to Harper Allen, June 4, 1945, 1–2, Folder: "Allen, Harper," Box 1, Condon Papers; Condon to Emilie Condon, May 7, 1945, 4, Folder: "Condon, Emilie Honzik, #11," Box 45, Condon Papers; "S-1 Section O.S.R.D., Scientific Research & Development Personnel, Westinghouse Research Laboratory Personnel Assigned to University of California Radiation Laboratory, 2-1-45 to 7-21-45," July 21, 1945, Folder: "Westinghouse Research Laboratory (Personnel Assigned to U. of Cal. Rad. Lab.)," Box 8, OSRD Records.

56. Condon to Emilie Condon, April 12, 1945, 2, Folder: "Condon, Emilie Honzik, #10," Box 45, Condon Papers.

CHAPTER 6. COLD WAR IN WASHINGTON

1. Condon to Emilie Condon, May 25, 1945, 4, May 31, 1945, 1–2, Folder: "Condon, Emilie Honzik, #12," Box 45, Condon Papers; "Science Bulletin," vol. 1, no. 1, n.d., 1, Folder: "American-Soviet Friendship, Inc., #1," Box 4, Condon Papers; Condon to Edwin S. Smith, May 26, 1944, Folder: "National Council of American-Soviet Friendship, Inc., #2," Box 91, Condon Papers; D. I. Vinogradoff to John W. White, April 20, 1945, Folder: "National Council of American-Soviet Friendship, Inc., #4," Box 91, Condon Papers; Condon to Smith, April 20, 1945, Folder: "American-Soviet Friendship, Inc., #3," Box 4, Condon Papers; "Atomic Scientist Defended," *New York Times*, July 23, 1947, 11; Condon to Emilie Condon, June 4, 1945, June 5, 1945, 2, Folder: "Trip to Russia, 1945," Box 117, Condon Papers.

2. Condon to Emilie Condon, June 9, 1945, Folder: "Trip to Russia, 1945," Box 117, Condon Papers; Robert S. Norris, *Racing for the Bomb: General Leslie R. Groves, The Manhattan Project's Indispensable Man* (South Royalton, VT: Steerforth Press, 2002), 458–59, 671; "Anniversary of the Academy of Sciences of the U.S.S.R.," *Science* 101 (June 15, 1945): 603.

3. Condon to Emilie Condon, June 20, 1945, 1, Folder: "Trip to Russia, 1945," Box 117, Condon Papers; Condon to Loeb, August 11, 1945, 1, Folder: "Condon, Edward Uhler, 26 Letters, 1934–52," Box 4, Loeb Papers; Smith to Henry A. Wallace, June 11, 1945, Folder: "Smith, Edwin S.," Box 112, Condon Papers.

4. Wallace to Harlow Shapley, April 18, 1945, Folder: "104251 [5 of 6]," Box 1031, Records of the Office of the Secretary, General Correspondence, 1903–50, Record Group 40, Records of the Department of Commerce, National Archives and Records Administration, College Park, MD (hereafter cited as Commerce Secretary's Correspondence); Shapley to Wallace, April 28, 1945, Folder: "104251 [2 of 6]," Box 1033, Commerce Secretary's Correspondence; Shapley to Wallace, August 1, 1945, Folder: "Wallace, Henry A.," Box 29A, Harlow Shapley Papers, University Archives, Harvard University, Cambridge, MA (hereafter cited as Shapley Papers).

5. Wallace to Harry S. Truman, October 5, 1945, 1, Briggs to the Secretary of Commerce, October 8, 1945, both in Folder: "104462–104482," Box 1086, Commerce Secretary's Correspondence. Shapley had also recommended to Wallace the physicist George Harrison, the dean of science at MIT, and John Tate, professor of physics at the University of Minnesota and the longtime editor of the *Physical Review*. Shapley to Wallace, August 1, 1945, Folder: "Wallace, Henry A.," Box 29A, Shapley Papers.

6. Rexmond C. Cochrane, *Measures for Progress: A History of the National Bureau of Standards* (Washington, DC: US Department of Commerce, 1966), 44, 61–62, 338n118, 400, 434; Gano Dunn to the Visiting Committee, July 5, 1945, 5, Frank B. Jewett to Dunn, July 6, 1945, Dunn to Jewett, August 29, 1945, all in Folder: "Standards, Bureau of, (1945–51)," Box 108, Vannevar Bush Papers, Manuscript Division, Library of Congress, Washington, DC (hereafter cited as Bush Papers); Dunn to Briggs, August 29, 1945, Folder: "Dunn/APV," Box 15, NBS Director's Correspondence; Shapley to Wallace, August 6, 1945, Folder: "Wallace, Henry A.," Box 29A, Shapley Papers.

7. Smith to Wallace, June 11, 1945, Folder: "Smith, Edwin S.," Box 112, Condon Papers; Condon to Emilie Condon, June 9, 1945, Folder: "Trip to Russia, 1945," Box 117, Condon Papers.

8. US Congress, House Subcommittee on Departments of State, Justice, Commerce, and the Judiciary Appropriations, *Department of Commerce Appropriation Bill for 1947: Hearings before the Subcommittee of the Committee on Appropriations*, 79th Cong., 2nd sess., January 29, 1946, 178–79. The Office of Research and Inventions became the Office of Naval Research in the summer of 1946. Harvey M. Sapolsky, *Science and the Navy: The History of the Office of Naval Research* (Princeton: Princeton University Press, 1990), 24, 27.

9. Robert M. Hutchins to Wallace, September 6, 1945, Folder: "104251 [7 of 7]," Box 1058, Commerce Secretary's Correspondence; Alice Kimball Smith, *A Peril and a Hope: The Scientists' Movement in America, 1945–47* (Chicago: University of Chicago

Press, 1965), 93–94; Condon to Emilie Condon, September 20, 1945, 2, September 21, 1945, 2, Folder: "Condon, Emilie Honzik, #13," Box 45, Condon Papers.

10. Condon to Emilie Condon, September 22, 1945, 1, 3, Folder: "Condon, Emilie Honzik, #13," Box 45, Condon Papers.

11. Briggs to Dunn, October 4, 1945, Folder: "Standards, Bureau of (1945–51)," Box 108, Bush Papers; Wallace to Truman, October 5, 1945, Folder: "104462–104482," Box 1086, Commerce Secretary's Correspondence.

12. Condon to Ridenour, October 20, 1945, 1, Folder: "Atomic Energy, #20," Box 7, Condon Papers; Condon to Emilie Condon, October 16, 1945, Folder: "Condon, Emilie Honzik, #14," Box 45, Condon Papers.

13. Condon to Emilie Condon, October 17, 1945, October 18, 1945, October 19, 1945, 1, October 22, 1945, Folder: "Condon, Emilie Honzik, #14," Box 45, Condon Papers; John Morton Blum, ed., *The Price of Vision: The Diary of Henry A. Wallace, 1942–1946* (Boston: Houghton Mifflin, 1973), 498; Hewlett and Anderson, *New World*, chaps. 12–13.

14. Compton to Condon, June 22, 1948, 2, Folder: "Compton, Karl Taylor, #5," Box 19, Condon Papers; Wallace, "Monday, October 15, 1945," 4, Diary of Henry A. Wallace (Microfilm), vol. 36, September 4, 1945–October 31, 1945, University Archives, University of Iowa, Ames, IA (hereafter cited as Wallace Diary); "Named to Be Director of Bureau of Standards," *New York Times*, October 30, 1945, 13; Smith to Condon, November 1, 1945, Condon to Smith, November 8, 1945, both in Folder: "WEC, #6," Box 126, Condon Papers; Cochrane, *Measures for Progress*, 435.

15. Cochrane, *Measures for Progress*, 607–11. On the bureau's early history, see Gustavus A. Weber, *The Bureau of Standards: Its History, Activities, and Organization* (Baltimore: Johns Hopkins University Press, 1925).

16. Briggs to the Secretary of Commerce, April 5, 1945, Folder: "Commerce: Program Committee," Box 19, NBS Director's Correspondence. On the bureau's wartime work, see Cochrane, *Measures for Progress*, chaps. 6–7.

17. William L. Brown, "H. A. Wallace and the Development of Hybrid Corn," *Annals of Iowa* 47 (Fall 1983): 167–79; A. Richard Crabb, *The Hybrid-Corn Makers: Prophets of Plenty* (New Brunswick, NJ: Rutgers University Press, 1947), chap. 10. Also on Wallace's background, see, for example, Edward L. Schapsmeier and Frederick H. Schapsmeier, *Henry A. Wallace of Iowa: The Agrarian Years, 1910–1940* (Ames: Iowa State University Press, 1968); Don S. Kirschner, "Henry A. Wallace as Farm Editor," *American Quarterly* 17 (Summer 1965): 187–202; and the essays on Wallace in the fall 1983 issue of *Annals of Iowa*.

18. The radio address is printed in its entirety in Henry A. Wallace, "The Value of Scientific Research to Agriculture," *Science* 77 (May 19, 1933): 475–80 (quote, 479). See also A. Hunter Dupree, *Science in the Federal Government: A History of Policies*

and Activities to 1940 (Cambridge, MA: Belknap Press of Harvard University Press, 1957), chap. 8; and Carroll W. Pursell Jr., "The Administration of Science in the Department of Agriculture, 1933–1940," *Agricultural History* 42 (July 1968): 231–40.

19. Brinkley, *End of Reform*, 148–54.

20. Mansel G. Blackford, *A History of Small Business in America* (New York: Twayne, 1991), 74–76. See also Jim F. Heath, "American War Mobilization and the Use of Small Manufacturers, 1939–1943," *Business History Review* 46 (Autumn 1972): 295–319; and Barton J. Bernstein, "The Debate on Industrial Reconversion: The Protection of Oligopoly and Military Control of the Economy," *American Journal of Economics and Sociology* 26 (April 1967): 159–72.

21. "Wallace Explains Full Employment," *New York Times*, September 6, 1945, 40; Henry A. Wallace, *Sixty Million Jobs* (New York: Reynal and Hitchcock, 1945), 123.

22. See Jonathan J. Bean, *Beyond the Broker State: Federal Policies toward Small Business, 1936–1961* (Chapel Hill: University of North Carolina Press, 1996), chap. 1.

23. Nelson R. Kellogg, "Gauging the Nation: Samuel Wesley Stratton and the Invention of the National Bureau of Standards" (PhD diss., Johns Hopkins University, 1991), 278–85; Cochrane, *Measures for Progress*, 224–25; Wallace, *Sixty Million Jobs*, 123–25.

24. Briggs became acting director after the death of George Burgess in July 1932. He served in that temporary capacity until the spring of 1933, when the president made the appointment permanent. "Dr. Briggs to Head Standards Bureau," *Washington Post*, March 28, 1933, 16. On Briggs's background and career, see Edward R. Landa and John R. Nimmo, "The Life and Scientific Contributions of Lyman J. Briggs," *Soil Science Society of America Journal* 67 (May–June 2003): 681–93.

25. Lewis Wood, "Wallace Is Sworn in at Gay Ceremony," *New York Times*, March 3, 1945, 21; Briggs to Philip M. Hauser, April 17, 1945, 2, Folder: "PRB-Plans and Procedures," Box 8, Records of the Office of the Director, Office Files of C. N. Coates, 1949–67, Record Group 167, Records of the National Institute of Standards and Technology, National Archives and Records Administration, College Park, MD (hereafter cited as Coates Papers); "Agenda for Meeting of Visiting Committee, Office of Secretary of Commerce, 9 a.m., June 22, 1945," n.d., 2, Folder: "Standards, Bureau of, (1945–51)," Box 108, Bush Papers.

26. Hauser, "Major Problems of the Department," August 10, 1945, 2, Folder 2, Box 11, Philip M. Hauser Papers, University Archives, University of Chicago, Chicago, IL. One major initiative Wallace spearheaded focused on the declassification and dissemination of wartime scientific and technical data through the Office of Declassification and Technical Services, renamed the Office of Technical Services in 1946. See Robert K. Stewart, "The Office of Technical Services: A New Deal Idea in the Cold War," *Knowledge* 15 (September 1993): 44–77, esp. 45–56.

27. "Wallace to Shift Commerce Bureau," *New York Times*, September 21, 1945, 15; Edward T. Folliard, "Truman Asks for 78% Increase in Commerce Dept. Budget," *Washington Post*, January 22, 1946, 5; "Truman Seeks $1,000,000 for Standards Bureau," *Washington Post*, March 9, 1946, 5.

28. Jewett to Dunn, July 28, 1942, Folder: "Standards, Bureau of (1942–44)," Box 108, Bush Papers; Loeb to Condon, November 27, 1945, Folder: "U.S. Dept. of Commerce, NBS—Congratulations on Appointment, #3," Box 122, Condon Papers; Donald H. Menzel to Shapley, July 30, 1945, 2–3, Folder: "PRB—Plans and Procedures," Box 8, Coates Papers; Shapley to Wallace, August 6, 1945, Folder: "Wallace, Henry A.," Box 29A, Shapley Papers.

29. Condon to John R. Steelman, October 20, 1949, 1, Folder: "Steelman, John Roy," Box 115, Condon Papers; William C. Foster to Wigner, January 9, 1948, Folder 6, Box 25, Wigner Papers. See also Margaret W. Rossiter, "Setting Federal Salaries in the Space Age," *Osiris* 7 (1992): 218–37.

30. William E. Deming to Hauser, August 9, 1945, Folder: "PRB—Plans and Procedures," Box 8, Coates Papers; Condon to Loeb, November 19, 1945, Folder: "Condon, Edward Uhler, 26 letters, 1934–52," Box 4, Loeb Papers; Loeb to Condon, November 27, 1945, Folder: "U.S. Dept. of Commerce, NBS—Congratulations on Appointment, #3," Box 122, Condon Papers.

31. "Frederick J. Bates," *Washington Post*, November 2, 1958, B2; Cochrane, *Measures for Progress*, 338n118, 571, 604, 608, 619; Breit to Condon, January 16, 1947, Folder 163 (Reel 3), Box 5, Breit Papers; "Condon Heads New Group on Atomic Physics," *Washington Post*, October 12, 1947, M1.

32. Cochrane, *Measures for Progress*, 403–8; "Interservice Radio Propagation Laboratory," *Technical News Bulletin of the National Bureau of Standards*, no. 344 (December 1945): 95, copy in Folder: "70301–413," Box 58, Monthly Bureau Reports to the Secretary of Commerce, 1943–46, Record Group 40, Records of the Department of Commerce, National Archives and Records Administration, College Park, MD (hereafter cited as Monthly Bureau Reports); Briggs to the Secretary of Commerce, May 9, 1945, 1–2, Folder: "70301–413," Box 58, Monthly Bureau Reports; Eugene C. Crittenden to the Director of the Bureau of Standards, November 8, 1945, Folder: "D/IDM," Box 2, NBS Director's Correspondence; Secretary of Commerce to Government Agencies, November 23, 1945, Folder: "DIV. XIV: Central Radio Propagation Lab," Box 110, Records of James H. Dellinger, Record Group 167, Records of the National Institute of Standards and Technology, National Archives and Records Administration, College Park, MD (hereafter cited as Dellinger Papers).

33. Wallace to Paul Porter, January 9, 1946, Folder: "67009/90–67009/104 [4 of 4]," Box 120, Commerce Secretary's Correspondence; Wallace to G. B. Meyers, January 10, 1946, Frame 730, Reel 38, Henry A. Wallace Papers (Microfilm), Manuscript Division,

Library of Congress, Washington, DC (hereafter cited as Wallace Papers); Condon to Division and Section Chiefs, April 19, 1946, 1, Folder: "D/IDM, Memo from D/U," Box 2, NBS Director's Correspondence; Condon to the Secretary of Commerce, June 25, 1946, 4–5, Folder: "510 Budget Estimates, (General), 1948," Box 198, Records of the Office of the Secretary, General Correspondence Relating to Appropriations, 1910–51, Record Group 40, Records of the Department of Commerce, National Archives and Records Administration, College Park, MD (hereafter cited as Commerce Appropriations Correspondence).

34. Breit to Condon, January 29, 1946, Folder: "D/AP," Box 1, NBS Director's Correspondence; Condon to the Secretary of Commerce, May 6, 1946, 3, Folder: "70301–417," Box 59, Monthly Bureau Reports; Condon to Menzel, March 4, 1946, Menzel to Condon, March 7, 1946, both in Folder: "D/AP," Box 1, NBS Director's Correspondence.

35. Cochrane, *Measures for Progress*, 607; Condon to Menzel, April 16, 1948, Folder: "Menzel, Donald H.," Box 86, Condon Papers; F. R. Cawley to Condon, August 19, 1947, Folder: "511.3 Amounts Approved by Secretary and Advice to Bureaus of Approved Amounts, 1949," Box 203, Commerce Appropriations Correspondence; US Congress, House Subcommittee on the Department of Commerce, *Department of Commerce Appropriation Bill for 1949: Hearings before the Subcommittee of the Committee on Appropriations*, 80th Cong., 2nd sess., January 20, 1948, 552; "Navy Increases Scientists' Pay to $14,000 a Year," *Chicago Daily Tribune*, October 15, 1947, 16. Two years later, in the fall of 1949, Congress passed legislation that established four hundred civil service positions with a salary range between $11,200 and $14,000. The Civil Service Commission selected personnel to fill 375 of the new slots, and the president handpicked candidates for the remaining 25 openings that came with the maximum salary of $14,000. Condon received a promotion to the highest pay grade in April 1950. "Notification of Personnel Action," October 28, 1949, Folder: "Steelman, John Roy," Box 115, Condon Papers; "Notification of Personnel Action," April 26, 1950, Folder: "U.S. Dept. of Commerce, National Bureau of Standards—Corning Glass Works, #4," Box 122, Condon Papers; Jerry Kluttz, "Civil Service Discloses 400 High-Pay Jobs," *Washington Post*, April 26, 1950, 1, 8.

36. Condon to Charles Sawyer, July 15, 1949, Folder: "67009/5 [6 of 6]," Box 114, Commerce Secretary's Correspondence; News and Notes, *Science* 115 (May 23, 1952): 564; Cochrane, *Measures for Progress*, 404–5, 558.

37. Condon to Foster, February 13, 1947, 1–2, Folder: "U.S. Department of Commerce, Under Secretary," Box 122, Condon Papers.

38. "Dr. DuBridge Installed as Caltech President in Imposing Ceremony," *Los Angeles Times*, November 13, 1946, A1; DuBridge to Condon, January 14, 1946, 2, Folder: "Atomic Energy, U.S. Senate Special Committee on Atomic Energy, #16," Box 10,

Condon Papers; George F. Hussey to Condon, April 3, 1946, Folder: "D/IGI, Part 1, Nov. 1945–June 1946," Box 3, NBS Director's Correspondence; Loeb to William R. Haseltine, January 3, 1946, Folder: "Letters Written by Loeb, January–March 1946," Box 19, Loeb Papers; Jerry Kluttz, "NOL Working in Terms of Push-Button War," *Washington Post*, January 13, 1946, M5.

39. Jerry Kluttz, "Condon Would Expand Bureau of Standards," *Washington Post*, August 20, 1946, 3; Condon to the Secretary of Commerce, June 25, 1946, 1–2, Folder: "510, Budget Estimates (General), 1948," Box 198, Commerce Appropriations Correspondence.

40. Donald C. Swain, "The Rise of a Research Empire: NIH, 1930–1950," *Science* 138 (December 14, 1962): 123; Wallace, "Monday, January 21, 1946," 1, vol. 37, November 1, 1945–January 21, 1946, Wallace Diary; Wallace to Truman, January 23, 1946, Frames 906–7, Truman to Wallace, January 24, 1946, Frame 911, both in Reel 38, Wallace Papers.

41. Gerald Gross, "Atom Research Laboratory Planned Here," *Washington Post*, August 16, 1946, 3; Condon to DuBridge, May 13, 1946, DuBridge to Condon, May 15, 1946, both in Folder: "D/AG, Jan.–June," Box 1, NBS Director's Correspondence; Richard G. Hewlett and Francis Duncan, *Atomic Shield, 1947–1952*, vol. 2 of *A History of the United States Atomic Energy Commission* (University Park: Pennsylvania State University Press, 1969), 224–25; Allan A. Needell, "Nuclear Reactors and the Founding of Brookhaven National Laboratory," *Historical Studies in the Physical Sciences* 14, no. 1 (1983): 95–105.

42. US Bureau of the Budget, *The Budget of the United States Government for the Fiscal Year Ending June 30, 1948* (Washington, DC: US Government Printing Office, 1947), 420; Condon to the Secretary of Commerce, June 25, 1946, 4–5, Folder: "510, Budget Estimates (General), 1948," Box 198, Commerce Appropriations Correspondence; US Congress, House Subcommittee on the Department of Commerce, *Department of Commerce Appropriation Bill for 1948: Hearings before the Subcommittee of the Committee on Appropriations*, 80th Cong., 1st sess., March 12, 1947, 287, 356–57; Cawley to Condon, August 16, 1946, September 6, 1946, Cawley to Morrow, December 13, 1946, all in Binder: "Appropriations, 1948," Box 168, Commerce Appropriations Correspondence.

43. Hewlett and Duncan, *New World*, 651–55; US Congress, House Subcommittee, *Department of Commerce Appropriation Bill for 1948*, 291–92; Needell, "Nuclear Reactors and the Founding of Brookhaven National Laboratory," 111–19; "Drs. E. U. Condon and D. W. Bronk to Aid Atomic Research," *New York Times*, June 6, 1947, 46.

44. Cawley to Condon, August 19, 1947, Folder: "511.3, Amounts Approved by Secretary and Advice to Bureaus of Approved Amounts, 1949," Box 203, Commerce Appropriations Correspondence; Melvyn Leffler, *A Preponderance of Power: National*

Security, the Truman Administration, and the Cold War (Stanford: Stanford University Press, 1992), chaps. 1–3. See also Alonzo L. Hamby, "Henry A. Wallace, the Liberals, and Soviet-American Relations," *Review of Politics* 30 (April 1968): 153–69; and Mark L. Kleinman, *A World of Hope, a World of Fear: Henry A. Wallace, Reinhold Niebuhr, and American Liberalism* (Columbus: Ohio State University Press, 2000).

45. Michael J. Hogan, *A Cross of Iron: Harry S. Truman and the Origins of the National Security State, 1945–1954* (Cambridge: Cambridge University Press, 1998), chap. 3; Aaron L. Friedberg, *In the Shadow of the Garrison State: America's Anti-Statism and Its Cold War Grand Strategy* (Princeton: Princeton University Press, 2000), 40–47, 81–97; W. Averell Harriman to James E. Webb, September 15, 1947, 1–2, Binder: "Appropriations, 1949," Box 200, Commerce Appropriations Correspondence; Cawley to Condon, August 19, 1947, Folder: "511.3, Amounts Approved by Secretary and Advice to Bureaus of Approved Amounts, 1949," Box 203, Commerce Appropriations Correspondence.

46. Condon to the Secretary of Commerce, June 25, 1946, 5, Folder: "510, Budget Estimates (General), 1948," Box 198, Commerce Appropriations Correspondence; US Congress, House Subcommittee, *Department of Commerce Appropriation Bill for 1948*, 291; Condon, "Monthly Activity Report, July 1946," August 8, 1946, 1, Folder: "70301–420," Box 59, Monthly Bureau Reports; Peter J. T. Morris, *The American Synthetic Rubber Research Program* (Philadelphia: University of Pennsylvania Press, 1989), 9–13, 102.

47. "Committee to Fix Policy on Rubber," *New York Times*, September 9, 1945, 39; Condon to Wallace, September 18, 1946, Wallace to Steelman, September 18, 1946, both in Folder: "67009 (Part 8) [2 of 5]," Box 112, Commerce Secretary's Correspondence; Steelman to the Secretary of Commerce, October 11, 1946, Folder: "Director's File, 1945 & 1946," Box 1, NBS Director's Correspondence. Undersecretary of Commerce Alfred Schindler replaced Wallace on an interim basis until Harriman took over on October 7. Condon reviewed the navy's plans and the bureau's response with Schindler, who then briefed Harriman. Condon to Alfred Schindler, October 1, 1946, Folder: "Director's File, 1945 & 1946," Box 1, NBS Director's Correspondence; Schindler to Steelman, October 4, 1946, Folder: "82218/14–82218/14-A [3 of 4]," Box 519, Commerce Secretary's Correspondence; Anthony Leviero, "Harriman Begins Commerce Task," *New York Times*, October 8, 1946, 1.

48. US Congress, House Subcommittee, *Department of Commerce Appropriation Bill for 1948*, 292; Cawley to Condon, August 19, 1947, Folder: "511.3, Amounts Approved by Secretary and Advice to Bureaus of Approved Amounts, 1949," Box 203, Commerce Appropriations Correspondence; US Congress, House Subcommittee, *Department of Commerce Appropriation Bill for 1949*, 543–45; "Truman Gets Rubber Bill," *New York Times*, April 1, 1948, 11. See also Morris, *American Synthetic Rubber Research Program*, 16–22.

49. "Working Fund Receipts by Advancing Agency, Fiscal Years 1949–1953," July 6, 1953 (attached to letter from Allen V. Astin to Mervin J. Kelly, July 7, 1953), Untitled Folder, Box 1, Allen V. Astin Papers, Manuscript Division, Library of Congress, Washington, DC (hereafter cited as Astin Papers); US Bureau of the Budget, *The Budget of the United States Government for the Fiscal Year Ending June 30, 1952* (Washington, DC: US Government Printing Office, 1951), 531; Condon to Martin D. Kamen, August 16, 1951, Folder: "Kamen, Martin, #2," Box 81, Condon Papers.

50. Cochrane, *Measures for Progress*, 212–14, 546–47; Kellogg, "Gauging the Nation," 187–99, 208–9, 234; US Congress, Senate Subcommittee of the Committee on Appropriations, *Departments of State, Justice, Commerce, and the Judiciary Appropriation Bill for 1949: Hearings before the Subcommittee of the Committee on Appropriations on H.R. 5607*, 80th Cong., 2nd sess., April 6, 1948, 602.

51. Dunn to the Visiting Committee, October 31, 1945, 2, Secretary of Commerce to the President of the Senate, n.d., 1, "Proposed Amendment to the Act Establishing the National Bureau of Standards," 1, all in Folder: "Dunn/APV," Box 15, NBS Director's Correspondence.

52. See Carroll W. Pursell Jr., "A Preface to Government Support of Research and Development: Research Legislation and the National Bureau of Standards, 1935–41," *Technology and Culture* 9 (April 1968): 145–64.

53. Cochrane, *Measures for Progress*, 558; Bush to Dunn, November 21, 1945, 1, Folder: "67009/5 [6 of 6]," Box 114, Commerce Secretary's Correspondence; Vannevar Bush, *Science, the Endless Frontier: A Report to the President on a Program of Postwar Scientific Research* (Washington, DC: US Government Printing Office, 1945), 14. See also Reingold, "Vannevar Bush's New Deal for Research," 302–3.

54. Bush quoted in G. Pascal Zachary, *Endless Frontier: Vannevar Bush, Engineer of the American Century* (New York: Free Press, 1997), 255.

55. Secretary of Commerce to the President of the Senate and Speaker of the House of Representatives, n.d. (attached to letter from Harriman to the Director of the Bureau of the Budget, January 31, 1947), 1, Folder "67009 (Part 8) [2 of 5]," Box 112, Commerce Secretary's Correspondence; Sawyer to the Director of the Bureau of the Budget, February 10, 1949 (plus attachments), Folder: "67009 (Part 11)–67009 (Part 12) [5 of 5]," Box 112, Commerce Secretary's Correspondence; M. Hale to C. V. Whitney, May 27, 1949, Folder: "67009 [1 of 6]," Box 113, Commerce Secretary's Correspondence; *An Act to Amend Section 2 of the Act of March 3, 1901 (31 Stat. 1449)*, Public Law 619, *U.S. Statutes at Large* 64, pt. 1 (1950–51): 371–73.

56. William Aspray and Michael Gunderloy, "Early Computing and Numerical Analysis at the National Bureau of Standards," *Annals of the History of Computing* 11 (Spring 1989): 3–4; Arnold N. Lowan, "The Computation Laboratory of the National Bureau of Standards," *Scripta Mathematica* 15 (March 1949): 33–34, 38–42; Briggs,

"Preface," October 10, 1945 (attached to letter from Arnold N. Lowan to Condon, January 14, 1946), Folder: "D/AG, 828C, Lowan," Box 1, NBS Director's Correspondence.

57. Deming to Condon, November 20, 1945, Folder: "D/AP," Box 1, NBS Director's Correspondence; John H. Curtiss to Division and Section Chiefs, July 28, 1946, Folder: "NBS Grad. School Courses," Box 110, Dellinger Papers; Thorvald A. Solberg to Condon, January 4, 1946, Folder: "D/AP," Box 1, NBS Director's Correspondence; US Congress, House Subcommittee, *Department of Commerce Appropriation Bill for 1948*, 311, 323.

58. Lowan to Briggs, January 14, 1946, Folder: "1946," Box 1, Records of Lyman J. Briggs, Record Group 167, Records of the National Institute of Standards and Technology, National Archives and Records Administration, College Park, MD (hereafter cited as Briggs Papers); Curtiss to the Mathematical Computing Advisory Committee of the Navy Department, October 23, 1946, 2, Folder: "174.9," Box 174, DuBridge Papers; Sapolsky, *Science and the Navy*, 24–25; Condon to Section and Division Chiefs, April 18, 1946, Folder: "D/IDM, Memo from D/U," Box 2, NBS Director's Correspondence; Curtiss to Condon, March 1, 1946, 1–2, Folder: "D/AG. Jan.–June," Box 1, NBS Director's Correspondence.

59. Briggs to the Office of the Chief of Ordnance, February 13, 1945, Folder: "Director's File, 1945 & 1946," Box 1, NBS Director's Correspondence; Cochrane, *Measures for Progress*, 388–99, appendix H; Kellogg, "Gauging the Nation," 423–26; "Agenda for Meeting of Visiting Committee, Office of Secretary of Commerce, 9 a.m., June 22, 1945," n.d., 4, Folder: "Standards, Bureau of (1945–51)," Box 108, Bush Papers.

60. Condon, "Memorandum to Division Chiefs," January 30, 1946, Folder: "D/IDM Memo from D/U," Box 2, NBS Director's Correspondence; Harry Diamond to Richard C. Tolman, June 3, 1946, 1, Folder 27, Box 3, Richard C. Tolman Papers, Archives and Special Collections, California Institute of Technology, Pasadena, CA (hereafter cited as Tolman Papers).

61. Willard H. Bennett to Loeb, March 26, 1946, May 22, 1946, 1, Folder: "Bennett, Willard Harrison, 1903–," Box 2, Loeb Papers; Loeb to Condon, April 4, 1946, 1, Folder: "Letters Written by Loeb, April–July 1946," Box 19, Loeb Papers; Jaques Cattell, ed., *American Men of Science: A Biographical Directory*, 8th ed. (Lancaster, PA: Science Press, 1949), *s.v.* "Bennett, Willard Harrison."

62. Condon, "Monthly Activity Report, December 1946," January 6, 1947, 1–2 (quote, 2), Folder: "70301/425, December 1946," Box 59, Monthly Bureau Reports; Condon, "Monthly Activity Report, October 1947," November 5, 1947, 1, Folder: "October 1947, 70301–435," Box 61, Monthly Bureau Reports; Condon, "Monthly Activity Report, November 1948," December 8, 1948, 3, Folder: "Monthly Report, November 1948, 70301/448," Box 62, Monthly Bureau Reports; Cledo Brunetti and A. S. Khouri, "Printed Electronic Circuits," *Electronics* 19 (April 1946): 104–8.

63. US Bureau of the Budget, *The Budget of the United States Government for the Fiscal Year Ending June 30, 1946* (Washington, DC: US Government Printing Office, 1945), A49; Curtiss to the Mathematical Computing Advisory Committee of the Navy Department, October 23, 1946, 2–3, 5, Folder: "174.9," Box 174, DuBridge Papers.

64. Curtiss to the Mathematical Computing Advisory Committee of the Navy Department, October 23, 1946, 1, 3–4, Folder: "174.9," Box 174, DuBridge Papers; Condon to Cawley, September 6, 1946, Binder: "Appropriations, 1948," Box 168, Commerce Appropriations Correspondence; John H. Curtiss, "The National Applied Mathematics Laboratories of the National Bureau of Standards: A Progress Report Covering the First Five Years of Its Existence," *Annals of the History of Computing* 11 (Summer 1989): 72.

65. Aspray and Gunderloy, "Early Computing and Numerical Analysis at the National Bureau of Standards," 4; Director of the Bureau of the Census to Condon, March 28, 1946, Condon to the Director of the Bureau of the Census, April 15, 1946, Director of the Bureau of the Census to Condon, April 18, 1946, Director of the Bureau of Standards to the Director of the Bureau of the Census, April 26, 1946, all in Folder: "D/IGI, Part 1, Nov. 1945–June 1946," Box 3, NBS Director's Correspondence; Nancy Stern, *From ENIAC to UNIVAC: An Appraisal of the Eckert-Mauchly Computers* (Bedford, MA: Digital Press, 1981), 102–4.

66. Condon to the Secretary of Commerce, July 8, 1946, 2, Folder: "70301–419," Box 59, Monthly Bureau Reports; Diamond to Tolman, June 3, 1946, 1, Folder 27, Box 3, Tolman Papers; Diamond and Curtiss to Condon, September 4, 1946, 1, Folder: "PRB-Plans and Procedures," Box 8, Coates Papers; Stern, *From ENIAC to UNIVAC*, 105–6.

67. J. H. Curtiss, "A Federal Program in Applied Mathematics," *Science* 107 (March 12, 1948): 259. Initial plans for the computation laboratory did not pan out given the absence of a stable, long-term funding source and a shift in research priorities away from the production of mathematical tables. "This group has always been supported on a temporary basis, first as a WPA project, then by OSRD funds, then by Navy money administered by the Bureau of Standards," Condon wrote Eugene Wigner, who served on the bureau's visiting committee, in June 1948. "While the military departments are now able and willing to support applied mathematics[,] their interest is more and more in analytical problem solving and in the development of electronic digital computers, and less and less in the kind of table work that this group does." Bleak prospects for some relief in the budget for fiscal year 1949 prompted Condon to cut the workforce and initiate plans for a transfer of the computation laboratory and its remaining personnel to Washington. Wigner and John A. Wheeler to Condon, June 14, 1948, Condon to Wigner, June 17, 1948, both in Folder: "DIG, 1–6," Box 5, NBS Director's Correspondence; Condon to Frederic R. Coudert, July 9, 1948, Folder: "D/IG, 7–12," Box 5, NBS Director's Correspondence; Notes, *Mathematical Tables and Other*

Aides to Computation 27 (July 1949): 494; Curtiss, "National Applied Mathematics Laboratories of the National Bureau of Standards," 73.

68. Theron B. Morrow to the Secretary of Commerce, July 21, 1948, August 3, 1948, Folder: "67009 (Part 10) [4 of 5]," Box 112, Commerce Secretary's Correspondence; Aspray and Gunderloy, "Early Computing and Numerical Analysis at the National Bureau of Standards," 5–6; US Congress, House Subcommittee of the Committee on Appropriations, *Department of Commerce Appropriation for 1952: Hearings before the Subcommittee of the Committee on Appropriations*, 82nd Cong., 1st sess., April 10, 1951, 503–4; Arthur L. Norberg, *Computers and Commerce: A Study of Technology and Management at Eckert-Mauchly Computer Company, Engineering Research Associates, and Remington Rand, 1946–1957* (Cambridge, MA: MIT Press, 2005), 205–13; Stern, *From ENIAC to UNIVAC*, 1, 106, 112–15, 146–49.

69. US Congress, House Subcommittee, *Department of Commerce Appropriation Bill for 1947*, 178–79; Curtiss, "Federal Program in Applied Mathematics," 260.

70. US Congress, Senate Subcommittee of the Committee on Appropriations, *Departments of State, Justice, Commerce, and the Judiciary Appropriations for 1951: Hearings before the Subcommittee of the Committee on Appropriations*, Part 2, 81st Cong., 2nd sess., May 18, 1950, 1546; Cochrane, *Measures for Progress*, 399–403, 616; Diamond to Tolman, June 3, 1946, 1, Folder 27, Box 3, Tolman Papers; Condon, "Monthly Activity Report, November 1946," December 6, 1946, 3, Folder: "70301/424, November 1946," Box 59, Monthly Bureau Reports; Sawyer to Robert Crooner, August 24, 1949, Folder: "83555–83586," Box 609, Commerce Secretary's Correspondence; "Rocket Station Wins Approval," *Los Angeles Times*, October 26, 1949, 6; Secretary of Commerce to Frederick J. Lawton, September 13, 1950, Folder: "D/AGL," Box 9, NBS Director's Correspondence; "Missile Study Set at Closed Hospital," *Los Angeles Times*, June 1, 1951, A1.

71. Condon, "Bureau Order, No. 50–11, Changes in Organization," May 18, 1950, 2, Folder: "Memoranda, Interoffice, 1950–66," Box 5, Astin Papers; Curtiss, "National Applied Mathematics Laboratories of the National Bureau of Standards," 73; Condon to Lawrence, December 26, 1950, Frame 042295, Reel 19, Lawrence Papers.

72. Condon to Emilie Condon, July 29, 1950, 1–2, Folder: "Condon, Emilie Honzik, #19," Box 45, Condon Papers; Leffler, *Preponderance of Power*, 369–74, 399–403; Hogan, *Cross of Iron*, 304–11; Marshall Andrews, "Condon Quits to Take Post in Glass Firm," *Washington Post*, August 11, 1951, 9; "Memorandum for Record," September 26, 1951, Folder: "Standards, Bureau of (1945–51)," Box 108, Bush Papers; "Cosmos Club, Proposal for Membership or Associate Privileges," December 8, 1950, Folder: "1948–1950," Box 1, Briggs Papers; Condon to Sawyer, August 6, 1951, 1, Folder: "American Physical Soc., #4," Box 4, Condon Papers; "Astin Named to Direct Bureau of Standards," *Washington Post*, May 21, 1952, 1.

73. *Biennial Report, 1953 and 1954, National Bureau of Standards, Miscellaneous Publication 213* (Washington, DC: US Government Printing Office, 1954), 10, 124, copy preserved in the library at the National Institute of Standards and Technology, Gaithersburg, MD; "Funds Advanced from Other Agencies, Fiscal Year 1953," n.d., Folder: "RF 40, Department of Commerce, National Bureau of Standards, 1–10," Box 3, Briefing Handbooks, 1952–53, Record Group 40, Records of the Department of Commerce, National Archives and Records Administration, College Park, MD.

CHAPTER 7. RECESSIONAL

1. See US Congress, House Committee on Un-American Activities, Special Subcommittee on National Security, *Report to the Full Committee of the Special Subcommittee on National Security of the Committee on Un-American Activities*, 80th Cong., 2nd sess., March 1, 1948.

2. Jessica Wang, "Science, Security, and the Cold War: The Case of E. U. Condon," *Isis* 83 (June 1992): 242. See also Michael A. Day, "E. U. Condon: Science, Religion, and the Politics of World Peace," *Physics in Perspective* 10 (March 2008): 4–55; Smith, *Peril and a Hope*; and Jessica Wang, *American Science in an Age of Anxiety, Scientists, Anticommunism, and the Cold War* (Chapel Hill: University of North Carolina Press, 1999).

3. Compton to Jewett, April 2, 1948, 2, Folder: "Jewett, Frank B., 1932–1949," Box 126, Compton Papers.

4. Hewlett and Anderson, *New World*, 435–41, 449–52; Condon to the Secretary of Commerce, June 25, 1946, 1, Folder: "510 Budget Estimates (General), 1948," Box 198, Commerce Appropriations Correspondence.

5. US Congress, House Committee, *Report to the Full Committee of the Special Subcommittee on National Security of the Committee on Un-American Activities*, 1; "Memorandum of Conversation," March 15, 1948, Folder: "Con–CZ, Misc., 1946–48," Box 236, W. Averell Harriman Papers, Manuscript Division, Library of Congress, Washington, DC (hereafter cited as Harriman Papers); William S. White, "Soviet Spy Links Laid to Dr. Condon, High Federal Aide," *New York Times*, March 2, 1948, 1.

6. Wang, "Science, Security, and the Cold War," 246; Condon, Handwritten Notes, November 27, 1947, 1–4, Folder: "U.S. Dept. of Commerce Loyalty Board, #2," Box 121, Condon Papers; Cochrane, *Measures for Progress*, 558; Compton to Condon, April 13, 1954, Condon to Compton, April 16, 1954, both in Folder: "Compton, K. T., #6," Box 19, Condon Papers.

7. James L. Morrill to James R. Killian, October 15, 1947, Morrill to Compton, October 15, 1947, Compton to Morrill, October 27, 1947, all in Folder 2, Box 150, Compton Papers; Compton to Harriman, November 20, 1947, December 30, 1947, Folder: "Harmon-Harrington, 1940–1956," Box 105, Compton Papers; Harriman to Compton, No-

vember 26, 1947, January 6, 1948, Folder: "CL–Com, Miscellaneous, 1946–48," Box 235, Harriman Papers; Condon to John T. Tate, January 2, 1948, Condon to Morrill, January 2, 1948, both in Folder: "University of Minnesota," Box 124, Condon Papers; Condon to Hansen, January 1, 1948, 2, Folder 42, Box 4, Hansen Papers.

8. "Memorandum to Undersecretary Foster from the Secretary," March 26, 1948, Folder: "Misc. Corresp., Sp–Sz, 1946–48," Box 253, Harriman Papers; Jewett to Compton, March 29, 1948, 3–4, Folder: "Jewett, Frank B., 1932–1949," Box 126, Compton Papers.

9. News and Notes, *Science* 103 (May 17, 1946): 623; Wang, "Science, Security, and the Cold War," 257; Condon to Henry Eyring, December 21, 1950, 1–3, Folder: "Eyring, Henry," Box 69, Condon Papers; Leffler, *Preponderance of Power*, 398–406.

10. Condon to His Family, August 5, 1951, 1–2, Condon to Emilie Condon, August 7, 1951, 1, both in Folder: "Condon, Emilie Honzik, #20," Box 46, Condon Papers; Jesse T. Littleton to Condon, July 12, 1951, Condon to Sawyer, August 7, 1951, Condon to the President, August 8, 1951, all in Folder: "Corning Glass Works, Employment, #2," Box 49, Condon Papers.

11. Margaret B. W. Graham and Alec T. Shuldiner, *Corning and the Craft of Innovation* (New York: Oxford University Press, 2001), 242–45; Condon to His Family, August 5, 1951, 3–4, Folder: "Condon, Emilie Honzik, #20," Box 46, Condon Papers; Condon to Robert D. Huntoon, April 8, 1952, Folder: "Handbook of Physics, Book 4, #1," Box 38, Condon Papers; Condon to William C. Decker, December 25, 1952, Folder: "Corning Glass Works, Employment, #2," Box 49, Condon Papers; Condon to Eugene C. Sullivan, January 30, 1953, 1–3, Folder: "Corning Glass Works, Foundation, Science Fellows, #1," Box 49, Condon Papers.

12. Condon to Sullivan, January 30, 1953, 3–4, Folder: "Corning Glass Works, Foundation, Science Fellows, #1," Box 49, Condon Papers; Condon to N. H. Furman, April 2, 1954, 1, Folder: "Furman, N. H.," Box 74, Condon Papers.

13. "Probers Issue Subpoena for Dr. Condon," *Washington Post*, August 20, 1952, 1; Willard Edwards, "House Hearings on Chicago Reds to Start Today," *Chicago Tribune*, September 2, 1952, A5; "Condon Given Clearance to See Secret Defense Data," *Washington Post*, October 19, 1954, 2; Charles S. Thomas to Condon, December 30, 1954, Folder: "U.S. Department of Defense, Secretary of the Navy, #1," Box 122, Condon Papers; Condon to Nathan David, July 26, 1955, 1, 3–5, Folder: "David, Nathan H., #2," Box 60, Condon Papers; Condon to Amory Houghton, January 4, 1954, Folder: "Houghton, Amory Jr., #2," Box 77, Condon Papers; Condon to Wayne and Marie Thornton, November 29, 1954, 1–3, Folder: "Move to Berkeley, 1954, #1," Box 88, Condon Papers.

14. Condon to Wayne and Marie Thornton, November 29, 1954, 1, Folder: "Move to Berkeley, 1954, #1," Box 88, Condon Papers; Condon to Emilie Condon, January 17, 1955, 2, January 20, 1955, 1, January 30, 1955, 1–4, Folder: "Condon, Emilie Honzik,

#24," Box 46, Condon Papers; Condon to Henry T. Heald, March 1, 1955, 1–4, Folder: "New York University," Box 94, Condon Papers; Condon to Houghton, March 1, 1955, Folder: "Corning Glass Works, 1955, #1," Box 46, Condon Papers.

15. Heald to Condon, March 15, 1955, Folder: "New York University," Box 94, Condon Papers. See Roger L. Geiger, *Research and Relevant Knowledge: American Research Universities since World War II* (New York: Oxford University Press, 1993); and Peter J. Westwick, *The National Labs: Science in an American System, 1947–1974* (Cambridge, MA: Harvard University Press, 2003). For representative case studies, see, for example, Rebecca S. Lowen, *Creating the Cold War University: The Transformation of Stanford* (Berkeley: University of California Press, 1997); Amy Sue Bix, "'Backing into Sponsored Research': Physics and Engineering at Princeton University," *History of Higher Education Annual* 13 (1993): 9–52; Christophe Lécuyer, "The Making of a Science-Based Technological University: Karl Compton, James Killian, and the Reform of MIT, 1930–1957," *Historical Studies in the Physical and Biological Sciences* 23, no. 1 (1992): 153–80; and Joanne Abel Goldman, "National Science in the Nation's Heartland: The Ames Laboratory and Iowa State University, 1942–1965," *Technology and Culture* 41 (July 2000): 435–59.

16. Charles W. Ufford to Condon, December 15, 1954, Condon to Ufford, December 18, 1954, January 15, 1955, 1, Ufford to Condon, January 19, 1955, all in Folder: "University of Pennsylvania," Box 124, Condon Papers. On Varian and the origins of NMR, see Timothy Lenoir and Christophe Lécuyer, "Instrument Makers and Discipline Builders: The Case of Nuclear Magnetic Resonance," *Perspectives on Science* 3, no. 3 (1995): 276–345.

17. Condon to Russell Varian, March 8, 1955, Folder: "Corning Glass Works, 1955, #1," Box 46, Condon Papers; Condon to William H. Armistead, March 19, 1955, 3–4, Folder: "Corning Glass Works, 1955, #2," Box 46, Condon Papers.

18. Condon to Armistead, June 6, 1955, 1–3, Folder: "CGW, Research Progress Meetings," Box 52, Condon Papers; Condon to Hugh Odishaw, June 26, 1955, Folder: "Handbook of Physics, First Edition, Hugh Odishaw, #3," Box 37, Condon Papers.

19. Armistead and Condon to Decker, June 14, 1955, Folder: "CGW, Research Progress Meetings," Box 52, Condon Papers; Condon to Felix Bloch, June 14, 1955, Armistead to R. H. Sands, June 23, 1955, Condon to Sands, June 23, 1955, all in Folder: "CGW, Contract Research with the University of California, Berkeley, #4," Box 48, Condon Papers; Condon to Armistead, March 4, 1955, 2–4, Folder: "Corning Glass Works, 1955, #2," Box 46, Condon Papers; Armistead to Joseph A. Pask, March 28, 1955, Pask to Armistead, March 28, 1955, Armistead to Sproul, April 1, 1955, all in Folder: "Corning Glass Works, Foundation, Science Fellows, #1," Box 49, Condon Papers; Sproul to Armistead, May 4, 1955, Folder: "University of California, Berkeley, #3," Box 123, Condon Papers.

20. Condon to Armistead, August 5, 1955, Folder: "CGW, Contract Research with the University of California, Berkeley, #5," Box 48, Condon Papers; "Draft Statement of Purpose for [University of California] Research Contract," July 22, 1955, "Cooperative Fundamental Research Program on Structure and Properties of Glass," September 6, 1955, both in Folder: "CGW, Contract Research with the University of California, Berkeley, #2," Box 48, Condon Papers; F. H. Knight to G. E. Lynn, October 5, 1955, Folder: "CGW, Contract Research with the University of California, Berkeley, #3," Box 48, Condon Papers.

21. T. H. Hazlett to Knight, August 31, 1955, Folder: "CGW, Contract Research with the University of California, Berkeley, #5," Box 48, Condon Papers; Condon to Armistead, September 17, 1955, 1, Folder: "Corning Glass Works, 1956, #3," Box 46, Condon Papers; Knight to Hazlett, September 13, 1955, Hazlett to Knight, September 21, 1955, Condon to Armistead, October 31, 1955, all in Folder: "CGW, Contract Research with the University of California, Berkeley, #3," Box 48, Condon Papers.

22. Knight to Robert F. Kerley, November 3, 1955, Folder: "CGW, Contract Research with the University of California, Berkeley, #3," Box 48, Condon Papers; Condon to Armistead, January 2, 1956, 3, January 11, 1956, Folder: "Corning Glass Works, 1956, #3," Box 46, Condon Papers; Morrough P. O'Brien to Clark Kerr, November 9, 1955, Folder: "472, Raytheon Manufacturing Co., Proposed Contract with, 1956," Box 44, Records of the Office of the Chancellor (CU-149), University Archives, University of California, Berkeley, CA; Condon to Armistead, January 12, 1956, 2, Armistead to Condon, January 19, 1956, both in Folder: "Corning Glass Works, 1956, #2," Box 46, Condon Papers.

23. Graham and Shuldiner, *Corning and the Craft of Innovation*, 453n6; Condon to Emilie Condon, January 17, 1955, 2, Folder: "Condon, Emilie Honzik, #24," Box 46, Condon Papers; George E. Pake to Ethan A. H. Shepley, March 22, 1955, Folder P, Box 3, Records of the Office of the Chancellor, Series 2, Files of Ethan Allen Hitchcock Shepley, 1953–57, University Archives, Washington University, St. Louis, MO; Condon to Barry Commoner, March 16, 1955, Commoner to Condon, April 4, 1956, both in Folder: "Commoner, Barry," Box 18, Condon Papers; Condon to Shepley, March 16, 1956, Folder: "Shepley, Ethan A. H.," Box 111, Condon Papers.

24. Pake to Condon, April 12, 1956, May 4, 1956, 2, Folder: "Pake, George E., #1," Box 100, Condon Papers; Shepley to Condon, May 3, 1956, Condon to Shepley, May 7, 1956, both in Folder: "Washington University," Box 125, Condon Papers; News Release, University of Colorado (Joint Institute for Laboratory Astrophysics), October 22, 1962, Folder: "University of Colorado, Joint Institute for Laboratory Astrophysics, #1," Box 124, Condon Papers; Condon to Richard E. Norberg, February 1, 1963, Folder: "Norberg, Richard E., #1," Box 98, Condon Papers; Condon to Pake, August 8, 1963, 1, September 23, 1963, Folder: "Pake, George E., #4," Box 100, Condon Papers; Martin

Weil, "Atomic Scientist Edward Condon Dies," *Washington Post*, March 27, 1974, C4. Emilie Condon died later that year, on October 31. "Mrs. Edward Condon," *New York Times*, November 4, 1974, 40.

25. Rabi to Condon, February 4, 1970, 1, Folder: "Princeton University, #5," Box 104, Condon Papers.

26. Westinghouse announced plans to reactivate the atom smasher in 1946. It operated until October 1958. "Westinghouse Program for Fundamental Research," *Journal of Applied Physics* 17 (December 1946): 1129; Harry G. Gail to David Speer, January 25, 1967, 2, Folder 13, Box 177, Westinghouse Records.

27. See Hewlett and Duncan, *Atomic Shield*, chap. 7; Richard G. Hewlett and Jack M. Holl, *Atoms for Peace and War, 1953–1961: Eisenhower and the Atomic Energy Commission*, vol. 3 of *A History of the United States Atomic Energy Commission* (Berkeley: University of California Press, 1989), chap. 7; Brian Balogh, *Chain Reaction: Expert Debate and Public Participation in American Commercial Nuclear Power, 1945–1975* (Cambridge: Cambridge University Press, 1991); and Richard G. Hewlett and Francis Duncan, *Nuclear Navy, 1946–1962* (Chicago: University of Chicago Press, 1974). On Rickover, see Francis Duncan, *Rickover and the Nuclear Navy: The Discipline of Technology* (Annapolis: Naval Institute Press, 1990).

28. Heilbron and Seidel, *Lawrence and His Laboratory*, 192–99; Hewlett and Duncan, *Atomic Shield*, 222–27, 251–53. See also Angela N. H. Creager, "The Industrialization of Radioisotopes by the U.S. Atomic Energy Commission," in *The Science-Industry Nexus: History, Policy, Implications*, ed. Karl Grandin, Nina Wormbs, and Sven Widmalm (Sagamore Beach, MA: Science History Publications/Watson, 2004).

29. Hewlett and Duncan, *Nuclear Navy*, 97–100; "Westinghouse Plans to Coordinate Activities for Atomic Research," *Iron Age* 159 (April 17, 1947): 127; "Westinghouse Forms Atomic Power Division," *Nucleonics* 3 (November 1948): 74; "Westinghouse Names Heads of Atomic Power Group," *Nucleonics* 4 (February 1949): 80; "Westinghouse Names New Executives for Atomic Power Division," *Iron Age* 163 (June 2, 1949): 130; Leland Johnson and Daniel Schaffer, *Oak Ridge National Laboratory: The First Fifty Years* (Knoxville: University of Tennessee Press, 1994), 36–37, 39–40, 52.

30. See Bean, *Beyond the Broker State*, chap. 5; and Mark R. Wilson, *Destructive Creation: American Business and the Winning of World War II* (Philadelphia: University of Pennsylvania Press, 2016), 77–88.

31. *Physics Today* published the full text of Condon's address. See E. U. Condon, "Some Thoughts on Science in the Federal Government," *Physics Today* 5 (April 1952): 6–13 (quote, 9).

32. See Geiger, *Research and Relevant Knowledge*, 298–99; and Mowery and Rosenberg, *Technology and the Pursuit of Economic Growth*, 150, 258.

33. See Harold C. Livesay, "Entrepreneurial Dominance in Businesses Large and

Small, Past and Present," *Business History Review* 63 (Spring 1989): 3–4; Morris, "Resource Networks," chap. 7; and Eric S. Hintz, "The Post-Heroic Generation: American Independent Inventors, 1900–1950," *Enterprise and Society* 12 (December 2011): 732–48.

34. Kevles, *Physicists*, chap. 23; Hounshell, "Evolution of Industrial Research in the United States," 41–46. For representative case studies, see, for example, George Wise, "Science at General Electric," *Physics Today* 37 (December 1984): 52–61; Scott G. Knowles and Stuart W. Leslie, "'Industrial Versailles': Eero Saarinen's Corporate Campuses for GM, IBM, and AT&T," *Isis* 92 (March 2001): 1–33; Margaret B. W. Graham, *RCA and the VideoDisc: The Business of Research* (Cambridge: Cambridge University Press, 1986); and John W. Servos, "Changing Partners: The Mellon Institute, Private Industry, and the Federal Patron," *Technology and Culture* 35 (April 1994): 221–57.

35. See Glen R. Asner, "The Linear Model, the U.S. Department of Defense, and the Golden Age of Industrial Research," in *Science-Industry Nexus*, ed. Grandin, Wormbs, and Widmalm; and David Kaiser, "Cold War Requisitions, Scientific Manpower, and the Production of American Physicists after World War II," *Historical Studies in the Physical and Biological Sciences* 33, no. 1 (2002): 131–59.

36. "In the Laboratories," *Science* 124 (October 26, 1956): 826; "Westinghouse Spurs Research," *Electrical World* 146 (October 8, 1956): 23–24. Also on postwar R&D at Westinghouse, see Glen R. Asner, "The Cold War and American Industrial Research" (PhD diss., Carnegie Mellon University, 2006), chaps. 8–9.

37. See Hounshell, "Evolution of Industrial Research in the United States," 47–56; Margaret B. W. Graham, "Corporate Research and Development: The Latest Transformation," *Technology in Society* 7, no. 2–3 (1985): 179–95, esp. 189–93; and Nathan Rosenberg and Richard R. Nelson, "The Roles of Universities in the Advance of Industrial Technology," in *Engines of Innovation*, ed. Rosenbloom and Spencer.

BIBLIOGRAPHY

MANUSCRIPT COLLECTIONS

Personal Papers

Astin, Allen V. Papers. Manuscript Division, Library of Congress, Washington, DC.

Barton, Henry A. Papers. Niels Bohr Library and Archives, American Institute of Physics, College Park, MD.

Birge, Raymond T. Papers. University Archives, University of California, Berkeley, CA.

Bitter, Francis. Papers. Institute Archives and Special Collections, Massachusetts Institute of Technology, Cambridge, MA.

Bleakney, Walker. Papers. Rare Books and Special Collections, Princeton University, Princeton, NJ.

Breit, Gregory. Papers (Microfilm). Niels Bohr Library and Archives, American Institute of Physics, College Park, MD.

Bush, Vannevar. Papers. Manuscript Division, Library of Congress, Washington, DC.

Condon, Edward U. Papers. American Philosophical Society, Philadelphia, PA.

DuBridge, Lee A. Papers. Archives and Special Collections, California Institute of Technology, Pasadena, CA.

Epstein, Paul S. Papers. Archives and Special Collections, California Institute of Technology, Pasadena, CA.

Erikson, Henry A. Papers. Archives and Special Collections, University of Minnesota, Minneapolis, MN.

Hansen, William W. Papers. Special Collections and University Archives, Stanford University, Stanford, CA.

Harnwell, Gaylord P. Papers. University Archives, University of Pennsylvania, Philadelphia, PA.

Harriman, W. Averell. Papers. Manuscript Division, Library of Congress, Washington, DC.

Hauser, Philip M. Papers. University Archives, University of Chicago, Chicago, IL.

Hutchisson, Elmer. Papers. Niels Bohr Library and Archives, American Institute of Physics, College Park, MD.

Lawrence, Ernest O. Papers (Microfilm). University Archives, University of California, Berkeley, CA.

Loeb, Leonard B. Papers. University Archives, University of California, Berkeley, CA.

Morse, Philip M. Papers. Institute Archives and Special Collections, Massachusetts Institute of Technology, Cambridge, MA.

Oppenheimer, J. Robert. Papers. Manuscript Division, Library of Congress, Washington, DC.

Rabi, Isidor I. Papers. Manuscript Division, Library of Congress, Washington, DC.

Shapley, Harlow. Papers. University Archives, Harvard University, Cambridge, MA.

Stephens, William E. Papers. University Archives, University of Pennsylvania, Philadelphia, PA.

Tolman, Richard C. Papers. Archives and Special Collections, California Institute of Technology, Pasadena, CA.

Tuve, Merle A. Papers. Manuscript Division, Library of Congress, Washington, DC.

Van Vleck, John H. Papers. Niels Bohr Library and Archives, American Institute of Physics, College Park, MD.

Wallace, Henry A. Diary (Microfilm). University Archives, University of Iowa, Ames, IA.

Wallace, Henry A. Papers (Microfilm). Manuscript Division, Library of Congress, Washington, DC.

Webster, David L. Papers (SCO131), Series D. Special Collections and University Archives, Stanford University, Stanford, CA.

Wigner, Eugene P. Papers. Rare Books and Special Collections, Princeton University, Princeton, NJ.

Institutional Records

Biographical Faculty Files. Records of the Office of Public Affairs, University Archives, Washington University, St. Louis, MO.

Biographical Material Files. Niels Bohr Library and Archives, American Institute of Physics, College Park, MD.

Board of Regents. Minutes. University Archives Digital Conservancy, Archives and Special Collections, University of Minnesota, Minneapolis, MN.

Bowman, John G. Papers. Administrative Files, 1921–45. Archives and Special Collections, University of Pittsburgh, Pittsburgh, PA.

Briefing Handbooks, 1952–53. Record Group 40, Records of the Department of Commerce, National Archives and Records Administration, College Park, MD.

Briggs, Lyman J. Records. Record Group 167, Records of the National Institute of Standards and Technology, National Archives and Records Administration, College Park, MD.

Deans' Reports to the Chancellor. Archives and Special Collections, University of Pittsburgh, Pittsburgh, PA.

Dellinger, James H. Records. Record Group 167, Records of the National Institute of Standards and Technology, National Archives and Records Administration, College Park, MD.

Faculty Files. University Archives, Rare Books and Special Collections, Princeton University, Princeton, NJ.

Fitzgerald, Rufus H. Papers. Administrative Files, 1945–55. Archives and Special Collections, University of Pittsburgh, Pittsburgh, PA.

McCormick, Samuel B. Papers. Administrative Files, 1904–20. Archives and Special Collections, University of Pittsburgh, Pittsburgh, PA.

McMillan, Edwin M. Historical Research Papers. Series 11, Subseries A and C. Record Group 434, Records of the United States Department of Energy, National Archives and Records Administration, San Bruno, CA.

Monthly Bureau Reports to the Secretary of Commerce, 1943–46. Record Group 40, Records of the Department of Commerce, National Archives and Records Administration, College Park, MD.

Personnel Records. Archives and Special Collections, University of Minnesota, Minneapolis, MN.

Records of the College of Engineering and Science. University Archives, Carnegie Mellon University, Pittsburgh, PA.

Records of the Department of Physics. University Archives, Rare Books and Special Collections, Princeton University, Princeton, NJ.

Records of the Department of Terrestrial Magnetism. Carnegie Institution for Science, Washington, DC.

Records of the Division of Research in the Natural Sciences. Archives and Special Collections, University of Pittsburgh, Pittsburgh, PA.

Records of the Graduate School. Publications. Archives and Special Collections, University of Pittsburgh, Pittsburgh, PA.

Records of the Office of the Chancellor (CU-149). University Archives, University of California, Berkeley, CA.

Records of the Office of the Chancellor, Series 2. Files of Ethan Allen Hitchcock Shepley, 1953–57. University Archives, Washington University, St. Louis, MO.

Records of the Office of the Director. Director's Correspondence, 1923–63. Record Group 167, Records of the National Institute of Standards and Technology, National Archives and Records Administration, College Park, MD.

Records of the Office of the Director. Office Files of C. N. Coates, 1949–67. Record Group 167, Records of the National Institute of Standards and Technology, National Archives and Records Administration, College Park, MD.

Records of the Office of the President, 1914–58 (CU-5), Series 2. University Archives, University of California, Berkeley, CA.

Records of the Office of the President, 1930–59 (Karl T. Compton and James R. Killian). Institute Archives and Special Collections, Massachusetts Institute of Technology, Cambridge, MA.

Records of the Office of the President, Series 15 (Harold W. Dodds). University Archives, Princeton University, Princeton, NJ.

Records of the Office of the President, 1930–44 (Thomas S. Gates). University Archives, Rare Books and Special Collections, University of Pennsylvania, Philadelphia, PA.

Records of the Office of the President, 1945–55 (George W. McClelland). University Archives, University of Pennsylvania, Philadelphia, PA.

Records of the Office of the President. Special Problem Folders, 1899–1958 (CU-5), Series 4. University Archives, University of California, Berkeley, CA.

Records of the Office of the Secretary. General Correspondence, 1903–50. Record Group 40, Records of the Department of Commerce, National Archives and Records Administration, College Park, MD.

Records of the Office of the Secretary. General Correspondence Relating to Appropriations, 1910–51. Record Group 40, Records of the Department of Commerce, National Archives and Records Administration, College Park, MD.

Records of the Office of the Secretary. Subject Files of Under Secretary of Commerce William C. Foster, 1946–48. Record Group 167, Records of the National Institute of Standards of Technology, National Archives and Records Administration, College Park, MD.

Records of the Westinghouse Electric Corporation. Thomas and Katherine Detre Library and Archives, Senator John Heinz History Center, Pittsburgh, PA.

S-1 Files. Lyman J. Briggs Alphabetical Files. Record Group 227, Records of the Office of Scientific Research and Development, National Archives and Records Administration, College Park, MD.

S-1 Materials. Manhattan Engineer District Biography Files. Record Group 227, Records of the Office of Scientific Research and Development, National Archives and Records Administration, College Park, MD.

Westinghouse Quarterly Research Reports. Accession No. 2017.0312. Thomas and Katherine Detre Library and Archives, Senator John Heinz History Center, Pittsburgh, PA.

Miscellaneous Documents

The Catalogue of Princeton University, 1928–1929. University Archives, Rare Books and Special Collections, Princeton University, Princeton, NJ.

The Catalogue of Princeton University, 1931–1932. University Archives, Rare Books and Special Collections, Princeton University, Princeton, NJ.

Condon, Edward U. "What Next? The Story of One American Physicist." Unpublished manuscript, n.d. Niels Bohr Library and Archives, American Institute of Physics, College Park, MD.

"A Conference on Problems of Modern Physics: Opening of the Charles Benedict Stuart Laboratory of Applied Physics, June 19 and 20, 1942." Copy of conference program preserved at the Indiana State Library, Indianapolis, IN.

"Union Switch & Signal Company," February 12, 1934. Accession No. 1810, Folder 7, Box 1130. Manuscripts and Archives, Hagley Museum and Library, Wilmington, DE.

Westinghouse Electric and Manufacturing Company. *55th Annual Report, 1940.* Copy preserved at the Library of Congress, Washington, DC.

Oral Histories

Coltman, John W. Interview by Thomas C. Lassman, June 29–30, 2004. Transcript, Niels Bohr Library and Archives, American Institute of Physics, College Park, MD.

Condon, Edward U. Interview by Charles Weiner, October 17, 1967. Transcript, session 1. Niels Bohr Library and Archives, American Institute of Physics, College Park, MD.

White, Milton G. Interview by Charles Weiner, February 27, 1973. Transcript, session 2. Niels Bohr Library and Archives, American Institute of Physics, College Park, MD.

PUBLISHED AND SECONDARY SOURCES

"Abstracts of Papers Presented at the Washington Meeting of the National Academy of Sciences." *Science* 81 (May 3, 1935): 416–26.

An Act to Amend Section 2 of the Act of March 3, 1901 (31 Stat. 1449). Public Law 619. *U.S. Statutes at Large* 64 (1950–51): 371–73.

"Advisory Council on Applied Physics." *Review of Scientific Instruments* 6 (December 1935): 383–84.

"Advisory Council on Applied Physics of the American Institute of Physics, Report of Meeting, October 28, 1936." *Journal of Applied Physics* 8 (February 1937): 98–106.

"Aeronautical College Assured for U.C.L.A." *Los Angeles Times*, June 10, 1943, A1.

Aitken, Hugh G. J. *Syntony and Spark: The Origins of Radio*. New York: John Wiley, 1976.

"American Cyanamid." *Fortune* 22 (September 1940): 66–71, 102–6.

"American Institute of Physics." *Review of Scientific Instruments* 7 (August 1936): 321.

"The American Institute of Physics." *Science* 74 (November 20, 1931): 508–9.

Andrews, Marshall. "Condon Quits to Take Post in Glass Firm." *Washington Post*, August 11, 1951, 9.

"Anniversary of the Academy of Sciences of the U.S.S.R." *Science* 101 (June 15, 1945): 603.

Asner, Glen R. "The Cold War and American Industrial Research." PhD diss., Carnegie Mellon University, 2006.

Asner, Glen R. "The Linear Model, the U.S. Department of Defense, and the Golden Age of Industrial Research." In *The Science-Industry Nexus: History, Policy, Implications*, edited by Karl Grandin, Nina Wormbs, and Sven Widmalm, 3–30. Sagamore Beach, MA: Science History Publications/Watson, 2004.

Aspray, William, and Michael Gunderloy. "Early Computing and Numerical Analysis at the National Bureau of Standards." *Annals of the History of Computing* 11 (Spring 1989): 3–12.

Assmus, Alexi. "The Americanization of Molecular Physics." *Historical Studies in the Physical and Biological Sciences* 23, no. 1 (1992): 1–34.

"Astin Named to Direct Bureau of Standards." *Washington Post*, May 21, 1952, 1.

Aston, F. W. *Mass Spectra and Isotopes*. 2nd ed. New York: Longmans, Green, 1942.

"The Atomic-Physics Observatory of the Carnegie Institution of Washington." *Science* 86 (July 23, 1937): 74–75.

"Atomic Scientist Defended." *New York Times*, July 23, 1947, 11.

"Atom Smasher." *Chicago Daily Tribune*, July 29, 1937, 9.

"Atom Smashing Equipment." *Journal of Applied Physics* 8 (November 1937): 758.

"At Westinghouse." *Time*, February 12, 1940, 44–45.

Balogh, Brian. *Chain Reaction: Expert Debate and Public Participation in American Commercial Nuclear Power, 1945–1975.* Cambridge: Cambridge University Press, 1991.

Barringer, E. C. "Five Interests Make Two-Thirds of Country's Iron and Steel." *Iron Trade Review* 85 (August 15, 1929): 386.

"The Bartol Research Foundation." *Science* 65 (June 17, 1927): 589.

Barton, Henry A. "Advisory Council on Applied Physics." *Review of Scientific Instruments* 6 (December 1935): 383–84.

Barton, Henry A. "A Conference on Applied Physics." *Review of Scientific Instruments* 6 (February 1935): 31–35.

Barton, Henry A. "New Journal of Applied Physics." *Review of Scientific Instruments* 7 (November 1936): 399.

Barton, Henry A. "Report of Conference on Applied Physics." *Review of Scientific Instruments* 7 (March 1936): 113–23.

Barton, William S. "Our Progress in the World of Science." *Los Angeles Times*, March 17, 1940, I16.

Bass, Lawrence W. "Solving Food Manufacturing Problems by Research Institute Method." *Food Industries* 2 (September 1930): 406–9.

Bean, Jonathan J. *Beyond the Broker State: Federal Policies toward Small Business, 1936–1961.* Chapel Hill: University of North Carolina Press, 1996.

Bernstein, Barton J. "The Debate on Industrial Reconversion: The Protection of Oligopoly and Military Control of the Economy." *American Journal of Economics and Sociology* 26 (April 1967): 159–72.

Biennial Report, 1953 and 1954, National Bureau of Standards, Miscellaneous Publication 213. Washington, DC: US Government Printing Office, 1954.

"Big Steel Bulb to Aid Science." *Boston Globe*, April 13, 1937, 13.

"The Biochemical Research Foundation of the Franklin Institute." *Science* 83 (January 10, 1936): 26.

Bird, Kai, and Martin J. Sherwin. *American Prometheus: The Triumph and Tragedy of J. Robert Oppenheimer.* New York: Knopf, 2005.

Bix, Amy Sue. "Backing into Sponsored Research: Physics and Engineering at Princeton University." *History of Higher Education Annual* 13 (1993): 9–52.

Blackford, Mansel G. *A History of Small Business in America.* New York: Twayne, 1991.

Blakeslee, Howard W. "Herbert Hoover Jr. Uses Atom-Throwers to Find Oil." *Atlanta Constitution*, February 29, 1940, 16.

Bleakney, W. Focusing and Separation of Charged Particles. US Patent 2,221,467, filed December 27, 1938, and issued November 12, 1940.

Bloch, F. "Leonard Isaac Schiff, 1915–1971." *Biographical Memoirs*, 54:301–23. Washington, DC: National Academy Press, 1983.

Blum, John Morton, ed. *The Price of Vision: The Diary of Henry A. Wallace, 1942–1946*. Boston: Houghton Mifflin, 1973.

Boot, Henry A. H., and John T. Randall. "Historical Notes on the Cavity Magnetron." *IEEE Transactions on Electron Devices* ED-23 (July 1976): 724–29.

Braun, Ernest. "Mechanical Properties of Solids." In *Out of the Crystal Maze: Chapters from the History of Solid-State Physics*, edited by Lillian Hoddeson et al., 317–58. New York: Oxford University Press, 1992.

Braun, Ernest. "Selected Topics from the History of Semiconductor Physics and Its Applications." In *Out of the Crystal Maze: Chapters from the History of Solid-State Physics*, edited by Lillian Hoddeson et al., 443–88. New York: Oxford University Press, 1992.

Breit, G., and E. U. Condon. "Interaction between Protons as Indicated by Scattering Experiments." Abstract. *Physical Review* 49 (June 1, 1936): 866.

Breit, G., E. U. Condon, and R. D. Present. "Theory of Scattering of Protons by Protons." *Physical Review* 50 (November 1, 1936): 825–45.

Brinkley, Alan. *The End of Reform: New Deal Liberalism in Recession and War*. New York: Vintage, 1995.

Brown, William L. "H. A. Wallace and the Development of Hybrid Corn." *Annals of Iowa* (Fall 1983): 167–79.

Brunetti, Cleo, and A. S. Khouri. "Printed Electronic Circuits." *Electronics* 19 (April 1946): 104–8.

Bryant, John H. "Microwave Technology and Careers in Transition: The Interests and Activities of Visitors to the Sperry Gyroscope Company's Klystron Plant in 1939–1940." *IEEE Transactions on Microwave Theory and Techniques* 38 (November 1990): 1545–58.

"Bucher Promoted by Westinghouse." *New York Times*, February 24, 1938, 31.

Bush, Vannevar. *Science, the Endless Frontier: A Report to the President on a Program of Postwar Scientific Research*. Washington, DC: US Government Printing Office, 1945.

Cannadine, David. *Mellon: An American Life*. New York: Knopf, 2006.

Carlisle, Prince M. "Chemical Gains to Feature Show." *New York Times*, November 30, 1941, F1.

Carlson, W. Bernard. "Academic Entrepreneurship and Engineering Education: Dugald C. Jackson and the MIT-GE Cooperative Engineering Course, 1907–1932." *Technology and Culture* 29 (July 1988): 536–67.

Carlson, W. Bernard. *Innovation as a Social Process: Elihu Thomson and the Rise of General Electric, 1870–1900*. Cambridge: Cambridge University Press, 1991.

Cassen, B. Super Voltage X-Ray Tube US Patent 2,342,789, filed April 19, 1941, and issued February 29, 1944.

Cassen, B., and E. U. Condon. "On Nuclear Forces." *Physical Review* 50 (November 1, 1936): 846–49.

Cattell, Jaques, ed. *American Men of Science: A Biographical Directory*. 7th ed. Lancaster, PA: Science Press, 1944.

Cattell, Jaques, ed. *American Men of Science: A Biographical Directory*. 8th ed. Lancaster, PA: Science Press, 1949.

Cattell, J. McKeen, and Jaques Cattell, eds. *American Men of Science: A Biographical Directory*. 6th ed. New York: Science Press, 1938.

Chandler, Alfred D., Jr. *Strategy and Structure: Chapters in the History of the American Industrial Enterprise*. Cambridge, MA: MIT Press, 1962.

Childs, Herbert. *An American Genius: The Life of Ernest Orlando Lawrence*. New York: E. P. Dutton, 1968.

Chubb, L. W. "Exploring the Atom." *Scientific Monthly* 45 (September 1937): 285–87.

Chubb, L. W. "New Products from Research in the Electrical Industry." *Edison Electric Institute Bulletin* 5 (July 1937): 289–94.

Coben, Stanley. "The Scientific Establishment and the Transmission of Quantum Mechanics to the United States, 1919–32." *American Historical Review* 76 (April 1971): 442–66.

Cochrane, Rexmond C. *Measures for Progress: A History of the National Bureau of Standards*. Washington, DC: US Department of Commerce, 1966.

Coltman, John W. "Resonant Cavity Magnetron." *Westinghouse Engineer* 6 (November 1946): 172–75.

"Committee to Fix Policy on Rubber." *New York Times*, September 9, 1945, 39.

"Compton Chosen M.I.T. President." *New York Times*, March 13, 1930, 3.

"Conditioned Heated Air Induces Fever That Cures." *Science News Letter* 35 (February 11, 1939): 90.

Condon, Edward U. "Nuclear Motions Associated with Electron Transitions in Diatomic Molecules." *Physical Review* 32 (December 1928): 858–72.

Condon, Edward U. "Physics in Industry." *Science* 96 (August 21, 1942): 172–74.

Condon, Edward U. "Reminiscences of a Life in and out of Quantum Mechanics." In *Proceedings of the International Symposium on Atomic, Molecular, and Solid-State Theory and Quantum Biology*, edited by Per-Olav Löwdin, 7–22. New York: John Wiley, 1973.

Condon, Edward U. "Science and the National Welfare." *Science* 107 (January 2, 1948): 2–7.

Condon, Edward U. "Sharpshooting at the Atom." *Popular Mechanics* 74 (July 1940): 1–5, 143A–44A.

Condon, Edward U. "Some Thoughts on Science in the Federal Government." *Physics Today* 5 (April 1952): 6–13.

Condon, Edward U. "A Theory of Intensity Distribution in Band Systems." *Physical Review* 28 (December 1926): 1182–1201.

Condon, E. U., and P. M. Morse. *Quantum Mechanics*. New York: McGraw-Hill, 1929.

Condon, E. U., and G. H. Shortley. *The Theory of Atomic Spectra*. New York: Macmillan, 1935.

"Condon Given Clearance to See Secret Defense Data." *Washington Post*, October 19, 1954, 2.

"Condon Heads New Group on Atomic Physics." *Washington Post*, October 12, 1947, M1.

"Conference on Applied Nuclear Physics." *Science* 92 (July 26, 1940): 73–74.

"Consolidated Engineering Corp." *Commercial and Financial Chronicle* 161 (June 25, 1945): 2784.

Cornell, Thomas D. *Establishing Research Corporation: A Case Study of Patents, Philanthropy, and Organized Research in Early Twentieth-Century America*. Tucson, AZ: Research Corporation, 2004.

Cornell, Thomas D. "Merle Tuve and His Program of Nuclear Studies at the Department of Terrestrial Magnetism: The Early Career of a Modern American Physicist." PhD diss., Johns Hopkins University, 1986.

"The Corporation." *Fortune* 13 (March 1936): 59–67, 152–204.

Couvares, Francis G. *The Remaking of Pittsburgh: Class and Culture in an Industrializing City, 1877–1919*. Albany: State University of New York Press, 1984.

Crabb, A. Richard. *The Hybrid-Corn Makers: Prophets of Plenty*. New Brunswick, NJ: Rutgers University Press, 1947.

"Cracking the Atom." *Daily Journal of Commerce* [Portland, OR], November 18, 1936, 2.

Creager, Angela N. H. "The Industrialization of Radioisotopes by the U.S. Atomic Energy Commission." In *The Science-Industry Nexus: History, Policy, Implications*, edited by Karl Grandin, Nina Wormbs, and Sven Widmalm, 141–67. Sagamore Beach, MA: Science History Publications/Watson, 2004.

Crolius, F. J. "Pittsburgh as a Tonnage Producing Center." *Blast Furnace and Steel Plant* 11 (October 1923): 514–16.

Curtiss, J. H. "A Federal Program in Applied Mathematics." *Science* 107 (March 12, 1948): 257–62.

Curtiss, John H. "The National Applied Mathematics Laboratories of the National Bureau of Standards: A Progress Report Covering the First Five Years of Its Existence." *Annals of the History of Computing* 11 (Summer 1989): 69–98.

"The Cyclotron at Princeton." *Science*, Supplement 83 (March 13, 1936): 8.

"Cyclotron to Be Built by the Massachusetts Institute of Technology." *Science*, Supplement 88 (July 15, 1938): 6.

"The Cyrus Fogg Brackett Professorship of Physics at Princeton University." *Science* 76 (July 29, 1932): 96.

Dale, Ernest. *The Great Organizers*. New York: McGraw-Hill, 1960.

Davis, Watson. "Experimenting with Millions of Volts." *Science News Letter* 19 (May 23, 1931): 326–27.

Davis, Watson. "A New Plan for the Production and Transmission of Electrical Power." *Science*, Supplement 81 (January 11, 1935): 8.

Day, Michael A. "E. U. Condon: Science, Religion, and the Politics of World Peace." *Physics in Perspective* 10 (March 2008): 4–55.

Dennis, Michael Aaron. "Reconstructing Sociotechnical Order: Vannevar Bush and U.S. Science Policy." In *States of Knowledge: The Co-Production of Science and Social Order*, edited by Sheila Jasanoff, 225–53. London: Routledge, 2004.

"Describes Feats of Atom Smasher." *New York Times*, June 3, 1937, 27.

Devlin, Stanley. "Undiscounted Earnings and Prospects." *Magazine of Wall Street* 63 (December 3, 1938): 212–14.

"Devotes $2,000,000 to Cancer Research." *New York Times*, June 9, 1932, 13.

Dieterich-Ward, Allen. *Beyond Rust: Metropolitan Pittsburgh and the Fate of Industrial America*. Philadelphia: University of Pennsylvania Press, 2015.

"Director of Laboratories Named by Westinghouse." *New York Times*, March 12, 1948, 35.

"Disks Spinning in Vacuum May Replace Present Generators." *Science News Letter* 24 (September 23, 1933): 202.

Dobbs, Phillip. "An Industry in Itself, Pioneer Feels Full Weight of Business Letdown." *Magazine of Wall Street* 50 (June 11, 1932): 230–31, 251–52.

"Donner Buys Home in Switzerland." *New York Times*, October 2, 1938, 56.

"Dr. Briggs to Head Standards Bureau." *Washington Post*, March 28, 1933, 16.

"Dr. DuBridge Installed as Caltech President in Imposing Ceremony." *Los Angeles Times*, November 13, 1946, A1.

"Dr. R. E. Hellmund of Westinghouse." *New York Times*, May 18, 1942, 15.

"Drs. E. U. Condon and D. W. Bronk to Aid Atomic Research." *New York Times*, June 6, 1947, 46.

Duncan, Francis. *Rickover and the Nuclear Navy: The Discipline of Technology*. Annapolis: Naval Institute Press, 1990.

Duncan, Robert Kennedy. "Industrial Fellowships of the Mellon Institute." *Science* 39 (May 8, 1914): 672–78.

Dupree, A. Hunter. *Science in the Federal Government: A History of Policies and Activities to 1940*. Cambridge, MA: Belknap Press of Harvard University Press, 1957.

Dyche, H. E., and R. E. Hellmund. "The Pitt-Westinghouse Graduate Program." *Electrical Engineering* 53 (January 1934): 103–8.

"Editorial Office of the J.A.P." *Journal of Applied Physics* 8 (September 1937): 614.

Edwards, Willard. "House Hearings on Chicago Reds to Start Today." *Chicago Tribune*, September 2, 1952, A5.

"Electronic Unit to Sell Shares." *Los Angeles Times*, May 25, 1945, A6.

"Electrostatic Generator at the University of Pennsylvania." *Water Tower* 26 (November 1939): 4–6.

England, J. Merton. *A Patron for Pure Science: The National Science Foundation's Formative Years, 1945–1957*. Washington, DC: National Science Foundation, 1982.

Fagen, M. D., ed. *A History of Engineering and Science in the Bell System: The Early Years, 1875–1925*. Murray Hill, NJ: Bell Telephone Laboratories, 1975.

"Fellowships in Wartime." *Review of Scientific Instruments* 13 (December 1942): 544.

Fenton, Roy M. "Heavy Machinery Makers Compared." *Magazine of Wall Street* 60 (July 31, 1937): 480–81.

Fenton, Roy M. "Two Electrical Equipment Giants Compared." *Magazine of Wall Street* 60 (June 19, 1937): 314–16, 336.

"$15,000,000 Estate Will Go to Charity." *New York Times*, June 21, 1927, 28.

"The Fifth Washington Conference on Theoretical Physics." *Science* 89 (February 24, 1939): 180–82.

Financial News of the Week. *Annalist* 52 (August 3, 1938): 181.

"$500,000 Added to Princeton Science Fund." *New York Times*, October 7, 1928, 1.

Folliard, Edward T. "Truman Asks for 78% Increase in Commerce Dept. Budget." *Washington Post*, January 22, 1946, 5.

"40% of Westinghouse Production for Defense." *Steel* 108 (January 6, 1941): 3.

"Frederick J. Bates." *Washington Post*, November 2, 1958, B2.

Friedberg, Aaron L. *In the Shadow of the Garrison State: America's Anti-Statism and Its Cold War Grand Strategy*. Princeton: Princeton University Press, 2000.

Galambos, Louis. "Theodore N. Vail and the Role of Innovation in the Modern Bell System." *Business History Review* 66 (Spring 1992): 95–126.

Geiger, Roger L. *Research and Relevant Knowledge: American Research Universities since World War II*. New York: Oxford University Press, 1993.

Geiger, Roger L. "Science, Universities, and National Defense, 1945–1970." *Osiris* 7 (1992): 26–48.

Geiger, Roger L. *To Advance Knowledge: The Growth of American Research Universities, 1900–1940*. New York: Oxford University Press, 1986.

"General Electric Co. to Build Machinery of Big Atom Smasher." *Chicago Daily Tribune*, December 8, 1940, B5.

"G. E. Pendray, 86, Rocket Proponent." *New York Times*, September 20, 1987, 60.

"German Atom Expert Heads New Laboratory." *Science News Letter* 25 (March 24, 1934): 184.

"German Scholars Get Places Here." *New York Times*, January 28, 1934, 22.

Gintzon, Edward L. "The $100 Idea." *IEEE Spectrum* 12 (February 1975): 30–39.

Goldman, Joanne Abel. "National Science in the Nation's Heartland: The Ames Laboratory and Iowa State University, 1942–1965." *Technology and Culture* 41 (July 2000): 435–59.

Graham, Margaret B. W. "Corporate Research and Development: The Latest Transformation." *Technology in Society* 7, no. 2–3 (1985): 179–95.

Graham, Margaret B. W. *RCA and the VideoDisc: The Business of Research*. Cambridge: Cambridge University Press, 1986.

Graham, Margaret B. W., and Bettye H. Pruitt. *R&D for Industry: A Century of Technical Innovation at Alcoa*. Cambridge: Cambridge University Press, 1990.

Graham, Margaret B. W., and Alec T. Shuldiner. *Corning and the Craft of Innovation*. New York: Oxford University Press, 2001.

Gray, George W. "What Holds the World Together." *Harper's Monthly*, September 1937, 434–38.

Grondahl, L. O. "An Adventure in Research: Copper-Oxide Rectifiers and Their Applications." *American Physics Teacher* 4 (September 1936): 105–12.

Grondahl, L. O. "A New Type of Contact Rectifier." Abstract. *Physical Review* 27 (June 1926): 813.

Grondahl, L. O. "The Role of Physics in Modern Industry." *Science* 70 (August 23, 1929): 175–83.

Gross, Gerald R. "Atom Research Laboratory Planned Here." *Washington Post*, August 16, 1946, 3.

Groves, Leslie R. *Now It Can Be Told: The Story of the Manhattan Project*. New York: Harper, 1962.

Guerlac, Henry E. *Radar in World War II, Sections A–C*. Volume 8 of *The History of Physics, 1800–1950*. New York: Tomash Publishers and the American Institute of Physics, 1987.

Hamby, Alonzo L. "Henry A. Wallace, the Liberals, and American-Soviet Relations." *Review of Politics* 30 (April 1968): 153–69.

Hamby, Alonzo L. "Sixty Million Jobs and the People's Revolution: The Liberals, the New Deal, and World War II." *Historian* 30 (August 1968): 578–98.

Hamor, W. A. "Pittsburgh as a Chemical Research Center." *Journal of Industrial and Engineering Chemistry* 14 (September 1922): 764–71.

Hamor, W. A., and A. S. Keller. "Natural Resources and Manufactures of Western Pennsylvania [part 1]." *Chemical and Metallurgical Engineering* 34 (July 1927): 426–30.

Hamor, W. A., and A. S. Keller. "Natural Resources and Manufactures of Western Pennsylvania [part 2]." *Chemical and Metallurgical Engineering* 34 (August 1927): 497–99.

Hart, David M. *Forged Consensus: Science, Technology, and Economic Policy in the United States, 1921–1953*. Princeton: Princeton University Press, 1998.

Haxby, R. O., et al. "Photo-Fission of Uranium and Thorium." *Physical Review* 58 (July 1, 1940): 92.

Haxby, R. O., et al. "Photo-Fission of Uranium and Thorium." *Physical Review* 59 (January 1, 1941): 57–62.

Hazlett, T. Lyle, and William W. Hummel. *Industrial Medicine in Western Pennsylvania, 1850–1950*. Pittsburgh: University of Pittsburgh Press, 1958.

Heath, Jim F. "American War Mobilization and the Use of Small Manufacturers, 1939–1943." *Business History Review* 46 (Autumn 1972): 295–319.

Heilbron, J. L., and Robert W. Seidel. *Lawrence and His Laboratory*. Volume 1 of *A History of the Lawrence Berkeley Laboratory*. Berkeley: University of California Press, 1989.

Here and There. *Journal of Applied Physics* 8 (January 1937): 72–74.

Hewlett, Richard G., and Oscar E. Anderson Jr. *The New World, 1939–1946*. Volume 1 of *A History of the United States Atomic Energy Commission*. University Park: Pennsylvania State University Press, 1962.

Hewlett, Richard G., and Francis Duncan. *Atomic Shield, 1947–1952*. Volume 2 of *A History of the United States Atomic Energy Commission*. University Park: Pennsylvania State University Press, 1969.

Hewlett, Richard G., and Francis Duncan. *Nuclear Navy, 1946–1962*. Chicago: University of Chicago Press, 1974.

Hewlett, Richard G., and Jack M. Holl. *Atoms for Peace and War, 1953–1961: Eisenhower and the Atomic Energy Commission*. Volume 3 of *A History of the United States Atomic Energy Commission*. Berkeley: University of California Press, 1989.

Hintz, Eric S. "The Post-Heroic Generation: American Independent Inventors, 1900–1950." *Enterprise and Society* 12 (December 2011): 732–49.

Hipple, John A. "Gas Analysis with the Mass Spectrometer." *Journal of Applied Physics* 13 (September 1942): 551–59.

Hipple, John A. "Portable Mass Spectrometer." *Nature* 150 (July 25, 1942): 111–12.

Hoch, Paul. "The Development of the Band Theory of Solids, 1933–1960." In *Out of the Crystal Maze: Chapters from the History of Solid-State Physics*, edited by Lillian Hoddeson et al., 182–235. New York: Oxford University Press, 1992.

Hoddeson, Lillian. "The Discovery of the Point Contact Transistor." *Historical Studies in the Physical Sciences* 12, no. 1 (1981): 41–76.

Hoddeson, Lillian. "The Emergence of Basic Research in the Bell Telephone System, 1875–1915." *Technology and Culture* 22 (July 1981): 512–44.

Hoddeson, Lillian. "The Entry of the Quantum Theory of Solids into the Bell Telephone Laboratories: A Case Study of the Industrial Application of Fundamental Science, 1925–40." *Minerva* 18 (Autumn 1980): 422–47.

Hoddeson, Lillian. "Research on Crystal Rectifiers during World War II and the Invention of the Transistor." *History and Technology* 11, no. 2 (1994): 121–30.

Hoddeson, Lillian, et al. *Critical Assembly: A Technical History of Los Alamos during the Oppenheimer Years, 1943–1945.* Cambridge: Cambridge University Press, 1993.

Hoerr, John P. *And the Wolf Finally Came: The Decline of the American Steel Industry.* Pittsburgh: University of Pittsburgh Press, 1988.

Hogan, Michael J. *A Cross of Iron: Harry S. Truman and the Origins of the National Security State, 1945–1954.* Cambridge: Cambridge University Press, 1998.

Holton, Gerald. "Striking Gold in Science: Fermi's Group and the Recapture of Italy's Place in Physics." *Minerva* 12 (April 1974): 159–98.

Hoover, H., Jr., and H. Washburn. "Analysis of Hydrocarbon Gas Mixtures by Mass Spectrometry." *California Oil World and Petroleum Industry* 34 (November 1941): 21–22.

Hoover, H., Jr., and H. Washburn. "A Preliminary Report on the Application of Mass Spectrometry to Problems in the Petroleum Industry." *Petroleum Technology* 3 (May 1940): 1–7.

Hounshell, David A. "The Evolution of Industrial Research in the United States." In *Engines of Innovation: U.S. Industrial Research at the End of an Era*, edited by Richard S. Rosenbloom and William J. Spencer, 13–85. Boston: Harvard Business School Press, 1996.

Hounshell, David A., and John Kenly Smith Jr. *Science and Corporate Strategy: DuPont R&D, 1902–1980.* Cambridge: Cambridge University Press, 1988.

"Huge Power Plan Gets Federal Aid." *New York Times*, December 5, 1933, 41.

Hughes, Thomas P. *Networks of Power: Electrification in Western Society, 1880–1930.* Baltimore: Johns Hopkins University Press, 1983.

Hutchisson, E. "Conference on Industrial Physics." *Review of Scientific Instruments* 6 (December 1935): 381–82.

"In the Laboratories." *Science* 124 (October 26, 1956): 826.

"Interservice Radio Propagation Laboratory." *Technical News Bulletin of the National Bureau of Standards*, no. 344 (December 1945): 95–96.

"James Bliss Austin Heads United States Steel Research Laboratory." *Bulletin of the American Ceramic Society* 26 (January 15, 1947): 2–5.

Johnson, Leland, and Daniel Schaffer. *Oak Ridge National Laboratory: The First Fifty Years.* Knoxville: University of Tennessee Press, 1994.

Jones, Vincent C. *Manhattan: The Army and the Atomic Bomb.* In *United States Army in World War II, Special Studies.* Washington, DC: US Army Center of Military History, 1985.

"Jones-Laughlin Steel to Be Reorganized." *New York Times*, December 6, 1922, 2.

Kaempffert, Waldemar. "Science in the News." *New York Times*, February 9, 1941, D5.

Kaiser, David. "Cold War Requisitions, Scientific Manpower, and the Production of American Physics after World War II." *Historical Studies in the Physical and Biological Sciences* 33, no. 1 (2002): 131–59.

Karasek, F. W. "Harold Washburn: MS Pioneer." *Research/Development* 23 (June 1972): 26–33.

Kargon, Robert H. "The New Era: Science and American Individualism in the 1920s." In *The Maturing of American Science: A Portrait of Science in Public Life Drawn from the Presidential Addresses of the American Association for the Advancement of Science,* edited by Robert H. Kargon, 1–29. Washington, DC: American Association for the Advancement of Science, 1974.

Kargon, Robert H., and Scott G. Knowles. "Knowledge for Use: Science, Higher Learning, and America's New Industrial Heartland, 1880–1915." *Annals of Science* 59, no. 1 (2002): 1–20.

Kellogg, Nelson R. "Gauging the Nation: Samuel Wesley Stratton and the Invention of the National Bureau of Standards." PhD diss., Johns Hopkins University, 1991.

Kerst, Donald W. "Development of the Betatron and Applications of High Energy Betatron Radiations." *American Scientist* 35 (January 1948): 57–84.

Kevles, Daniel J. "The National Science Foundation and the Debate over Postwar Research Policy, 1942–1945: A Political Interpretation of *Science, the Endless Frontier.*" *Isis* 68 (March 1977): 5–26.

Kevles, Daniel J. *The Physicists: The History of a Scientific Community in Modern America.* New York: Knopf, 1978.

Kingdon, K. K., et al. "Fission of the Separated Isotopes of Uranium." *Physical Review* 57 (April 15, 1940): 749.

Kintner, Samuel M. "The Engineer and Research." *Journal of Engineering Education* 19 (April 1929): 806–11.

Kintner, Samuel M. "Making Research Profitable." *Manufacturing Industries* 14 (December 1927): 415–18.

Kirschner, Don S. "Henry A. Wallace as Farm Editor." *American Quarterly* 17 (Summer 1965): 187–202.

Kleinberg, S. J. *The Shadow of the Mills: Working Class Families in Pittsburgh, 1870–1907.* Pittsburgh: University of Pittsburgh Press, 1989.

Kleinman, Daniel Lee. "Layers of Interests, Layers of Influence: Business and the

Genesis of the National Science Foundation." *Science, Technology, and Human Values* 19 (Summer 1994): 259–82.

Kleinman, Mark L. *A World of Hope, a World of Fear: Henry A. Wallace, Reinhold Niebuhr, and American Liberalism.* Columbus: Ohio State University Press, 2000.

Kline, Ronald R. "Construing 'Technology' as 'Applied Science': Public Rhetoric of Scientists and Engineers in the United States, 1880–1945." *Isis* 86 (June 1995): 194–221.

Kline, Ronald R. *Steinmetz: Engineer and Socialist.* Baltimore: Johns Hopkins University Press, 1992.

Kline, Ronald R., and Thomas C. Lassman. "Competing Research Traditions in American Industry: Uncertain Alliances between Engineering and Science at Westinghouse Electric, 1886–1935." *Enterprise and Society* 6 (December 2005): 601–45.

Kluttz, Jerry. "Civil Service Discloses 400 High-Pay Jobs." *Washington Post*, April 26, 1950, 1, 8.

Kluttz, Jerry. "Condon Would Expand Bureau of Standards." *Washington Post*, August 20, 1946, 3.

Kluttz, Jerry. "NOL Working in Terms of Push-Button War." *Washington Post*, January 13, 1946, M5.

Knowles, Scott G., and Stuart W. Leslie. "'Industrial Versailles': Eero Saarinen's Corporate Campuses for GM, IBM, and AT&T." *Isis* 92 (March 2001): 1–33.

Kohler, Robert E. *Partners in Science: Foundations and Natural Scientists, 1900–1945.* Chicago: University of Chicago Press, 1991.

Kragh, Helge. *Quantum Generations: A History of Physics in the Twentieth Century.* Princeton: Princeton University Press, 1999.

Krasik, Sidney. "The Klystron: Radar-Receiver-Oscillator." *Westinghouse Engineer* 6 (November 1946): 176–79.

Krause, Paul. *The Battle for Homestead, 1880–1892: Politics, Culture, and Steel.* Pittsburgh: University of Pittsburgh Press, 1992.

"Laboratories of the Gulf Research & Development Corporation." *Review of Scientific Instruments* 6 (November 1935): 338–41.

Landa, Edward R., and John R. Nimmo. "The Life and Scientific Contributions of Lyman J. Briggs." *Soil Science Society of America Journal* 67 (May–June 2003): 681–93.

Lassman, Thomas C. "Putting the Military Back into the History of the Military-Industrial Complex: The Management of Technological Innovation in the U.S. Army, 1945–1960." *Isis* 106 (March 2015): 94–120.

Lécuyer, Christophe. "The Making of a Science-Based Technological University:

Karl Compton, James Killian, and the Reform of MIT, 1930–1957." *Historical Studies in the Physical and Biological Sciences* 23, no. 1 (1992): 153–80.

Lécuyer, Christophe. "MIT, Progressive Reform, and 'Industrial Service,' 1890–1920." *Historical Studies in the Physical and Biological Sciences* 26, no. 1 (1995): 35–88.

Lécuyer, Christophe. "What Do Universities Really Owe Industry? The Case of Solid State Electronics at Stanford." *Minerva* 43 (March 2005): 51–71.

Leffler, Melvyn P. *A Preponderance of Power: National Security, the Truman Administration, and the Cold War.* Stanford: Stanford University Press, 1992.

Lenoir, Timothy, and Christophe Lécuyer. "Instrument Makers and Discipline Builders: The Case of Nuclear Magnetic Resonance." *Perspectives on Science* 3, no. 3 (1995): 276–345.

Leslie, Stuart W. *The Cold War and American Science: The Military-Industrial-Academic Complex at MIT and Stanford.* New York: Columbia University Press, 1993.

Leviero, Anthony. "Harriman Begins Commerce Task." *New York Times*, October 8, 1946, 1, 17.

"Life Goes On." *Fortune* 9 (January 1934): 42–47, 86–93.

Livesay, Harold C. "Entrepreneurial Dominance in Businesses Large and Small, Past and Present." *Business History Review* 63 (Spring 1989): 1–21.

Livingston, M. Stanley, and John P. Blewett. *Particle Accelerators.* New York: McGraw-Hill, 1962.

Longenecker, Charles. "Jones & Laughlin Steel Corporation in the Service of the Country for Almost a Century." *Blast Furnace and Steel Plant* 29 (August 1941): 858–93.

Lowan, Arnold N. "The Computation Laboratory of the National Bureau of Standards." *Scripta Mathematica* 15 (March 1949): 33–63.

Lowen, Rebecca S. *Creating the Cold War University: The Transformation of Stanford.* Berkeley: University of California Press, 1997.

"Makes $200,000 Gift for Princeton Chair." *New York Times*, October 2, 1927, 16.

Martin, Joseph D., and Michel Janssen. "Beyond the Crystal Maze: Twentieth-Century Physics from the Vantage Point of Solid State Physics." *Historical Studies in the Natural Sciences* 45 (November 2015): 631–40.

"Mass Spectrographic." *Industrial and Engineering Chemistry, Analytical Edition* 13 (October 15, 1941): 749–50.

"Mass Spectrometer Analysis of Gasoline." *Petroleum Engineering* 14 (April 1943): 44.

"Mass Spectrometer for Gas Analysis." *Review of Scientific Instruments* 13 (May 1942): 238–39.

McRoberts, D. C. "Contributors to Rubber Compounding Progress: The New Jersey Zinc Co.'s Personnel and Laboratory Facilities." *India Rubber World* 92 (July 1935): 31–34.

"Measure the Forces between Cores of Hydrogen Atoms." *Science News Letter* 29 (March 14, 1936): 167.

"The Mellon-Owned Oil Company." *Barron's* 15 (May 6, 1935): 14.

Meyerson, Seymour. "Reminiscences of the Early Days of Mass Spectrometry in the Petroleum Industry." *Organic Mass Spectrometry* 21 (April 1986): 197–208.

"The Middle Atlantic Region." *Index* 11 (May 1931): 93–98.

"Mightiest Atom Smasher at East Pittsburgh, PA." *Life*, August 30, 1937, 36–37.

Millman, S., ed. *A History of Engineering and Science in the Bell System: Communications Sciences, 1925–1980*. Murray Hill, NJ: AT&T Bell Laboratories, 1984.

Miranti, Paul J. "Corporate Learning and Quality Control at the Bell System, 1877–1929." *Business History Review* 79 (Spring 2005): 39–72.

"Missile Study Set at Closed Hospital." *Los Angeles Times*, June 1, 1951, A1.

Morris, Peter J. T. *The American Synthetic Rubber Research Program*. Philadelphia: University of Pennsylvania Press, 1989.

Morris, Susan W. "Resource Networks: Industrial Research in Small Enterprises, 1860–1930." PhD diss., Johns Hopkins University, 2003.

Morrow, J. B. "Research Displaces Rule-of-Thumb." *Coal Age* 32 (November 1927): 244–45.

Morse, Philip M. *In at the Beginnings: A Physicist's Life*. Cambridge, MA: MIT Press, 1977.

"Most U.S. Steel Corp. Subsidiary Plants Are Clustered about Pittsburgh." *Steel* 96 (March 25, 1935): 21.

Mowery, David C., and Nathan Rosenberg. *Technology and the Pursuit of Economic Growth*. Cambridge: Cambridge University Press, 1989.

"Mrs. Edward Condon." *New York Times*, November 4, 1974, 40.

"Named to Be Director of Bureau of Standards." *New York Times*, October 30, 1945, 13.

"National Defense." *Review of Scientific Instruments* 12 (May 1941): 283.

"The National Defense Research Committee." *Science* 92 (November 22, 1940): 484–86.

"Navy Increases Scientists' Pay to $14,000 a Year." *Chicago Daily Tribune*, October 15, 1947, 16.

Needell, Allan A. "Nuclear Reactors and the Founding of Brookhaven National Laboratory." *Historical Studies in the Physical Sciences* 14, no. 1 (1983): 93–122.

Neumann, Tracy. *Remaking the Rust Belt: The Postindustrial Transformation of North America*. Philadelphia: University of Pennsylvania Press, 2016.

"New Appointments." *Journal of Applied Physics* 10 (April 1939): 267.

"New 'Atom Smasher' Shown." *New York Times*, February 4, 1938, 6.

"New Hydrogen Atom Is Found to Be Natural." *Washington Post*, April 29, 1934, M1, M3.

"New Method of Splitting Uranium Atoms Discovered." *Los Angeles Times,* July 2, 1940, 9.

"New Research Program at Pittsburgh Plate Glass." *Glass Industry* 18 (December 1937): 411–14.

"New Way to Split Uranium and Release Energy Found." *Science News Letter* 38 (July 6, 1940): 3–4.

"New Weapons Foreseen." *New York Times,* December 7, 1948, 3.

News and Notes. *Science* 103 (May 17, 1946): 621–27.

News and Notes. *Science* 103 (May 24, 1946): 651–53.

News and Notes. *Science* 115 (May 23, 1952): 562–66.

Nier, Alfred O., et al. "Further Experiments on Fission of Separated Uranium Isotopes." *Physical Review* 57 (April 15, 1940): 748.

Nier, Alfred O., et al. "Nuclear Fission of Separated Uranium Isotopes." *Physical Review* 57 (March 15, 1940): 546.

Norberg, Arthur L. *Computers and Commerce: A Study of Technology and Management at Eckert-Mauchly Computer Company, Engineering Research Associates, and Remington Rand, 1946–1957.* Cambridge, MA: MIT Press, 2015.

Norris, Robert S. *Racing for the Bomb: General Leslie R. Groves, the Manhattan Project's Indispensable Man.* South Royalton, VT: Steerforth Press, 2002.

Notes. *Mathematical Tables and Other Aides to Computation* 27 (July 1949): 494–97.

Notes and News. *American Mathematical Monthly* 38 (November 1931): 545–46.

"Opportunities in Expanding Industries." *Magazine of Wall Street* 57 (March 14, 1936): 651–52, 672.

"Organizes New Products Division." *Steel* 101 (August 16, 1937): 35.

Osgood, T. H. "Physics in 1940." *Journal of Applied Physics* 12 (February 1941): 84–99.

Owens, Larry. "The Counterproductive Management of Science in the Second World War: Vannevar Bush and the Office of Scientific Research and Development." *Business History Review* 68 (Winter 1994): 515–76.

Owens, Larry. "MIT and the Federal 'Angel': Academic R&D and Federal-Private Cooperation before World War II." *Isis* 81 (June 1990): 189–213.

Pang, Alex Soojung-Kim. "Edward Bowles and Radio Engineering at MIT, 1920–1940." *Historical Studies in the Physical and Biological Sciences* 20, no. 2 (1990): 313–37.

Parton, James. "Pittsburg." *Atlantic Monthly* 21 (January 1868): 17–36.

Passer, Harold C. *The Electrical Manufacturers, 1875–1900: A Study of Competition, Entrepreneurship, Technical Change, and Economic Growth.* Cambridge, MA: Harvard University Press, 1953.

"Personalities in Research." *Scientific American* 156 (January 1937): 3.

"Personalities in Science." *Scientific American* 165 (July 1941): 3.

"Physics Symposium at Cornell." *Review of Scientific Instruments* 7 (September 1936): 361.

"Pittsburgh as an Industrial Center." *Industrial and Engineering Chemistry* 14 (June 20, 1936): 246–48.

"Pittsburgh to Be Headquarters for U.S. Steel Corporation." *Blast Furnace and Steel Plant* 25 (December 1937): 1317–18.

Plant and Product of the Mesta Machine Company. Pittsburgh: Mesta Machine Company, 1919.

"Portable Atom-Sorter for Industrial Uses." *Science News Letter* 40 (July 5, 1941): 9.

Potter, Robert D. "Some Papers Read at the New York Meeting of the American Physical Society." *Science* 91 (March 1, 1940): 10, 12.

"Princeton Seniors in Class Day Events." *New York Times*, June 17, 1930, 20.

"Princeton to Have Two New Chairs." *New York Times*, May 31, 1927, 20.

"Probers Issue Subpoena for Dr. Condon." *Washington Post*, August 20, 1952, 1.

"The Production of Synthetic Rubber." *Science*, Supplement 98 (August 20, 1943): 8–9.

"Prof. Harry S. Hower." *New York Times*, October 11, 1941, 17.

Pursell, Carroll W., Jr. "The Administration of Science in the Department of Agriculture, 1933–1940." *Agricultural History* 42 (July 1968): 231–40.

Pursell, Carroll W., Jr. "A Preface to Government Support of Research and Development: Research Legislation and the National Bureau of Standards, 1935–41." *Technology and Culture* 9 (April 1968): 145–64.

Rabkin, Yakov M. "Technological Innovation in Science: The Adoption of Infrared Spectroscopy by Chemists." *Isis* 78 (March 1987): 31–54.

"Radar Countermeasures." *Electronics* 19 (January 1946): 92–97.

"Recent Appointments." *Review of Scientific Instruments* 14 (November 1943): 347–48.

"Record-Breaking Magnetic Field Employed in New Mass Spectrometer." *Electronics* 13 (May 1940): 40, 44.

"Record Expansion for Westinghouse." *New York Times*, March 11, 1941, 35.

Reich, Leonard S. "Irving Langmuir and the Pursuit of Science and Technology in the Corporate Environment." *Technology and Culture* 24 (April 1981): 199–221.

Reich, Leonard S. *The Making of American Industrial Research: Science and Business at GE and Bell, 1876–1926.* Cambridge: Cambridge University Press, 1985.

Reingold, Nathan. "Choosing the Future: The U.S. Research Community, 1944–1946." *Historical Studies in the Physical and Biological Sciences* 25, no. 2 (1995): 301–28.

Reingold, Nathan. "Vannevar Bush's New Deal for Research: Or the Triumph of the Old Order." *Historical Studies in the Physical and Biological Sciences* 17, no. 2 (1987): 299–344.

"Research Fellowship Appointments Announced." *Review of Scientific Instruments* 9 (June 1938): 206.

"Research in Pittsburgh." *Industrial and Engineering Chemistry* 14 (July 20, 1936): 281–84.

"Revolutionary Method of Power Transmission Urged." *Science News Letter* 26 (December 29, 1934): 408.

Rider, Robin E. "Alarm and Opportunity: Emigration of Mathematicians and Physicists to Britain and the United States, 1933–1945." *Historical Studies in the Physical Sciences* 15, no. 1 (1984): 107–76.

Roberts, O. L. "Quantitative Analysis by Mass Spectrometry." *Petroleum Engineer* 14 (May 1943): 109–11, 114–16.

"Robot Chemist to Aid Testing of Butadiene." *Chicago Daily Tribune*, August 8, 1943, A7.

"Rocket Station Wins Approval." *Los Angeles Times*, October 26, 1949, 6.

Rosenberg, Nathan, and Richard R. Nelson. "The Roles of Universities in the Advance of Industrial Technology." In *Engines of Innovation: U.S. Industrial Research at the End of an Era*, edited by Richard S. Rosenbloom and William J. Spencer, 87–109. Boston: Harvard Business School Press, 1996.

Rosenof, Theodore. "The Economic Ideas of Henry A. Wallace, 1933–1948." *Agricultural History* 41 (April 1967): 143–53.

Rossiter, Margaret W. "Setting Federal Salaries in the Space Age." *Osiris* 7 (1992): 218–37.

Russo, Arturo. "Fundamental Research at Bell Laboratories: The Discovery of Electron Diffraction." *Historical Studies in the Physical Sciences* 12, no. 1 (1981): 117–60.

"Samuel M. Kintner, Engineer, 64, Dead." *New York Times*, September 29, 1936, 27.

"Samuel Montgomery Kintner." *Journal of Applied Physics* 8 (February 1937): 117–21.

Sapolsky, Harvey M. *Science and the Navy: The History of the Office of Naval Research*. Princeton: Princeton University Press, 1990.

Savage, Kirk. "Monuments of a Lost Cause: The Postindustrial Campaign to Commemorate Steel." In *Beyond the Ruins: The Meanings of Deindustrialization*, edited by Jefferson Cowie and Joseph Heathcott, 237–56. Ithaca: Cornell University Press, 2003.

Schapsmeier, Edward L., and Frederick H. Schapsmeier. *Henry A. Wallace of Iowa: The Agrarian Years, 1910–1940*. Ames: Iowa State University Press, 1968.

Schapsmeier, Edward L., and Frederick H. Schapsmeier. *Prophet in Politics: Henry A. Wallace and the War Years, 1940–1965*. Ames: Iowa State University Press, 1970.

Schatz, Ronald W. *The Electrical Workers: A History of Labor at General Electric and Westinghouse, 1923–1960*. Urbana: University of Illinois Press, 1983.

Schweber, S. S. "The Empiricist Temper Regnant: Theoretical Physics in the United States, 1920–1950." *Historical Studies in the Physical and Biological Sciences* 17, no. 1 (1986): 55–98.

Scientific News and Notes. *Science* 52 (October 29, 1920): 404.

Scientific Notes and News. *Science* 63 (February 5, 1926): 161–62.

Scientific Notes and News. *Science* 69 (April 19, 1929): 420–24.

Scientific News and Notes. *Science* 72 (September 26, 1930): 316.

Scientific News and Notes. *Science* 86 (August 27, 1937): 193.

Scientific News and Notes. *Science* 89 (March 3, 1939): 195–97.

Scientific News and Notes. *Science* 92 (November 8, 1940): 424–27.

Scientific Notes and News. *Science* 93 (June 13, 1941): 564–66.

Scientific News and Notes. *Science* 98 (October 15, 1943): 340–42.

Scientific Notes and News. *Science* 101 (April 27, 1945): 427–29.

"Scientists Pool Resources in \$60,000 Research Plan." *Science News Letter* 36 (August 19, 1939): 120.

"Scientists Unleash Largest Atom-Attacking Machine." *Science News Letter* 19 (December 2, 1933): 364–66.

"Second in Electrical Equipment." *Barron's* 16 (December 28, 1936): 14.

Seely, Bruce. "Research, Engineering, and Science in American Engineering Colleges, 1900–1960." *Technology and Culture* 34 (April 1993): 344–86.

Segrè, Emilio. *Enrico Fermi: Physicist.* Chicago: University of Chicago Press, 1970.

Seidel, Robert W. "Accelerating Science: The Postwar Transformation of the Lawrence Radiation Laboratory." *Historical Studies in the Physical Sciences* 13, no. 2 (1983): 375–400.

Seitz, Frederick. *The Modern Theory of Solids.* New York: McGraw-Hill, 1940.

Seitz, Frederick. *On the Frontier: My Life in Science.* New York: AIP Press, 1994.

Seitz, Frederick, and T. A. Read. "Theory of the Plastic Properties of Solids. I." *Journal of Applied Physics* 12 (February 1941): 100–118.

Serber, Robert. *The Los Alamos Primer: The First Lectures on How to Build an Atomic Bomb.* Edited by Richard Rhodes. Berkeley: University of California Press, 1992.

Servos, John W. "Changing Partners: The Mellon Institute, Private Industry, and the Federal Patron." *Technology and Culture* 35 (April 1994): 221–57.

Servos, John W. "Engineers, Businessmen, and the Academy: The Beginnings of Sponsored Research at the University of Michigan." *Technology and Culture* 37 (October 1996): 721–62.

Servos, John W. "The Industrial Relations of Science: Chemical Engineering at MIT, 1900–1939." *Isis* 71 (December 1980): 531–49.

Servos, John W. "Mathematics and the Physical Sciences in America, 1880–1930." *Isis* 77 (December 1986): 611–29.

Seth, Suman. *Crafting the Quantum: Arnold Sommerfeld and the Practice of Theory, 1890–1926*. Cambridge, MA: MIT Press, 2010.

Shenstone, A. G. "E. P. Adams, Princeton Physicist." *Science* 125 (February 22, 1957): 339.

"Silk Belt Gathers Huge Electric Charge." *Science News Letter* 20 (September 19, 1931): 184.

"The Sixth Annual Summer Symposium in Theoretical Physics at the University of Michigan." *Science* 77 (March 31, 1933): 320.

Skinner, Charles E., and R. W. E. Moore. "The New Westinghouse Research Building." *Electrical World* 71 (June 1, 1918): 1132–33.

Slepian, J. X-Ray Tube. US Patent 1,645,304, filed April 1, 1922, and issued October 11, 1927.

Smith, Alice Kimball. *A Peril and a Hope: The Scientists' Movement in America, 1945–47*. Chicago: University of Chicago Press, 1965.

Smith, George David. *From Monopoly to Competition: The Transformations of Alcoa, 1888–1986*. Cambridge: Cambridge University Press, 1988.

Smith, M. W. "The Importance of Research and Development in Maintaining Technical Progress." *Electrical Engineering* 57 (December 1938): 484–88.

Solomon, Ernest, and Louis C. Rubin. "Mass Spectrometric Gas Analysis." *American Gas Association Monthly* 27 (October 1945): 461–63.

Sopka, Katherine Russell. *Quantum Physics in America, 1920–1935*. New York: Arno Press, 1980.

"Source of Rare Hydrogen Isotope Made at Princeton." *Science News Letter* 27 (March 23, 1935): 179.

Stephan, Karl. "Experts at Play: Magnetron Research at Westinghouse, 1930–1934." *Technology and Culture* 42 (October 2000): 737–49.

Stern, Nancy. *From ENIAC to UNIVAC: An Appraisal of the Eckert-Mauchly Computers*. Bedford, MA: Digital Press, 1981.

"Stettinius Quits $100,000-a-Year Steel Job to Give Full Time to National Defense." *New York Times*, June 5, 1940, 1, 45.

Stewart, Irvin. *Organizing Scientific Research for War: The Administrative History of the Office of Scientific Research and Development*. Boston: Little, Brown, 1948.

Stewart, Robert K. "The Office of Technical Services: A New Deal Idea in the Cold War." *Knowledge* 15 (September 1993): 44–77.

Stine, C. M. A. "The Place of Fundamental Research in an Industrial Research Organization." *Transactions of the American Institute of Chemical Engineers* 32 (June 29, 1936): 127–37.

"Story of Research: The Resnatron." *Westinghouse Engineer* 6 (March 1946): 47.

"The Structure of Solids." *Science*, Supplement 87 (March 4, 1938): 10.

"The Summer Symposium on Theoretical Physics at the University of Michigan." *Science* 83 (June 5, 1936): 544.

"Supervoltage X-Ray Tube Produces 'Hard' Rays." *Science News Letter* 45 (March 11, 1944): 169.

Swain, Donald C. "The Rise of a Research Empire: NIH, 1930–1950." *Science* 138 (December 14, 1962): 1233–37.

"Symposia on Theoretical Physics and Chemical Kinetics." *Science* 69 (April 19, 1929): 420.

Tarr, Joel A., ed. *Devastation and Renewal: The Environmental History of Pittsburgh and Its Region.* Pittsburgh: University of Pittsburgh Press, 2003.

"Ten Million Volts." *Princeton Alumni Weekly* 32 (January 8, 1932): 291.

"Ten Will 'Explore' into Pure Science." *New York Times*, December 19, 1937, 49.

"3 Germans to Join Carnegie Faculty." *New York Times*, September 13, 1933, 8.

Tillotson, E. W. "Researches in Glass at Mellon Institute." *Chemical and Metallurgical Engineering* 34 (April 1927): 232.

"To Direct Research Work." *Iron Age* 119 (June 9, 1927): 1682.

"Top Changes Made in Research Group." *New York Times*, October 2, 1944, 27.

"Trade-In Plan Set Up for Mass Spectrometers." *Chemical and Engineering News* 41 (August 19, 1963): 23.

Travis, Anthony S. *Dyes Made in America, 1915–1980: The Calco Chemical Company, American Cyanamid, and the Raritan River.* Jerusalem: Hexagon Press and the Sidney M. Edelstein Center for the History and Philosophy of Science, Technology, and Medicine, Hebrew University, 2004.

"Truman Gets Rubber Bill." *New York Times*, April 1, 1948, 11.

"Truman Seeks $1,000,000 for Standards Bureau." *Washington Post*, March 9, 1946, 5.

Turchetti, Simone. "The Invisible Businessman: Nuclear Physics, Patenting Practices, and Trading Activities in the 1930s." *Historical Studies in the Physical and Biological Sciences* 37, no. 1 (2006): 153–72.

"U.C.L.A. Fills Dean's Chair." *Los Angeles Times*, September 23, 1944, A1.

"United Celebrates Golden Anniversary." *Iron and Steel Engineer* 28 (December 1951): 124–27.

University and Educational Notes. *Science* 67 (May 25, 1928): 530.

US Bureau of the Budget. *The Budget of the United States Government for the Fiscal Year Ending June 30, 1946.* Washington, DC: US Government Printing Office, 1945.

US Bureau of the Budget. *The Budget of the United States Government for the Fiscal Year Ending June 30, 1948.* Washington, DC: US Government Printing Office, 1947.

US Bureau of the Budget. *The Budget of the United States Government for the Fiscal*

Year Ending June 30, 1952. Washington, DC: US Government Printing Office, 1951.

US Congress. House of Representatives. Subcommittee of the Committee on Appropriations. *Department of Commerce Appropriation for 1952: Hearings before the Subcommittee of the Committee on Appropriations.* 82nd Cong., 1st sess., April 10, 1951.

US Congress. House of Representatives. Subcommittee of the Committee on Departments of State, Justice, Commerce, and the Judiciary Appropriations. *Department of Commerce Appropriation Bill for 1947: Hearings before the Subcommittee of the Committee on Appropriations.* 79th Cong., 2nd sess., January 29, 1946.

US Congress. House of Representatives. Subcommittee on the Department of Commerce. *Department of Commerce Appropriation Bill for 1948: Hearings before the Subcommittee of the Committee on Appropriations.* 80th Cong., 1st sess., March 12, 1947.

US Congress. House of Representatives. Subcommittee on the Department of Commerce. *Department of Commerce Appropriation Bill for 1949: Hearings before the Subcommittee of the Committee on Appropriations.* 80th Cong., 2nd sess., January 20, 1948.

US Congress. House of Representatives. Subcommittee on Un-American Activities, Special Subcommittee on National Security. *Report to the Full Committee of the Special Subcommittee on National Security of the Committee on Un-American Activities.* 80th Cong., 2nd sess., March 1, 1948.

US Congress. Senate. Subcommittee of the Committee on Appropriations. *Departments of State, Justice, Commerce, and the Judiciary Appropriation Bill for 1949: Hearings before the Subcommittee of the Committee on Appropriations on H.R. 5607.* 80th Cong., 2nd sess., April 6, 1948.

US Congress. Senate. Subcommittee of the Committee on Appropriations. *Departments of State, Justice, Commerce, and the Judiciary Appropriations for 1951: Hearings before the Subcommittee of the Committee on Appropriations. Part 2.* 81st Cong., 2nd sess., May 18, 1950.

Usselman, Steven W. "From Novelty to Utility: George Westinghouse and the Business of Innovation in the Age of Edison." *Business History Review* 66 (Summer 1992): 251–304.

Van de Graaff, R. J., et al. "The Electrostatic Production of High Voltage for Nuclear Investigations." *Physical Review* 43 (February 1, 1933): 149–57.

Van Voorhis, M. G. "Industrial Fellowship System Expands in Temple of Research." *National Petroleum News* 29 (June 23, 1937): 50–52.

Wall, Joseph Frazier. *Andrew Carnegie.* New York: Oxford University Press, 1970.

Wallace, Henry A. *Sixty Million Jobs.* New York: Reynal and Hitchcock, 1945.

Wallace, Henry A. "The Value of Scientific Research to Agriculture." *Science* 77 (May 19, 1933): 475–80.

Wallace, Henry A. "Why S. 1850 Is the Better Bill." *Science* 103 (June 21, 1946): 724–25, 729–30.

"Wallace Explains Full Employment." *New York Times*, September 6, 1945, 40.

"Wallace to Shift Commerce Bureau." *New York Times*, September 21, 1945, 15.

Walther, H. "Internal Friction in Solids." *Scientific Monthly* 41 (September 1935): 275–77.

Wang, Jessica. *American Science in an Age of Anxiety: Scientists, Anticommunism, and the Cold War*. Chapel Hill: University of North Carolina Press, 1999.

Wang, Jessica. "Liberals, the Progressive Left, and the Political Economy of Postwar American Science: The National Science Foundation Debate Revisited." *Historical Studies in the Physical and Biological Sciences* 26, no. 1 (1995): 139–66.

Wang, Jessica. "Science, Security, and the Cold War: The Case of E. U. Condon." *Isis* 83 (June 1992): 238–69.

Warren, Kenneth. *Big Steel: The First Century of the United States Steel Corporation, 1901–2001*. Pittsburgh: University of Pittsburgh Press, 2001.

Warren, Kenneth. *Triumphant Capitalism: Henry Clay Frick and the Industrial Transformation of America*. Pittsburgh: University of Pittsburgh Press, 1996.

Weart, Spencer R. "The Physics Business in America, 1919–1940: A Statistical Reconnaissance." In *The Sciences in the American Context: New Perspectives*, edited by Nathan Reingold, 295–358. Washington, DC: Smithsonian Institution Press, 1979.

Weart, Spencer R. "The Solid Community." In *Out of the Crystal Maze: Chapters from the History of Solid-State Physics*, edited by Lillian Hoddeson et al., 617–69. New York: Oxford University Press, 1992.

Weber, Gustavus A. *The Bureau of Standards: Its History, Activities, and Organization*. Baltimore: Johns Hopkins University Press, 1925.

Weber, Michael. *Don't Call Me Boss: David L. Lawrence, Pittsburgh's Renaissance Mayor*. Pittsburgh: University of Pittsburgh Press, 1988.

Weil, Martin. "Atomic Scientist Edward Condon Dies." *Washington Post*, March 27, 1974, C4.

Weiner, Charles. "A New Site for the Seminar: The Refugees and American Physics in the Thirties." In *The Intellectual Migration: Europe and America, 1930–1960*, edited by Donald Fleming and Bernard Bailyn, 190–234. Cambridge, MA: Belknap Press of Harvard University Press, 1969.

Weiner, Charles. "1932: Moving into the New Physics." *Physics Today* 25 (May 1972): 40–49.

Wells, W. H. "Production of High Energy Particles." *Journal of Applied Physics* 9 (November 1938): 677–89.

Wells, W. H., et al. "Design and Preliminary Performance Tests of the Westinghouse Electrostatic Generator." *Physical Review* 58 (July 15, 1940): 162–73.

"Westinghouse Air Brake." *Fortune* 15 (March 1937): 115–19, 180–92.

"Westinghouse Electric." *Fortune* 17 (February 1938): 43–57, 146, 149.

"Westinghouse Forms Atomic Power Division." *Nucleonics* 3 (November 1948): 74.

"Westinghouse Forms X-Ray Company." *Radiology* 15 (September 1930): 408–9.

"Westinghouse Names Heads of Atomic Power Group." *Nucleonics* 4 (February 1949): 80.

"Westinghouse Names New Executives for Atomic Power Division." *Iron Age* 163 (June 2, 1949): 130.

"Westinghouse Plans to Coordinate Activities for Atomic Research." *Iron Age* 159 (April 17, 1947): 127.

"Westinghouse Program for Fundamental Research." *Journal of Applied Physics* 17 (December 1946): 1129.

"Westinghouse Research Fellows." *Science* 95 (June 5, 1942): 571.

"Westinghouse Research Fellowships." *Journal of Applied Physics* 9 (January 1938): 34–35.

"Westinghouse Research Fellowships." *Journal of Applied Physics* 10 (May 1939): 319–20.

"The Westinghouse Research Fellowships." *Science* 86 (December 31, 1937): 605–6.

"Westinghouse Research Fellowships." *Science* 91 (April 12, 1940): 351.

"Westinghouse Research Fellowships Granted." *Journal of Applied Physics* 11 (May 1940): 340.

"Westinghouse Research Grant." *Journal of Applied Physics* 9 (January 1938): 35.

"Westinghouse Spurs Research." *Electrical World* 146 (October 8, 1956): 23–24.

"Westinghouse Will Endow Scientists." *Los Angeles Times*, December 20, 1937, 5.

Westwick, Peter J. *The National Labs: Science in an American System, 1947–1974*. Cambridge, MA: Harvard University Press, 2003.

"Where Radar Frequencies Come From." *Westinghouse Engineer* 6 (January 1946): 4–5.

White, Langdon. "The Iron and Steel Industry of the Pittsburgh District." *Economic Geography* 4 (April 1928): 115–39.

White, William S. "Soviet Spy Links Laid to Dr. Condon, High Federal Aide." *New York Times*, March 2, 1948, 1.

"Why Science Seeks to Smash Atom." *Washington Post*, August 15, 1937, A6.

Wikoff, Alan G. "Some Research of General Interest in Progress at the Mellon Institute." *Chemical and Metallurgical Engineering* 28 (April 9, 1923): 625–32.

Wilson, Mark R. *Destructive Creation: American Business and the Winning of World War II*. Philadelphia: University of Pennsylvania Press, 2016.

Wise, George. "Science at General Electric." *Physics Today* 37 (December 1984): 52–61.

Wise, George. *Willis R. Whitney, General Electric, and the Origins of U.S. Industrial Research*. New York: Columbia University Press, 1985.

Wood, Lewis. "Wallace Is Sworn in at Gay Ceremony." *New York Times*, March 3, 1945, 21.

Woodbury, Robert O. *Battlefronts of Industry: Westinghouse in World War II*. New York: John Wiley, 1948.

Zachary, G. Pascal. *Endless Frontier: Vannevar Bush, Engineer of the American Century*. New York: Free Press, 1997.

INDEX

Note: EC stands for Edward Condon. Page numbers followed by n and a number refer to endnotes. Page numbers in *italics* indicate illustrative material.

Calco Chemical Company, 98

California Institute of Technology
(Caltech), 5, 9, 11, 18, 20, 27, 35, 64,
66, 153

Canfield, L. D., 96

Capt, James, 165–66

Carnegie, Andrew, 25

Carnegie Institute of Technology (CIT),
26, 76–77, 197n2

Carnegie Institution of Washington, 19,
38, 39, 47, 49, 77–78, 79, 84

Carothers, Wallace, 52. *See also* E. I. du
Pont de Nemours and Company

Cassen, Benedict, 20, 33, 34, 86, 118–19,
224n20

Central Radio Propagation Laboratory
(CRPL), 151–52, 168. *See also* Bureau
of Standards

Chadwick, James, 16

Charles Benedict Stuart Laboratory of
Applied Physics (Purdue University),
xv

Chicago Bridge and Iron Works, 50

Chubb, Lewis, *30*; becomes research
director at Westinghouse, 31; concerns
about publicity of nuclear physics
program, 41–42, 48; recruits EC, 57,
60; rejects EC's proposal for a new
corporate laboratory at Berkeley, 128;
supports nuclear physics research,
34–35, 40–42, 44, 51, 112–13; supports
semiconductor research, 75–76

Civil Service Commission, 234n35

Clinton Engineer Works, 124. *See also*
Oak Ridge

Coast and Geodetic Survey, 165

Cockroft, John, 13, 16

Cold War: and anticommunism, 157,
170–73, 174–75; and academic and

industrial R&D, xvi, 175, 181–83; and
federal R&D establishment, xviii, 181

Coltman, John, 120

Compton, Karl, *10*; on EC's nomination
to head Bureau of Standards,
142; on EC's scientific ability and
temperament, 13, 170–71; microwave
radar, 116–17, 118; recruits EC to
Princeton University, 9–12; seeks
employment for EC at University of
Minnesota, 172; supports publication
of EC's first textbook, 13–14

Condon, Caroline Marie (daughter of
EC), 4

Condon, Edward (EC), *3, 20, 105, 143*
academic career: accepts
appointments at Washington
University and University of
Colorado, 178; conducts theoretical
physics research at Princeton
and University of Minnesota,
12–15; education at Berkeley, 2–4;
encourages and conducts nuclear
physics research at Princeton,
18–21; postdoctoral research in
Germany, 4–5; resigns from
Princeton to join Westinghouse,
61; seeks appointments at New
York University and University of
Pennsylvania, 175–76
career at Bell Telephone Laboratories,
5–9
career at Corning Glass Works,
173–75, 176–77
career at National Bureau of
Standards. *See* Bureau of Standards
career at Westinghouse Electric
and Manufacturing Company:
accepts position at, 54–55, 57,